Methods in Molecular Biology

Series Editor
John M. Walker
School of Life and Medical Sciences
University of Hertfordshire
Hatfield, Hertfordshire, UK

For further volumes:
http://www.springer.com/series/7651

For over 35 years, biological scientists have come to rely on the research protocols and methodologies in the critically acclaimed *Methods in Molecular Biology* series. The series was the first to introduce the step-by-step protocols approach that has become the standard in all biomedical protocol publishing. Each protocol is provided in readily-reproducible step-by step fashion, opening with an introductory overview, a list of the materials and reagents needed to complete the experiment, and followed by a detailed procedure that is supported with a helpful notes section offering tips and tricks of the trade as well as troubleshooting advice. These hallmark features were introduced by series editor Dr. John Walker and constitute the key ingredient in each and every volume of the *Methods in Molecular Biology* series. Tested and trusted, comprehensive and reliable, all protocols from the series are indexed in PubMed.

Cell-Secreted Vesicles

Methods and Protocols

Edited by

Seppo Vainio

GeneCellNano, Oulu University, Oulu, Finland

Humana Press

Editor
Seppo Vainio
GeneCellNano
Oulu University
Oulu, Finland

ISSN 1064-3745 ISSN 1940-6029 (electronic)
Methods in Molecular Biology
ISBN 978-1-0716-3202-4 ISBN 978-1-0716-3203-1 (eBook)
https://doi.org/10.1007/978-1-0716-3203-1

Cover Illustration Caption: A breast cancer cell (MCF-7) expressing an EGFP-HAS3 secreting EVs. Figure provided by Dr. Kirsi Rilla, University of Eastern Finland and Finnish Society of Extracellular Vesicles (FISEV).

This Humana imprint is published by the registered company Springer Science+Business Media, LLC, part of Springer Nature.
The registered company address is: 1 New York Plaza, New York, NY 10004, U.S.A.

Preface

This current Springer *Methods in Molecular Biology* volume targets the exciting but technically challenging field of cell-secreted extracellular vesicles (EVs). A wealth of names for EVs have been used. Their naming typically depends on the process their secretion is associated with. It is noteworthy that, even though EVs are secreted locally by cells, they can spread systemically and even between individuals. It is now evident that the EV signaling system is universal in nature (Vainio S. The Widespread Implications of a Nano-Bio Communication System. Rose Croix J. 13 82: 66–82, 2019).

One expected value of EVs is their capacity to cargo the same molecular classes that are used in diagnostics. These include DNA, RNA, proteins/enzymes, metabolites, lipids, and sugars in body fluids including blood. Due to this reason, EVs provide openings to develop novel diagnostic measures via identification of molecular disease signatures. Moreover, presence of EVs in body fluids that are not yet used in large-scale routine molecular diagnostics such as saliva, sweat, and urine offer good reasons to develop such non-invasive diagnostic technological solutions.

The identification of EV functionalities has made it possible also to target new problems in a scientifically sound way. In this regard, the capacity of EVs to cross various biological barriers such as the blood-brain barrier represents a major step forward to increase our understanding of central nervous system communication and diseases. Moreover, in mammals, the gut and its microbiome EVs can enter the systemic circulation and cross the placental barrier during pregnancy and offer transgenerational signaling to be analyzed in detail.

Traditionally, experimental medicine has much studied disease processes from the point of view of one or few receptor signal transduction pathways. Once the associated receptor ligand binding protein structure had been solved, assays for small molecules binding in drug screening protocols have become possible. EVs provide a more complex cellular regulatory signaling system. Due to this, more sophisticated analytic technologies are essentially needed for progress and therapeutic development. We have now begun to have the technological means for this.

Purification of EVs from biosamples and individual EV-based molecular content analysis are essentially needed for trustworthy data to accumulate. Solutions to the technical challenges of pure EV isolation have started to emerge as well. Biotechnological, biochemical, and genetic engineering of EV molecular content is central when addressing their roles in cells, including stem and progenitor cells, organoids, and in vivo. Such technologies are also becoming available. Moreover, capacities to make use of the wealth of man-made liposomes and nanomaterials for EV imaging, targeting, and compound delivery are also being advanced.

We can now conclude that the skills to target EVs from multiple viewpoints have developed during the recent years. As a result of this, sophisticated protocols have become available. Part I of the current protocols book focuses on methods of EV purification and analysis. In Part II, various ways to study EV functions are described. Finally, Part III focuses on some specific systems and models allowing for the study of EVs of different origin. The reader hopefully will find in the depicted technological protocols hands-on guidance to target their favorite EVs in a variety of biological systems. I hope that the arrays of methods displayed will provide the essentials to conduct systematic assays toward more in-depth understanding of the mode of assembly, secretion, and targeting of EVs to serve eventually as new therapeutic openings.

I would like to thank John Walker for giving me the opportunity to edit this book and all the authors for their contributions. I also wish to thank Dr. Ulla Saarela and Dr. Anatoliy Samoylenko from my team for their time and input on assembling this methods book.

Oulu, Finland ***Seppo Vainio***

Contents

Preface . . . . *v*
Contributors . . . . *ix*

Part I Purification and Analysis of Single or Multiple EVs

1 Plasmon-Enhanced Characterization of Single Extracellular Vesicles 3
Mi Ho Jeong, Taehwang Son, and Hyungsoon Im

2 Detection of Cell-Derived Exosomes Via Surface-Enhanced Raman Scattering Using Aggregated Silver Nanoparticles 15
Yang Li, Yunpeng Wang, Jinwei Tian, and Jian-An Huang

3 In Vivo Analysis of Heterogeneous Extracellular Vesicles Using a Red-Shifted Bioluminescence Resonance Energy Transfer Reporter Protein 23
Gloria I. Perez, Michael H. Bachmann, and Masamitsu Kanada

4 Characterization of Extracellular Vesicles by Transmission Electron Microscopy and Immunolabeling Electron Microscopy 33
Maribel Lara Corona, Ilse Hurbain, Graça Raposo, and Guillaume van Niel

5 Extracellular Vesicle Isolation by a Tangential-Flow Filtration-Based Large-Scale Purification Method 45
Runjie Yuan, Yijun Zhou, Gabriel F. Arias, and Dirk P. Dittmer

6 Metabolomics Analysis of Urinary Extracellular Vesicles by Nuclear Magnetic Resonance and Liquid Chromatography–Mass Spectrometry 57
Marta Martin-Lorenzo, Dolores Molero, and Gloria Alvarez-Llamas

7 Methodologies for Scalable Production of High-Quality Purified Small Extracellular Vesicles from Conditioned Medium 69
Gabriella Andriolo, Elena Provasi, Andrea Brambilla, Stefano Panella, Sabrina Soncin, Viviana Lo Cicero, Marina Radrizzani, Lucia Turchetto, and Lucio Barile

8 Automated On-Line Isolation and Fractionation Method for Subpopulations of Extracellular Vesicles 99
Susanne K. Wiedmer, Evgen Multia, Thanaporn Liangsupree, and Marja-Liisa Riekkola

Part II Targeting Cell Behavior Functions of EVs

9 A Novel Assay for Investigating the Role of Exosomes in Tumor Cell-Endothelial Cell Crosstalk 111
Yan Qiu, Wenli Jiang, and Ye Zeng

10 Nanoparticle (NP) Loading by Direct Incubation with Extracellular Vesicles-Secretor Cells: NP Encapsulation and Exosome Characterization ... 121
Ana Martín-Pardillos and Pilar Martín-Duque

11 Two Complementary Strategies to Quantitate Extracellular Vesicle Uptake Using Bioluminescence and Non-Lipidic Dyes ... 133
Víctor Toribio and María Yáñez-Mó

12 Integrin-Mediated Exosomal Homing to Organs ... 145
Eun Jeong Park and Motomu Shimaoka

13 Methods and Protocols for Using Extracellular Vesicles as Delivery Vehicles in Neuronal Research ... 159
Manjusha Vaidya, Nasif Sayeed, Caroline Hobson, Sandeep Sreerama, Jonhoi Smith, Riya Shah, and Kiminobu Sugaya

14 Quantitative Analysis of Extracellular Vesicle Release Using Artificial MicroRNAs ... 191
Albert Lu

PART III SYSTEMS AND MODELS TO STUDY EVs

15 Purification of Bacterial-Enriched Extracellular Vesicle Samples from Feces by Density Gradient Ultracentrifugation ... 211
Nadiya Byts, Olha Makieieva, Artem Zhyvolozhnyi, Genevieve Bart, Johanna Korvala, Jenni Hekkala, Sonja Salmi, Anatoliy Samoylenko, and Justus Reunanen

16 Isolation and Characterization of Extracellular Vesicles from Lymphocytes ... 227
Lujain Al-Ghadir and Zhi Chen

17 Biogenesis of Mesoporous Silica Nanoparticles Enclosed in Extracellular Vesicles by Mouse Renal Adenocarcinoma Cells ... 241
Feby Wijaya Pratiwi, Keerthanaa Balasubramanian Shanthi, Olha Makieieva, Zih An Chen, Artem Zhyvolozhnyi, Ilkka Miinalainen, Genevieve Bart, Anatoliy Samoylenko, and Si-Han Wu

18 Magnetic Separation of Cell-Secreted Vesicles with Tailored Magnetic Particles and Downstream Applications ... 257
Mireia Bernuz, Arnau Pallarès-Rusiñol, Rosanna Rossi, Carolina Fernández-Senac, Mercè Martí, and María Isabel Pividori

19 Cilia-Derived Extracellular Vesicles in *Caenorhabditis Elegans*: In Vivo Imaging and Quantification of Extracellular Vesicle Release and Capture ... 277
Adrià Razzauti, Teresa Lobo, and Patrick Laurent

20 Exosome-Based COVID-19 Vaccine ... 301
Jaeyoung Kim and Nikita Thapa

Index ... *313*

Contributors

LUJAIN AL-GHADIR • *Faculty of Biochemistry and Molecular Medicine, University of Oulu, Oulu, Finland*

GLORIA ALVAREZ-LLAMAS • *Immunology Department, IIS-Fundacion Jimenez Diaz, Madrid, Spain; RICORS2040, Fundacion Jimenez Diaz, Madrid, Spain; Biochemistry and Molecular Biology Department, Universidad Complutense, Madrid, Spain*

GABRIELLA ANDRIOLO • *Lugano Cell Factory, Istituto Cardiocentro Ticino, Ente Ospedaliero Cantonale, Lugano, Switzerland*

GABRIEL F. ARIAS • *Department of Biochemistry and Biophysics, The University of North Carolina, Chapel Hill, NC, USA*

MICHAEL H. BACHMANN • *Institute for Quantitative Health Science and Engineering (IQ), Department of Microbiology and Molecular Genetics, Michigan State University, East Lansing, MI, USA*

LUCIO BARILE • *Cardiovascular Theranostics, Istituto Cardiocentro Ticino, Laboratories for Translational Research, Ente Ospedaliero Cantonale, Bellinzona, Switzerland; Faculty of Biomedical Sciences, Università Svizzera italiana, Lugano, Switzerland*

GENEVIEVE BART • *Laboratory of Developmental Biology, Disease Networks Research Unit, Faculty of Biochemistry and Molecular Medicine, University of Oulu and Kvantum Institute, Oulu, Finland*

MIREIA BERNUZ • *Grup de Sensors i Biosensors, Departament de Química, Universitat Autònoma de Barcelona, Bellaterra, Spain; Biosensing and Bioanalysis Group, Institute of Biotechnology and Biomedicine, Bellaterra, Spain*

ANDREA BRAMBILLA • *Lugano Cell Factory, Istituto Cardiocentro Ticino, Ente Ospedaliero Cantonale, Lugano, Switzerland*

NADIYA BYTS • *Biocenter Oulu & Cancer and Translational Medicine Research Unit, Faculty of Medicine, University of Oulu, Oulu, Finland*

ZHI CHEN • *Faculty of Biochemistry and Molecular Medicine, University of Oulu, Oulu, Finland*

ZIH AN CHEN • *Department of Chemistry, National Taiwan University, Taipei, Taiwan*

VIVIANA LO CICERO • *Lugano Cell Factory, Istituto Cardiocentro Ticino, Ente Ospedaliero Cantonale, Lugano, Switzerland*

MARIBEL LARA CORONA • *Université de Paris, Institute of Psychiatry and Neuroscience of Paris (IPNP), Paris, France*

DIRK P. DITTMER • *Department of Microbiology and Immunology, The University of North Carolina, Chapel Hill, NC, USA*

JENNI HEKKALA • *Biocenter Oulu & Cancer and Translational Medicine Research Unit, Faculty of Medicine, University of Oulu, Oulu, Finland*

CAROLINE HOBSON • *Burnett School of Biomedical Sciences, College of Medicine, University of Central Florida, Orlando, FL, USA*

JIAN-AN HUANG • *Faculty of Medicine, Faculty of Biochemistry and Molecular Medicine, University of Oulu, Oulu, Finland*

ILSE HURBAIN • *Institut Curie, PSL Research University, CNRS, UMR144, Cell and Tissue Imaging Facility(PICT-IBiSA), Paris, France*

HYUNGSOON IM • *Center for Systems Biology, Massachusetts General Hospital, Boston, MA, USA*
MI HO JEONG • *Center for Systems Biology, Massachusetts General Hospital, Boston, MA, USA*
WENLI JIANG • *Institute of Biomedical Engineering, West China School of Basic Medical Sciences and Forensic Medicine, Sichuan University, Chengdu, China*
MASAMITSU KANADA • *Institute for Quantitative Health Science and Engineering (IQ), Department of Pharmacology and Toxicology, Michigan State University, East Lansing, MI, USA*
JAEYOUNG KIM • *CK Exogene, Inc., Seongnam, Gyeonggi do, Republic of Korea*
JOHANNA KORVALA • *Biocenter Oulu & Cancer and Translational Medicine Research Unit, Faculty of Medicine, University of Oulu, Oulu, Finland*
PATRICK LAURENT • *Laboratory of Neurophysiology, ULB Neuroscience Institute (UNI), Université Libre de Bruxelles (ULB), Bruxelles, Belgium*
YANG LI • *College of Pharmacy, Harbin Medical University, Harbin, Heilongjiang, China*
THANAPORN LIANGSUPREE • *Department of Chemistry, University of Helsinki, Helsinki, Finland*
TERESA LOBO • *Laboratory of Neurophysiology, ULB Neuroscience Institute (UNI), Université Libre de Bruxelles (ULB), Bruxelles, Belgium*
ALBERT LU • *Departament de Biomedicina, Unitat de Biologia Cel·lular, Facultat de Medicina i Ciències de la Salut, Centre de Recerca Biomèdica CELLEX, Institut d'Investigacions Biomèdiques August Pi i Sunyer, Universitat de Barcelona, Barcelona, Spain*
OLHA MAKIEIEVA • *Laboratory of Developmental Biology, Disease Networks Research Unit, Faculty of Biochemistry and Molecular Medicine, University of Oulu and Kvantum Institute, Oulu, Finland*
MERCÈ MARTÍ • *Biosensing and Bioanalysis Group, Institute of Biotechnology and Biomedicine, Bellaterra, Spain*
PILAR MARTÍN-DUQUE • *Instituto de Investigaciones Sanitarias de Aragón (IIS Aragón), Zaragoza, Spain; Ciber Bioingeniería y Biomateriales (CIBER-BBN), Instituto de Salud Carlos III, Madrid, Spain; Surgery Department, Medicine Medical School, University of Zaragoza, Zaragoza, Spain*
MARTA MARTIN-LORENZO • *Immunology Department, IIS-Fundacion Jimenez Diaz, Madrid, Spain*
ANA MARTÍN-PARDILLOS • *Instituto de Nanociencia y Materiales de Aragon (INMA), CSIC-Universidad de Zaragoza, Zaragoza, Spain; Department of Chemical Engineering and Environmental Technology (IQTMA), University of Zaragoza, Zaragoza, Spain; Instituto de Investigaciones Sanitarias de Aragón (IIS Aragón), Zaragoza, Spain*
ILKKA MIINALAINEN • *Biocenter Oulu, Department of Pathology, Oulu University Hospital, University of Oulu, Oulu, Finland*
DOLORES MOLERO • *CAI-NMR, Universidad Complutense, Madrid, Spain*
EVGEN MULTIA • *Department of Chemistry, University of Helsinki, Helsinki, Finland*
GUILLAUME VAN NIEL • *Institut Curie, PSL Research University, CNRS, UMR144, Cell and Tissue Imaging Facility(PICT-IBiSA), Paris, France; GHU Paris Psychiarie et Neurosciences, Hôpital Sainte Anne, Paris, France*

ARNAU PALLARÈS-RUSIÑOL • *Grup de Sensors i Biosensors, Departament de Química, Universitat Autònoma de Barcelona, Bellaterra, Spain; Biosensing and Bioanalysis Group, Institute of Biotechnology and Biomedicine, Bellaterra, Spain*
STEFANO PANELLA • *Cardiovascular Theranostics, Istituto Cardiocentro Ticino, Laboratories for Translational Research, Ente Ospedaliero Cantonale, Bellinzona, Switzerland*
EUN JEONG PARK • *Department of Molecular Pathobiology and Cell Adhesion Biology, Mie University Graduate School of Medicine, Tsu, Mie, Japan*
GLORIA I. PEREZ • *Institute for Quantitative Health Science and Engineering (IQ), College of Osteopathic Medicine, Michigan State University, East Lansing, MI, USA*
MARÍA ISABEL PIVIDORI • *Grup de Sensors i Biosensors, Departament de Química, Universitat Autònoma de Barcelona, Bellaterra, Spain; Biosensing and Bioanalysis Group, Institute of Biotechnology and Biomedicine, Bellaterra, Spain*
FEBY WIJAYA PRATIWI • *Laboratory of Developmental Biology, Disease Networks Research Unit, Faculty of Biochemistry and Molecular Medicine, University of Oulu and Kvantum Institute, Oulu, Finland*
ELENA PROVASI • *Lugano Cell Factory, Istituto Cardiocentro Ticino, Ente Ospedaliero Cantonale, Lugano, Switzerland*
YAN QIU • *Institute of Biomedical Engineering, West China School of Basic Medical Sciences and Forensic Medicine, Sichuan University, Chengdu, China*
MARINA RADRIZZANI • *Lugano Cell Factory, Istituto Cardiocentro Ticino, Ente Ospedaliero Cantonale, Lugano, Switzerland*
GRAÇA RAPOSO • *Institut Curie, PSL Research University, CNRS, UMR144, Cell and Tissue Imaging Facility(PICT-IBiSA), Paris, France*
ADRIÀ RAZZAUTI • *Laboratory of Neurophysiology, ULB Neuroscience Institute (UNI), Université Libre de Bruxelles (ULB), Bruxelles, Belgium*
JUSTUS REUNANEN • *Biocenter Oulu & Cancer and Translational Medicine Research Unit, Faculty of Medicine, University of Oulu, Oulu, Finland*
MARJA-LIISA RIEKKOLA • *Department of Chemistry, University of Helsinki, Helsinki, Finland*
ROSANNA ROSSI • *Grup de Sensors i Biosensors, Departament de Química, Universitat Autònoma de Barcelona, Bellaterra, Spain; Biosensing and Bioanalysis Group, Institute of Biotechnology and Biomedicine, Bellaterra, Spain*
SONJA SALMI • *Biocenter Oulu & Cancer and Translational Medicine Research Unit, Faculty of Medicine, University of Oulu, Oulu, Finland; Laboratory of Developmental Biology, Disease Networks Research Unit, Faculty of Biochemistry and Molecular Medicine, University of Oulu and Kvantum Institute, Oulu, Finland*
ANATOLIY SAMOYLENKO • *Laboratory of Developmental Biology, Disease Networks Research Unit, Faculty of Biochemistry and Molecular Medicine, University of Oulu and Kvantum Institute, Oulu, Finland*
NASIF SAYEED • *Burnett School of Biomedical Sciences, College of Medicine, University of Central Florida, Orlando, FL, USA*
RIYA SHAH • *Burnett School of Biomedical Sciences, College of Medicine, University of Central Florida, Orlando, FL, USA*
KEERTHANAA BALASUBRAMANIAN SHANTHI • *Laboratory of Developmental Biology, Disease Networks Research Unit, Faculty of Biochemistry and Molecular Medicine, University of Oulu and Kvantum Institute, Oulu, Finland*

MOTOMU SHIMAOKA • *Department of Molecular Pathobiology and Cell Adhesion Biology, Mie University Graduate School of Medicine, Tsu, Mie, Japan*
JONHOI SMITH • *Burnett School of Biomedical Sciences, College of Medicine, University of Central Florida, Orlando, FL, USA*
TAEHWANG SON • *Center for Systems Biology, Massachusetts General Hospital, Boston, MA, USA*
SABRINA SONCIN • *Lugano Cell Factory, Istituto Cardiocentro Ticino, Ente Ospedaliero Cantonale, Lugano, Switzerland*
SANDEEP SREERAMA • *Burnett School of Biomedical Sciences, College of Medicine, University of Central Florida, Orlando, FL, USA*
KIMINOBU SUGAYA • *Burnett School of Biomedical Sciences, College of Medicine, University of Central Florida, Orlando, FL, USA*
NIKITA THAPA • *CK Exogene, Inc., Seongnam, Gyeonggi do, Republic of Korea*
JINWEI TIAN • *Department of Cardiology, The Second Affiliated Hospital of Harbin Medical University; The Key Laboratory of Myocardial Ischemia, Ministry of Education, Harbin Medical University, Harbin, Heilongjiang, China*
VÍCTOR TORIBIO • *Departamento de Biología Molecular, Instituto Universitario de Biología Molecular (IUBM), Universidad Autónoma de Madrid (UAM), Madrid, Spain; Centro de Biología Molecular Severo Ochoa (CBMSO), Instituto de Investigación Sanitaria La Princesa (IIS-IP), Madrid, Spain*
LUCIA TURCHETTO • *Lugano Cell Factory, Istituto Cardiocentro Ticino, Ente Ospedaliero Cantonale, Lugano, Switzerland*
MANJUSHA VAIDYA • *Burnett School of Biomedical Sciences, College of Medicine, University of Central Florida, Orlando, FL, USA*
YUNPENG WANG • *College of Pharmacy, Harbin Medical University, Harbin, Heilongjiang, China*
SUSANNE K. WIEDMER • *Department of Chemistry, University of Helsinki, Helsinki, Finland*
SI-HAN WU • *Graduate Institute of Nanomedicine and Medical Engineering, Taipei Medical University, Taipei, Taiwan*
MARÍA YÁÑEZ-MÓ • *Departamento de Biología Molecular, Instituto Universitario de Biología Molecular (IUBM), Universidad Autónoma de Madrid (UAM), Madrid, Spain; Centro de Biología Molecular Severo Ochoa (CBMSO), Instituto de Investigación Sanitaria La Princesa (IIS-IP), Madrid, Spain*
RUNJIE YUAN • *Department of Microbiology and Immunology, The University of North Carolina, Chapel Hill, NC, USA*
YE ZENG • *Institute of Biomedical Engineering, West China School of Basic Medical Sciences and Forensic Medicine, Sichuan University, Chengdu, China*
YIJUN ZHOU • *Department of Microbiology and Immunology, The University of North Carolina, Chapel Hill, NC, USA*
ARTEM ZHYVOLOZHNYI • *Laboratory of Developmental Biology, Disease Networks Research Unit, Faculty of Biochemistry and Molecular Medicine, University of Oulu and Kvantum Institute, Oulu, Finland*

Part I

Purification and Analysis of Single or Multiple EVs

Chapter 1

Plasmon-Enhanced Characterization of Single Extracellular Vesicles

Mi Ho Jeong, Taehwang Son, and Hyungsoon Im

Abstract

Extracellular vesicles (EVs) represent heterogeneous populations of membrane-bound vesicles shed from almost all kinds of cells. Although superior to conventional methods, most newly developed EV sensing platforms still require a certain number of EVs, measuring bulk signals from a group of vesicles. A new analytical approach that enables single EV analysis could be extremely valuable for understanding EVs' subtypes, heterogeneity, and production dynamics during disease development and progression. Here, we describe a new nanoplasmonic sensing platform for sensitive single EV analysis. Termed nPLEX-FL (nanoplasmonic EV analysis with enhanced fluorescence detection), the system amplifies EVs' fluorescence signals using periodic gold nanohole structures, enabling sensitive, multiplexed analysis of single EVs.

Key words Surface plasmon resonance, Plasmon-enhanced fluorescence, Biosensing, Extracellular vesicles, Molecular analysis

1 Introduction

Fluorescence detection is a promising method for single EV characterization. Fluorescence imaging enables the visualization of individual EVs captured on a solid substrate [1]. Advances in nanoflow cytometry bring the resolution down to sub-100 nm for detecting small EVs. Multiplexed molecular profiling of single EVs, however, often requires sophisticated multi-step signal amplification strategies, such as branched DNAs [2, 3] or enzymatic reactions [4], otherwise undetected by weak signals [5]. This is because the majority population of EVs is small vesicles (<200 nm) with limited numbers of epitopes and surface areas for antibody labeling (i.e., weak fluorescence signals). Therefore, developing simple yet robust signal amplification strategies is critical to boost signals generated from individual EVs.

Mi Ho Jeong and Taehwang Son contributed equally with all other contributors.

Seppo Vainio (ed.), *Cell-Secreted Vesicles: Methods and Protocols*, Methods in Molecular Biology, vol. 2668,
https://doi.org/10.1007/978-1-0716-3203-1_1,

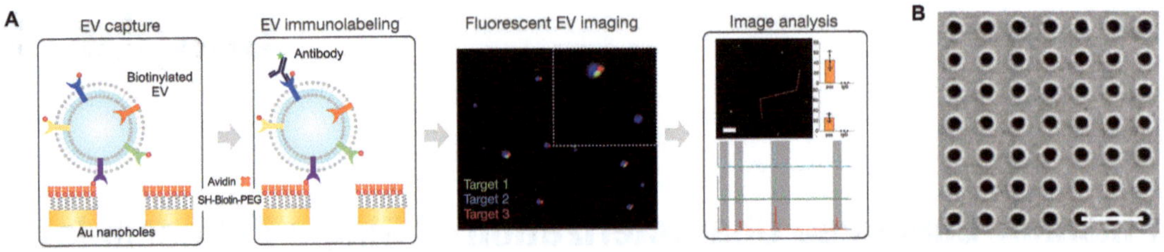

Fig. 1 Overview of nano-plasmonic EV analysis with enhanced fluorescence detection (nPLEX-FL). (**a**) Biotinylated EV are captured on avidin-coated gold nanohole surface. The captured EVs are immunostained by antibodies for target EV surface proteins, followed by fluorophore-conjugated secondary antibody labeling. The fluorescently labeled EVs are imaged by a fluorescence microscope, and images are analyzed to detect individual EVs and measure target protein levels from fluorescence intensities. (**b**). A scanning electron micrograph of gold nanoholes. The hole diameter is 200 nm, and the periodicity is 500 nm. Scale bar: 1 μm. (Reproduced from Ref. [5] with permission from John Wiley and Sons)

The nano-plasmonic EV analysis with enhanced fluorescence detection (nPLEX-FL) is based on plasmonic metallic nanostructures that amplify fluorescence signals of EVs captured on the nanostructure surface [5]. The approach is simple and does not require additional chemical reactions or amplification processes, as the enhancement is achieved simply by using a gold substrate with nanostructures instead of glass slides or other plain substrates. In particular, the nPLEX-FL sensor comprises periodic gold nanohole arrays in an optically opaque gold film. The periodic nanohole arrays support surface plasmon resonances extended in a long range, making them more suitable for EVs than other nanostructures supporting localized resonances [6]. The detection is done by capturing target EVs on the gold nanohole surface and labeling them using fluorescently conjugated antibodies (*see* Fig. 1). The underlying gold nanohole arrays then amplify EVs' fluorescence signals without any additional procedures, and thus the assay is compatible with most conventional fluorescence microscopes. To achieve maximum enhancement, it is recommended to use a fluorophore having a good overlap with the resonant wavelength of gold nanoholes (*see* Fig. 2).

In this chapter, we detail the nPLEX-FL sensing method, including EV isolation, chip fabrication, assay procedures, and image analysis.

2 Materials

The following materials are necessary to isolate EVs, biotinylate EVs, prepare nPLEX-FL chips, and detect EVs using a fluorescence microscope. All materials are prepared and stored at room temperature unless manufacturers recommend other storage conditions. For example, the superblock solution is recommended to store at 4 °C. Proper storage at recommended temperatures is critical for antibodies to preserve their affinity.

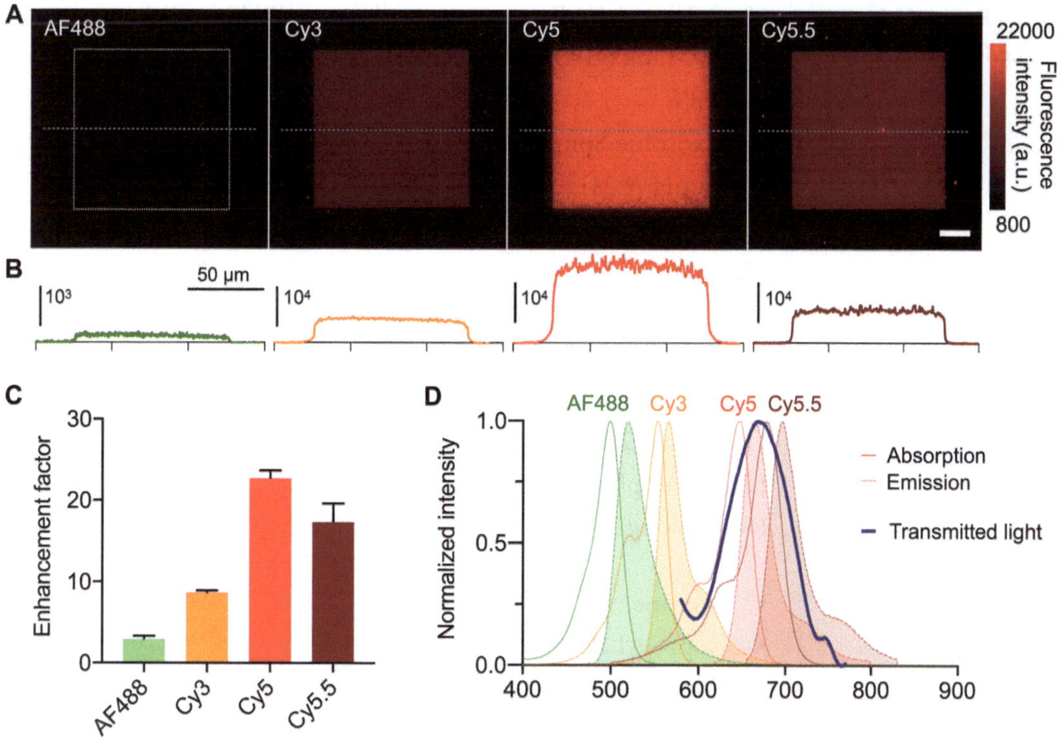

Fig. 2 Plasmon-enhanced fluorescence intensity in the nPLEX-FL system. (**a**). Four different fluorophore-conjugated streptavidin (streptavidin with AF488, Cy3, Cy5, or Cy5.5) coated on nPLEX-FL chips showed the distinguishing enhancement of fluorescence depending on the absorption and emission spectra. Scale bar: 20 μm. Au nanoholes are made in the 100 × 100 μm^2 sized square area highlighted by a white dashed box. (**b**) The cross-sectional intensity profiles along the blue dashed lines of images in A. (**c**) Enhancement factors of fluorescence intensity on nanohole area compared to flat Au area depending on the fluorophores. (**d**) The overlaid spectrum of plasmon-supported light transmission on periodic nanoholes. (Reproduced from Ref. [5] with permission from John Wiley and Sons)

2.1 Extracellular Vesicle Isolation

1. Fetal Bovine Serum, exosome-depleted (Gibco, A2720801) for EV isolation from cell culture (*see* **Note 1**).
2. Cell Strainer, 40 μm.
3. Syringe Filter, 0.22 μm.
4. Centricon Plus-70 Centrifugal Filter (Millipore Sigma, UFC701008).
5. Amicon Ultra-2 Centrifugal Filter (Millipore Sigma, UFC201024).
6. Sepharose CL-4B (Cytiva, 17,015,001).
7. 10 mL syringe.
8. 11 μm Nylon Net Filter, 25 mm (Millipore Sigma, NY1102500).
9. Syringe Tip Caps (Covidien, 8,881,682,085).
10. PBS (*see* **Note 2**).

2.2 Reagents for nPLEX-FL Assays

1. 0.2% and 1% Bovine serum albumin (BSA) in PBS.
2. EZ-Link Sulfo-NHS-LC-Biotin.
3. PBS.
4. Exosome Spin Columns MW 3000 (Invitrogen, 44-844-49).
5. BupH™ MES Buffered Saline.
6. Biotin PEG Thiol (NANOCS, PG2-BNTH-1 k).
7. Neutravidin.
8. Fixation/Permeabilization solution.
9. Antibodies.
10. Superblock™ Blocking Buffer (Thermo Scientific, 37,515).
11. Tween 20.
12. Fluorophore-conjugated streptavidin.
13. Prolong™ Gold Antifade Mountant.

2.3 Fabrication for nPLEX-FL Chips

- Acetone.
- Isopropyl alcohol.
- Test grade silicon wafer.
- Electron beam evaporator.

2.4 nPLEX Measurement System

1. Epi-fluorescence upright microscope.
2. Objective lens.
3. Fluorescence filter sets (DAPI, GFP, TXRED, Cy3, Cy5, Cy5.5).
4. Scientific sCMOS monochrome camera.
5. Deionized water.

2.5 Image Analysis

1. ImageJ Fiji Program

3 Methods

All procedures are performed at room temperature unless otherwise specified.

3.1 Extracellular Vesicle Isolation

1. For EV isolation from cell lines, change the cell culture medium with a new one with 1% exosome-depleted FBS and incubate for 48 h (*see* **Note 3**).
2. Collect the conditioned medium and pass through a 40 μm cell strainer to get rid of floating cells.

3. Centrifuge the conditioned medium at 300 × *g* for 10 min to get rid of cells and apoptotic bodies (*see* **Note 4**).
4. Load the processed medium onto a Centricon Plus-70 Centrifugal filter and centrifuge the filter with 3500 × *g* for 30 min at 4 °C to concentrate the medium. After centrifugation, invert the filter and centrifuge at 1000 × *g* for 2 min at 4 °C to recover EVs captured on the filter.
5. Add PBS to the concentrated EV solution to be a final volume of 1 mL.
6. Add 1 mL of the concentrate onto the top of the size exclusion chromatography (SEC) column (*see* **Note 5**) and then add 6 mL of PBS (*see* **Note 6**).
7. Discard the fractions 1–3 and collect the fractions 4 and 5 for EVs (*see* **Note** 7).
8. Concentrate the isolated EVs using an Amino Ultra-2-Centrifugal Filter and centrifuging at 3500 × *g* for 30 min at 4 °C. After centrifugation, invert the filter and centrifuge at 1000 × *g* for 2 min at 4 °C to recover captured EVs from the filter.
9. Aliquot an appropriate volume of EVs for nanoparticle tracking analysis (or other) to measure EV concentration and size distribution (*see* **Note 8**).

3.2 Biotinylation of Extracellular Vesicles

1. Dilute ~10^{10} EVs using 0.2% BSA in PBS to make a 100 μL solution as a final volume.
2. Add 3.3 μL of Sulfo-NHS-LC-Biotin 10 mM in PBS to the EV solution to make the final concentration of Sulfo-NHS-LC-Biotin of 330 μM and incubate for 1 h.
3. Remove unbound biotin with an Exosome Spin Column MW 3000.
4. Filter the biotinylated EVs using a 0.22 μm syringe filter to get rid of aggregated particles (*see* **Note 9**).

3.3 nPLEX-FL Chip Fabrication

1. Prepare or obtain a Si wafer with periodic nanohole patterns with 200 nm in diameter and 500 nm in periodicity (*see* Fig. 3, *see* **Notes 10** and **11**).
2. Clean the patterned Si wafer using acetone, followed by isopropyl alcohol for 5 min each.
3. Dry the cleaned wafer with nitrogen stream.
4. Deposit a 100 nm gold film with a 5-nm titanium adhesion layer using an electron-beam evaporator (*see* **Note 12**).
5. Inspect nanopatterns in the fabricated chip by scanning electron microscopy.

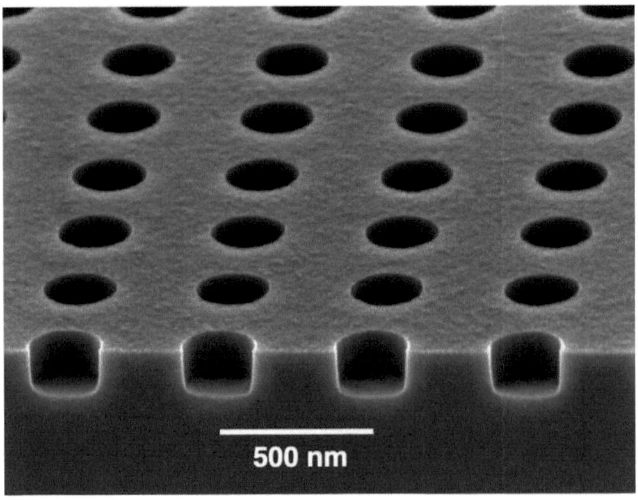

Fig. 3 A scanning electron micrograph of periodic nanohole patterns made in a Si wafer by interference lithography. The hole diameter is 200 nm, and the periodicity is 500 nm

3.4 nPLEX-FL Assay

1. Incubate an nPLEX-FL chip with 10 mM biotinylated thiol-PEG in deionized water overnight.
2. Wash the PEGylated chip with deionized water.
3. Apply 50 μg/mL of neutravidin in PBS with 0.2% bovine serum albumin on the chip surface and incubate for 1 h. Then, wash the chip to remove unbound neutravidin.
4. Apply biotinylated EVs on the chip surface and incubate for 30 min.
5. Wash the chip with PBS.
6. Apply fixation/permeabilization solution to captured EVs for 10 min followed by PBS washing.
7. Incubate the chip surface with Superblock for 20 min (*see* **Notes 13**).
8. Wash the chip with PBS.
9. Dilute primary antibodies in 1% BSA in PBST and incubate them on the chip surface for 20 min.
10. Wash the chip multiple times with PBST to wash out unbound antibodies.
11. Dilute fluorophore-conjugated secondary antibodies in 1% BSA in PBST and incubate them on the chip surface for 10 min (*see* **Note 14**).
12. Wash the chip multiple times with PBST to wash out unbound antibodies.
13. Label captured biotinylated EVs with Cy3-conjugated streptavidin.

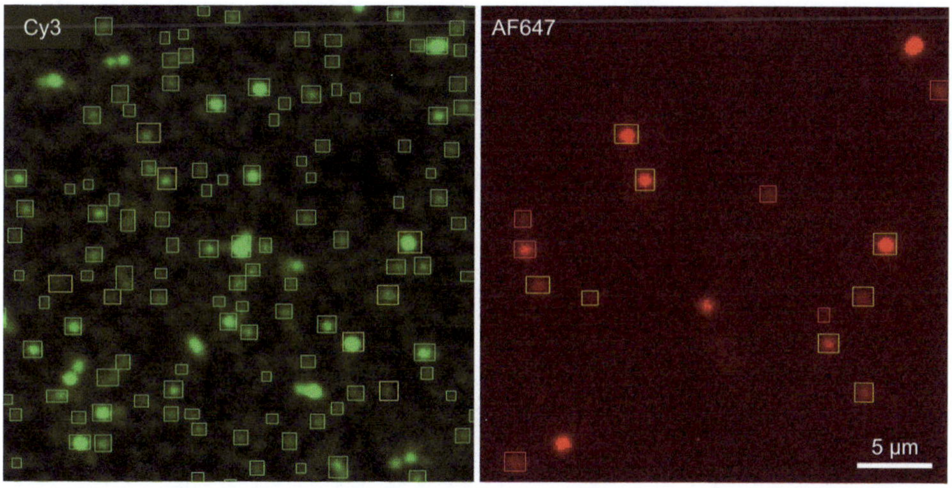

Fig. 4 A representative dual-channel nPLEX-FL fluorescence image. Left: biotinylated EVs are labeled by Cy3-conjugated streptavidin. Right: EVs are labeled by anti-EGFR primary antibodies, followed by AF647-conjugated secondary antibodies

14. Mount the chip on a glass slide.
15. Apply a mounting solution on the chip surface and cover the solution with a coverslip (*see* **Notes 15**).
16. Image labeled EVs using a multi-channel fluorescence microscope (*see* **Note 16**).

3.5 Image Analysis

1. Load multi-channel EV fluorescence images to ImageJ Fiji (*see* Fig. 4, *see* **Note 17**).
2. Assign each channel image into different colors and construct a stacked image under "Image - Color - Merge Channels."
3. Remove background signals using "rolling ball algorithm" under "Process - Substrate Background." (*see* **Note 18**).
4. Detect the EV locations based on the streptavidin channel using the ComDet plugin (*see* **Note 19**).
5. Optimize "Particle size" and "Intensity threshold" values to determine the best value to detect EVs (*see* Fig. 5, *see* **Note 20**).
6. Save the analysis result as a CVS file.
7. Collect the values at column "IntergrIntCh2" (AF488) of Ch1 (Cy3), which is the intensity at the EGFR channel of EVs detected by streptavidin (*see* **Note 21**).
8. Use IgG control experiment data to calculate the threshold. The threshold is defined by mean + 3 × standard deviation (*see* **Note 22**).
9. Count the number of marker-positive EVs above the threshold and calculate the co-localization rate with other markers.

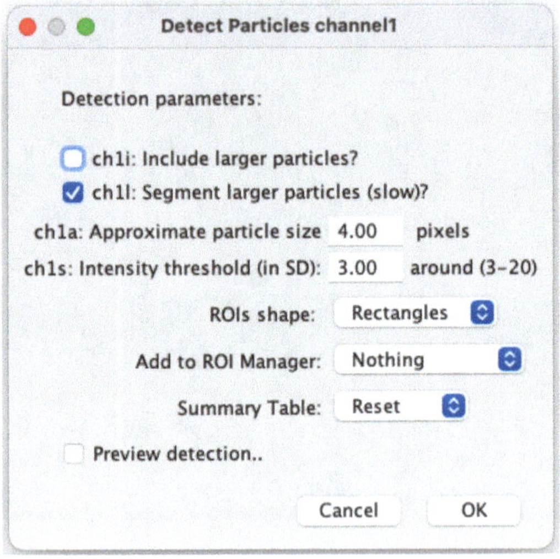

Fig. 5 ImageJ Comdet Plugin. Particle size 4 and intensity threshold 3 are used as a default setting

4 Notes

1. A common FBS solution already contains a large amount of EVs. Therefore, it is recommended to use the exosome-depleted version for EV isolation.
2. PBS could be purchased either by power (Thermo Scientific, 28,372) or stock solution (Thermo Scientific, 28,348). Dissolve the powder or dilute the stock solution using de-ionized (DI) water based on the protocol provided by the manufacturer. Other buffer preparation kits (e.g., BupH™ MES Buffered Saline Packs, Thermo Scientific, 28,390) are also commercially available and can be prepared in the same way unless otherwise specified by a manufacturer.
3. In the case of 150 mm culture dish, about 10 dishes are carried out to collect conditioned medium. It is typically done when the confluence of cells becomes about 70%.
4. After centrifugation, filtering through a 0.22 μm syringe filter can be done to separate small EVs (e.g., exosomes) from large ones (e.g., microvesicles).
5. Size exclusion chromatography columns are commercially available (IZON, qEVoriginal/70 nm) or custom-made. Briefly, put an 11 μm nylon net at the bottom of a 10 mL syringe and fill the syringe with Sepharose CL-4B. Add 5–10 mL of PBS to the syringe packed with sepharose for washing. Finally, add 3 mL of PBS and cap the syringe tip. Seal the syringe with a paraffin film and store it at 4 °C until used. For details, please refer to the section "Method" in [7].

6. The extra PBS is added to keep the column wet and continue the flow.
7. The fraction is 1 mL of solution coming through the size exclusion filter. The fraction is counted after adding the EV sample on the top of the column.
8. Multiple freeze-thawing makes several components in EVs degraded or makes lose the consistency of the assay result. Thus, aliquoting the EV sample is required for maintaining the fresh condition and consistent result.
9. This process is optional to isolate small EVs (e.g., exosomes) from large ones (e.g., microvesicles).
10. Different methods could be used to fabricate the plasmonic nanohole chips. The first-generation chips were fabricated using focused ion-beam milling [6]. The second-generation chips were fabricated using interference lithography followed by anisotropic Si etching to make nanoholes in a suspended silicon nitride membrane [8].
11. Here, we describe the simplest method that does not require sophisticated lithography tools. First, interference lithography on Si wafers was done through LumArray, Inc. (Somerville, MA, USA). Briefly, 200-nm thick, low-stress silicon nitride was first deposited by low-pressure chemical vapor deposition (LPCVD) onto 100 mm diameter Si wafers. The wafers were spin-coated with an anti-reflection coating (ARC) and a negative photoresist. The photoresist was patterned by exposing it with two orthogonal grating images using an Interference Lithography (IL) exposure tool. The holes are transferred through the underlying ARC by reactive ion etching in O_2, and the patterned ARC then acts as an etch mask to transfer the holes into the silicon nitride film by CF_4 reactive ion etching. The remaining ARC etch mask is then removed by a "piranha" cleaning [9].
12. If titanium is not available, chromium can also be used as an adhesion layer material.
13. This is to block the sensing surface with blocking proteins and reduce the non-specific binding of EVs. Other blocking materials, such as bovine serum albumin and fetal bovine serum, could also be used.
14. The available fluorescence filter sets in a fluorescence microscope should be considered in selecting fluorophores. As shown in Fig. 2, the nPLEX-FL chip amplifies fluorescence signals in a certain wavelength range. As a rule of thumb, a key or weak-signal marker is assigned to Cy5 (or AF647) to achieve the maximum signal enhancement.

15. A mounting solution is used to preserve the sample from drying. However, this is an irreversible process. If an additional reaction is necessary, a PBST solution could be applied instead of a mounting solution covered by a coverslip.
16. Optimal exposure time varies by several factors, including the choice of fluorophores, light source intensity, the efficiency of fluorescence filters, and camera sensitivity, among others. Therefore, it is recommended to test different exposure times and find one showing the highest signal-to-noise ratio.
17. In this study, we used ImageJ Fiji and its plugin. Other image analysis software (e.g., cell profiler) or custom-developed algorithms in Matlab, Python, etc., could also be used.
18. We use the Rolling ball radius of 20 for image analysis, but the optimized value could be different by the imaging condition.
19. The comet plugin and its instruction can be found at https://github.com/ekatrukha/ComDet.
20. We used particle size 4 and intensity threshold 3, which worked well under our experimental condition.
21. Here, we assume EVs are stained by Cy3-conjugated streptavidin and anti-EGFR antibody with AF488 to obtain EGFR signal levels (*see* Fig. 4). The streptavidin labeling can be replaced by antibody-based fluorescence labeling targeting antigens enriched in EVs, such as CD9, CD63, and CD81. The procedure first identifies streptavidin-positive EVs and calculates EGFR signals on those EVs.
22. Isotype IgG control experiments are done to set threshold values.

Acknowledgments

This work was supported by National Institute of Health grants (R21CA217662, R01GM138778) the Technology Innovation Program (20008829) funded by the Ministry of Trade, Industry, and Energy, Republic of Korea. M.H.J. is supported by the Basic Science Research Program through the National Research Foundation of Korea (NRF), funded by the Ministry of Education (NRF-2021R1A6A3A14039686).

References

1. Lee K, Fraser K, Ghaddar B, Yang K, Kim E, Balaj L, Chiocca EA, Breakefield XO, Lee H, Weissleder R (2018) Multiplexed profiling of single extracellular vesicles. ACS Nano 12:494–503
2. Wu D, Yan J, Shen X, Sun Y, Thulin M, Cai Y, Wik L, Shen Q, Oelrich J, Qian X, Dubois KL, Ronquist KG, Nilsson M, Landegren U, Kamali-Moghaddam M (2019) Profiling surface proteins on individual exosomes using a proximity barcoding assay. Nat Commun 10:3854
3. Shen W, Guo K, Adkins GB, Jiang Q, Liu Y, Sedano S, Duan Y, Yan W, Wang SE, Bergersen K, Worth D, Wilson EH, Zhong W

(2018) A single extracellular vesicle (EV) flow cytometry approach to reveal EV heterogeneity. Angew Chem Int Ed Engl 57:15675–15680
4. Yang KS, Ciprani D, O'Shea A, Liss AS, Yang R, Fletcher-Mercaldo S, Mino-Kenudson M, Fernández-Del Castillo C, Weissleder R (2021) Extracellular vesicle analysis allows for identification of invasive IPMN. Gastroenterology 160: 1345–1358.e11
5. Min J, Son T, Hong JS, Cheah PS, Wegemann A, Murlidharan K, Weissleder R, Lee H, Im H (2020) Plasmon-enhanced biosensing for multiplexed profiling of extracellular vesicles. Adv Biosyst 4:e2000003
6. Im H, Shao H, Park YI, Peterson VM, Castro CM, Weissleder R, Lee H (2014) Label-free detection and molecular profiling of exosomes with a nano-plasmonic sensor. Nat Biotechnol 32:490–495
7. Van Deun J, Jo A, Li H, Lin HY, Weissleder R, Im H, Lee H (2020) Integrated dual-mode chromatography to enrich extracellular vesicles from plasma. Adv Biosyst 4:e1900310
8. Yang KS, Im H, Hong S, Pergolini I, Del Castillo AF, Wang R, Clardy S, Huang CH, Pille C, Ferrone S, Yang R, Castro CM, Lee H, Del Castillo CF, Weissleder R (2017) Multiparametric plasma EV profiling facilitates diagnosis of pancreatic malignancy. Sci Transl Med 9: eaal3226
9. Schmidt HG (2022) Safe piranhas: a review of methods and protocols. ACS Chem Health Safety 29:54–61

Chapter 2

Detection of Cell-Derived Exosomes Via Surface-Enhanced Raman Scattering Using Aggregated Silver Nanoparticles

Yang Li, Yunpeng Wang, Jinwei Tian, and Jian-An Huang

Abstract

Exosomes are small extracellular vesicles that contain RNA, lipids, and proteins and can act as cellular messengers, carrying information to cells and tissues in the body. Thus, sensitive, label-free, and multiplexed analysis of exosomes may help in early diagnosis of important diseases. Here, we describe the process of pretreatment of cell-derived exosomes, preparation of surface-enhanced Raman scattering (SERS) substrates, and label-free SERS detection of exosomes using sodium borohydride aggregators. This method can enable the observation of SERS signals of exosomes that are clear and stable and have a good signal-to-noise ratio.

Key words Exosome, Cell sample, Label-free, SERS, Raman scattering, Silver nanoparticles

1 Introduction

Exosomes are membrane vesicles with a diameter of 30–150 nm and play key roles in life activities such as immune responses and intercellular signaling. Exosomes contain DNA, RNA, and proteins. Their molecular contents reflect the physiological activities of cells, which are potential biomarkers for early diagnosis and prognosis of cancer.

Surface-enhanced Raman spectroscopy (SERS) provides a sensitive, multiplexed, and label-free tool for detecting biomolecules owing to its narrow fingerprint peak information. Plasmonic metal nanostructures such as silver nanoparticles are usually used in label-free SERS methods. Under illumination from a resonant laser, the nanoparticles can induce a localized electromagnetic field called a hot spot. When targeted molecules are added to this hot spot, their Raman signals are significantly enhanced, which is promising in single-molecular analysis. The multiplexing capability of SERS

Seppo Vainio (ed.), *Cell-Secreted Vesicles: Methods and Protocols*, Methods in Molecular Biology, vol. 2668,
https://doi.org/10.1007/978-1-0716-3203-1_2,

offers a great potential for detecting rich molecules on and in exosomes, including membrane proteins, intra-vesicular nucleic acids, and proteins.

However, it is difficult to obtain a SERS signal for nucleic acids or proteins of biomolecules itself by common nanoparticle-based SERS technology [1–14]. For example, silver nanoparticles guided by calcium ions and positively charged silver sol particles reinforced by hydroxylamine hydrochloride failed to identify quadruple-stranded DNA structures known as i-motifs and G-quadruplexes [7–9]. In case of SERS detection of proteins, good SERS signals were difficult to obtain because of various charge levels carried by the protein surfaces [5]. Moreover, the SERS signals acquired by the traditional nanoparticle aggregation method are usually dominated by aromatic amino acid and backbone signals [15].

Here, we used sodium borohydride without Raman activity as an aggregator to prepare silver nanoparticles. Under the action of the aggregator, we detected fingerprint SERS spectra of cell-derived exosomes. This method can not only prevent exosome biomolecules from being oxidized by silver nanoparticles with the addition of excessive aggregators but also enhance SERS signals of different exosomes at the same time.

2 Materials

Prepare all solutions using ultrapure water (prepared by purifying deionized water, to attain a sensitivity of 18 MΩ-cmat 25 °C) and analytical grade reagents. Prepare and store all reagents at room temperature (unless indicated otherwise). Diligently follow all waste disposal regulations when disposing waste materials.

2.1 Preparation of Aggregator

1. 3 M hydrochloric acid solution: Dissolve 25 ml hydrochloric acid (12 M) in 75 mL water.
2. Sodium borohydride solution: Dissolve 0.0378 g sodium borohydride in 19.9 mL water.
3. Add 20 μL of 3 M hydrochloric acid solution to the sodium borohydride solution to prepare the aggregator (*see* **Note 1**).
4. Silver nitrate solution: Dissolve 0.033 g silver nitrate in 5 mL water.

2.2 Western Blotting

1. Cell culture media: Dulbecco's Modified Eagle Medium (DMEM) culture media.
2. SDS electrophoresis buffer: Mix 3.02 g Tris-base, 18.8 g glycine, and 1 g SDS in a volumetric flask and make up to 1 L with water.

3. Trans (10×): Mix 30.3 g Tris-base and 151.1 g glycine in a volumetric flask and make up to 1 L with water.
4. Transfer buffer: Mix 100 ml Trans and 200 mL methanol in a volumetric flask and make up to 1 L with water.
5. Tris-buffered saline (TBS; 10×): Mix 30.3 g Tris-base and 80 g NaCl in a volumetric flask and make up to 1 L with water.
6. TBS containing 0.1% Tween-20 (TBST).

3 Methods

Carry out all procedures at room temperature unless specified otherwise.

3.1 Sample Pretreatment

3.1.1 Preparation of Silver Nanoparticles

1. Dissolve 0.066 g sodium borohydride in 500 mL deionized water and centrifuge for 8 min (1650 rpm). Then pour in the silver nitrate solution and centrifuge for 16 min (2050 rpm) (*see* **Note 2**).
2. Centrifuge the solution at 5200 rpm for 20 min and discard the supernatant to obtain silver nanoparticles (*see* **Note 3**).
3. Store the product away from light (*see* **Note 4**) (*see* Fig. 1).

3.1.2 Cell Culture

1. Culture serum-free cells for tolerance (*see* **Note 5**).
2. Add cells in a cell culture flask and adjust the density of adherent cells to be above 70%.
3. Wash the cells with phosphate buffered saline (PBS) twice.
4. Culture in serum-free medium for 48 h. Store the resulting supernatant (the initial sample) at 4 °C (*see* **Note 6**).

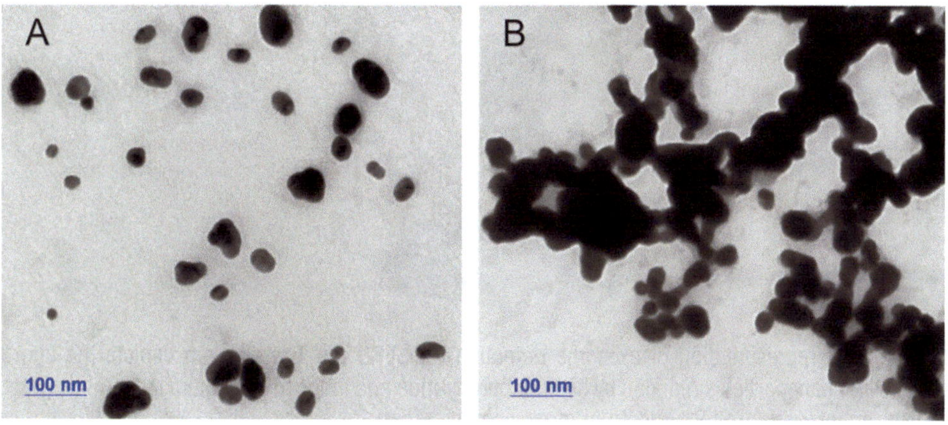

Fig. 1 TEM images of silver nanoparticles (**a**) and aggregated silver nanoparticles (**b**)

3.1.3 Exosome Isolation and Preparation

1. Isolate exosomes from the cell culture media through ultracentrifugation.
2. Centrifuge 4 mL of the initial sample at 300 × *g* for 10 min and isolate the supernatant.
3. Centrifuge at 2000 × *g* for 20 min and isolate the supernatant.
4. Centrifuge at 10000 × *g* for 30 min and isolate the supernatant.
5. Centrifuge at 100000 × *g* for 70 min, discard the supernatant, and re-dissolve the precipitate in 4 mL PBS.
6. Centrifuge at 100000 × *g* for 70 min, discard the supernatant, and re-dissolve the precipitate in 40 μL PBS (*see* **Note 7**).

Store the exosomes at −80 °C for long-term preservation (*see* **Note 8**).

3.2 Identification of Exosomes

1. Identify the isolated exosomes via transmission electron microscopy (TEM), nanoparticle tracking analysis (NTA), and Western blotting.
2. Carry out negative staining of the exosomes before TEM. Characterize the structure and morphology of the exosomes based on the TEM images. Further characterize the size distribution and concentration of exosomes by NTA (*see* **Note 9**).
3. Detect exosome marker proteins and identify exosomes via Western blotting. Commonly used exosomal membrane marker proteins include CD63, CD9, TSG101, and Alix.

3.3 Detection of Exosomes Using the SERS Method (see Fig. 2)

1. Dilute the exosome sample five-fold after pretreatment (*see* **Note 10**).
2. Mix 20 μL of the diluted exosome sample and 10 μL of silver nanoparticles (*see* **Note 11**) (*see* Fig. 3).

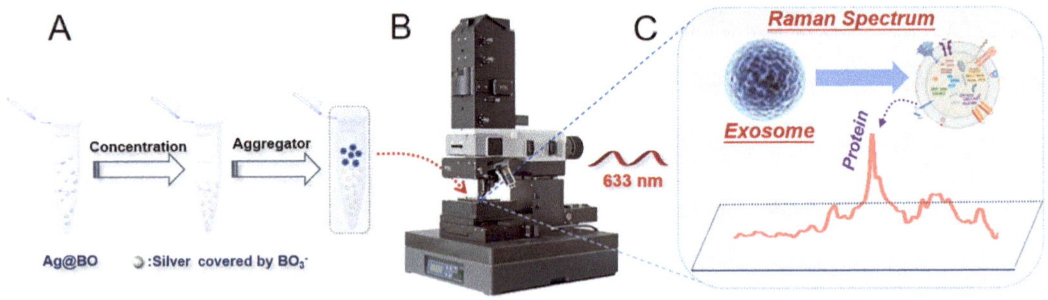

Fig. 2 Schematic representation of exosome detection via SERS. (**a**) The diagram depicts the aggregation process of silver nanoparticles reduced by sodium borohydride and induced by sodium ions; (**b**) diagram of the testing instrument; and (**c**) SERS spectrum of exosomes

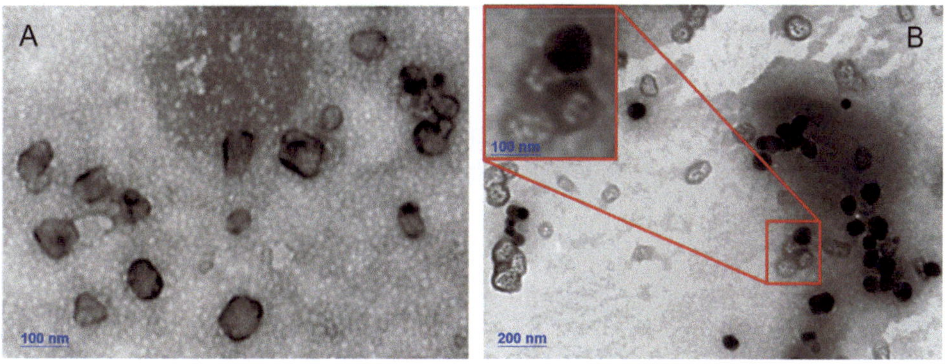

Fig. 3 TEM images of exosomes derived from mouse cardiomyocytes. (**a**) TEM image of exosomes; (**b**) TEM image of silver nanoparticles adsorbed on the surface of exosomes

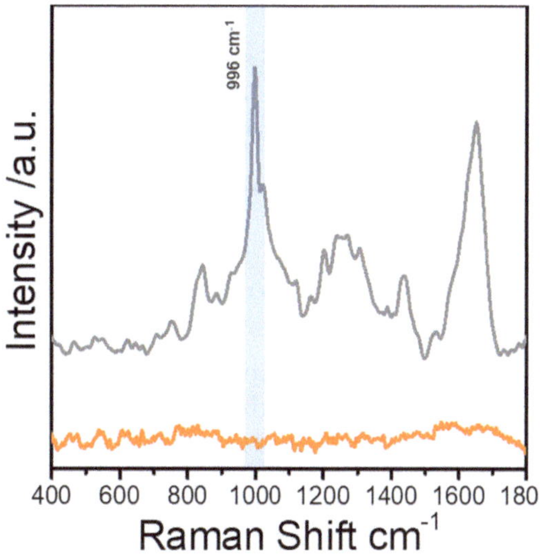

Fig. 4 SERS spectra of exosomes derived from mouse cardiomyocytes in different systems. Raman spectrum of exosomes (orange line); SERS spectrum of silver nanoparticles adsorbed on the surface of exosomes (gray line)

3. Add 5 μL of the newly formulated aggregator into the sample (*see* **Note 12**).
4. Measure SERS using a Laser confocal Raman spectrometer (WITec alpha300 RA, Germany) with an excitation wavelength of 532 nm. Set the parameters as follows: 10 s exposure time and 15 mW laser power. Conduct each test at least three times (*see* **Note 13**) (*see* Figs. 4 and 5).

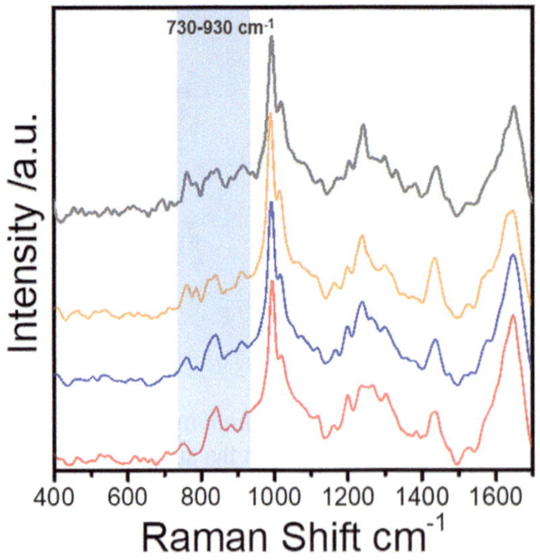

Fig. 5 SERS spectra of exosomes derived from human umbilical vein endothelial cells (gray line), mouse colon cancer cells (yellow line), mouse vascular endothelial smooth muscle cells (blue line), and mouse cardiomyocytes (red line)

4 Notes

1. We found that adding 3 M hydrochloric acid was more effective.
2. Sodium borohydride reacts with water to form borate ions, and then the silver nitrate solution is poured into the sodium borate solution. We found that silver nanoparticles that are produced by the reduction of silver nitrate by excess borate ions have a good detection effect on samples.
3. We found that the aggregated silver nanoparticle-detected samples showed good SERS signals.
4. Silver nanoparticles are best stored away from light and result in aggregation under long exposure to light conditions.
5. If the cells are intolerant to serum-free medium, a medium containing 5% exosome-free serum should be used.
6. Serum-free medium was added to secrete exosomes derived from cell samples under starvation conditions.
7. After centrifugation at 300 × *g*, the supernatant was collected to remove living cells in the culture medium; after centrifugation at 2000 × *g*, the supernatant was collected to remove dead cells in the culture medium; after centrifugation at 10000 × g, the supernatant was collected to remove cell fragments. After centrifugation at 100000 × *g*, the precipitate was re-dissolved in PBS and used to remove residual proteins.

8. Exosomes can be stored at 4 °C for a short period (1–2 days), but RNA significantly degrades. For later use, please store at −80 °C.
9. Phosphomolybdic staining of exosomes is required, otherwise exosomes cannot be observed under transmission electron microscopy.
10. The high concentration of exosomes mixed with enhanced substrate showed "agglomeration," and no SERS signal was observed.
11. Exosomes derived from mouse cardiomyocytes showed complete membrane structure with a diameter of approximately 80 nm (*see* Fig. 3a), and the enhanced nanoparticle could be well adsorbed on the surface of exosomes (*see* Fig. 3b).
12. We find that it is best to prepare this fresh each time.
13. High-laser power will lead to sample damage; if the power is too low, the sample signal strength will be weak. Multiple scanning is preferable for better signal-to-noise ratio.

The Raman spectrum of exosomes derived from mouse cardiomyocytes, (*see* Fig. 4, orange line) indicates that exosome signals were absent. With the addition of enhanced base, SERS signal with good signal-to-noise was detected under the action of the aggregator, and the characteristic peak of protein appeared at 996 cm^{-1} (*see* Fig. 4, gray line). The SERS signals of exosomes derived from human umbilical vein endothelial cells, mouse vascular endothelial smooth muscle cells, mouse colon cancer cells, and mouse cardiomyocytes were further detected. As shown in Fig. 5, the peak shapes were similar, and there were small changes in peak strength and peak position in the range of 730–930 cm^{-1}, which could be identified by statistical analysis.

Acknowledgments

HMU Marshal Initiative Funding (No.: HMUMIF-21012); National Natural Science Foundation for Youth (No.: 82202648); Jian-An Huang acknowledges the DigiHealth-project, a strategic profiling project at the University of Oulu that is supported by the Academy of Finland (project number 326291) and the University of Oulu.

References

1. Han X, Ye LB, Li BZ, Bo G, Cai WJ, Hong Z (2006) Expression, purification and characterization of the Hepatitis B virus entire envelope large protein in Pichia pastoris. Protein Expr Purif 49:168–175
2. Epstein SC, Huff AR, Winesett ES, Londergan CH, Charkoudian LKJNC (2019) Tracking

carrier protein motions with Raman spectroscopy. Nat Commun 10:2227

3. Jakubek RS, Handen J, White SE, Asher SA, Lednev IKJTTiAC (2018) Ultraviolet resonance Raman spectroscopic markers for protein structure and dynamics. Trends Analyt Chem 103:223–229
4. Li D, Xia L, Zhou Q, Wang L, Chen D, Gao X (2020) Label-free detection of miRNA using surface-enhanced Raman spectroscopy. Anal Chem 92:12769–12773
5. Xu L-J, Zong C, Zheng X-S, Hu P, Feng J-M, Ren BJAc (2014) Label-free detection of native proteins by surface-enhanced Raman spectroscopy using iodide-modified nanoparticles. Anal Chem 86:2238–2245
6. Zeng J, Dong M, Zhu B, Gao X, Chen D, Li YJAC (2021) Label-free detection of C–T mutations by surface-enhanced Raman spectroscopy using thiosulfate-modified nanoparticles. Anal Chem 93:1951–1956
7. Breuzard G, Millot J-M, Riou J-F, MJAc M (2003) Selective interactions of ethidiums with G-quadruplex DNA revealed by surface-enhanced Raman scattering. Anal Chem 75: 4305–4311
8. Ghodke HB, Krishnan R, Vignesh K, Kumar GP, Narayana C, Krishnan YJACIE (2007) The I-tetraplex building block: rational design and controlled fabrication of robust 1D DNA scaffolds through non-Watson–crick interactions. Angew Chem Int Ed Engl 46:2646–2649
9. Li Y, Han X, Yan Y, Cao Y, Xiang X, Wang S (2018) Label-free detection of tetramolecular i-motifs by surface-enhanced Raman spectroscopy. Anal Chem 90:2996–3000
10. Yang Y, Peng Y, Lin C, Long L, Hu J, He J (2021) Human ACE2-functionalized gold "virus-trap" nanostructures for accurate capture of SARS-CoV-2 and single-virus SERS detection. Nanomicro Lett 13:1–13
11. Yang S, Dai X, Stogin BB, Wong T-SJPotNAoS (2016) Ultrasensitive surface-enhanced Raman scattering detection in common fluids. Proc Natl Acad Sci U S A 113:268–273
12. Ando J, Asanuma M, Dodo K, Yamakoshi H, Kawata S, Fujita K (2016) Alkyne-tag SERS screening and identification of small-molecule-binding sites in protein. J Am Chem Soc 138: 13901–13910
13. Liu T-Y, Tsai K-T, Wang H-H, Chen Y, Chen Y-H, Chao Y-C (2011) Functionalized arrays of Raman-enhancing nanoparticles for capture and culture-free analysis of bacteria in human blood. Nat Commun 2:1–8
14. Zhou H, Yang D, Ivleva NP, Mircescu NE, Sr S, Niessner R (2015) Label-free in situ discrimination of live and dead bacteria by surface-enhanced Raman scattering. Anal Chem 87:6553–6561
15. Bao Y, Li Y, Ling L, Xiang X, Han X, Zhao B (2020) Label-Free and Highly Sensitive Detection of Native Proteins by Ag IANPs via Surface-Enhanced Raman Spectroscopy. Anal Chem 92:14325–14329

Chapter 3

In Vivo Analysis of Heterogeneous Extracellular Vesicles Using a Red-Shifted Bioluminescence Resonance Energy Transfer Reporter Protein

Gloria I. Perez, Michael H. Bachmann, and Masamitsu Kanada

Abstract

Current methods for characterizing the biodistribution of extracellular vesicles (EVs) are not sensitive enough to track EVs in vivo, despite significant advances over the past decade. Commonly used lipophilic fluorescent dyes are convenient, but lack specificity and yield inaccurate spatiotemporal images in the long-term tracking of EVs. In contrast, protein-based fluorescent or bioluminescent EV reporters have more accurately revealed their distribution in cells and mouse models. Here, we describe a red-shifted bioluminescence resonance energy transfer (BRET) EV reporter, PalmReNL, to analyze the trafficking of small EVs (<200 nm; sEVs) and medium/large EVs (>200 nm; m/lEVs) in mice. Its advantages are that (i) background signals in bioluminescence imaging (BLI) are negligible and (ii) the photons PalmReNL emits have spectral wavelengths longer than 600 nm and can more efficiently penetrate tissues than reporters emitting shorter wavelength light.

Key words Biodistribution, Exosomes, Microvesicles, Extracellular vesicles, sEVs, m/lEVs, Bioluminescence resonance energy transfer, Dual reporter

1 Introduction

Cells release a number of different extracellular vesicles (EVs) that are heterogeneous in size, constitution, and function [1]. Advances in EV research remain restricted by current technical limitations for separating heterogeneous EV populations. Therefore, authors are argued to use operational terms for EV subtypes based on physical characteristics of EVs, such as size and density. Currently, small EVs (sEVs) and medium/large EVs (m/lEVs) are well-accepted classifications according to the Minimal Information for Studies of Extracellular Vesicles (MISEV) 2018 guidelines [2]. Over the past decade, various EV imaging tools have been developed, and the

Seppo Vainio (ed.), *Cell-Secreted Vesicles: Methods and Protocols*, Methods in Molecular Biology, vol. 2668,
https://doi.org/10.1007/978-1-0716-3203-1_3,

studies involved have successfully characterized EV biodistribution in several mouse models [3–6]. However, previous studies on EV biodistribution have only focused on sEVs, but not m/lEVs.

To label EVs, various fluorescent and bioluminescent proteins have been broadly utilized. For in vivo imaging, luciferases with emissions above 600 nm are preferred, because lower wavelength photons are absorbed, mostly by hemoglobin molecules [7]. Luciferases that utilize energy-rich coelenterazine or derivatives as a substrate are ATP-independent, which allows for the monitoring of extracellular events. Bioluminescent EV reporters developed in previous studies were the outcome of using Gaussia luciferase (Gluc; peak emission, 470 nm) or NanoLuc (peak emission, 460 nm) for in vivo imaging [8]. More recently, Wu et al. reported a bioluminescence resonance energy transfer (BRET)-based EV reporter, called PalmGRET [9]. PalmGRET was produced by genetically fusing a palmitoylation signal sequence [10] to an eGFP-NanoLuc BRET reporter, GpNluc [11] (peak emission, 507 nm). Despite its green-shifted emission spectrum, most of its emitted bioluminescent light is absorbed by hemoglobin, with only the tail end of the distribution above 600 nm escaping absorption.

To overcome the limitation in studying EV biodistribution, we developed a novel EV reporter, PalmReNL, in which a palmitoylation signal sequence is genetically fused to one of the brightest red-shifted BRET reporters, Red-eNanoLantern [12] (peak emission, 581 nm), and delivered to cells by non-viral gene transfer using a transposon vector for stable reporter gene expression.

The present protocol provides detailed procedures to: establish PalmReNL-EV producer cells using the *Sleeping beauty* (SB) transposon system [13, 14]; isolate PalmReNL-sEVs and -m/lEVs from conditioned media using a cost-effective filtration method; visualize EVs in vivo in mouse models of disease, and analyze EVs ex vivo. While this protocol was optimized for EVs derived from 4 T1 mouse mammary carcinoma cells, it may be adapted to other tumor cell lines or to primary cells, such as mesenchymal stem cells, for investigating therapeutic EVs.

2 Materials

2.1 Cell Culture and Animals

1. Murine mammary carcinoma 4 T1 cells (ATCC Cat. No. CRL-2539) stably transfected with a Sleeping Beauty (SB) transposon expressing PalmReNL under control of the CAGGS promoter and selected for puromycin resistance [15].
2. DMEM supplemented with GlutaMAX, 10% (vol/vol) fetal bovine serum (FBS), 100 unit/mL penicilin–100 μg/mL streptomycin.

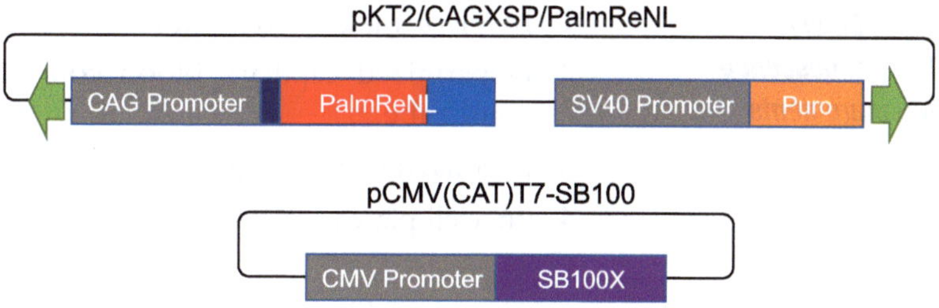

Fig. 1 *Sleeping beauty* (SB) transposon and transposase expression vector. pKT2/CAGXSP/PalmReNL is a SB transposon vector plasmid encoding the fusion gene PalmReNL, which consists of a palm-signal, tdTomato red fluorescent protein, and NanoLuc. This is expressed from the CAGGS promoter followed by a puromycin selection cassette, all of which are flanked by inverted SB transposon repeats. pCMV(CAT)T7-SB100 is an expression vector containing the gene for an improved SB transposase, SB100X, under control of the IE-CMV promoter [14]

3. TransIT-2020 transfection reagent (MIR5404, Mirus Bio, WI, USA), Lipofectamine 3000 (L3000008, Invitrogen / Thermo Fisher).
4. Tissue culture treated 100 mm dishes.
5. Eight-week-old female BALB/c mice (*see* **Note 1**).

2.2 Plasmids (see Fig. 1)

1. A *Sleeping Beauty* (SB) transposon vector for expressing PalmReNL and a puromycin resistance, pKT2/CAGXSP/PalmReNL (available from Addgene, #182970).
2. An improved SB transposase expression vector, pCMV (CAT)T7-SB100 [14] (available from Addgene, #34879).

2.3 EV Isolation and Characterization

1. EV-depleted FBS prepared by 18-h ultracentrifugation at 100,000 × *g*, 4 °C [16].
2. Sorvall WX+ Ultracentrifuge (Thermo Fisher).
3. Refrigerated microcentrifuge 5424 R (Eppendorf).
4. 0.2 μm PES membrane filters.
5. 50 nm membrane filters (WHA110603, Whatman), filter holders (SX0002500, EMD Millipore), a vacuum manifold (QIAvac 24 Plus, Qiagen).
6. 10 mL syringes.
7. ZetaView Multiple Parameter Particle Tracking Analyzer (Particle Metrix, Germany).
8. Bradford Protein Assay.

2.4 Ex Vivo Bioluminescence Measurements

1. In Vivo Imaging System, IVIS Lumina (PerkinElmer).
2. Heparinized capillary blood collection with conical bottom inner tube.
3. Furimazine (Fz; 25 μM).
4. 96-well plates, 384-well plates.

2.5 Fluorescence Microscopy

1. All-in-one Fluorescence Microscope BZ-X700 (KEYENCE).
2. Glass bottom chamber slides.

2.6 In Vivo Bioluminescence Imaging (BLI)

1. SAS3 anesthesia system (Summit Anesthesia Support).
2. EVAC 4 waste gas evacuation system (Universal Vaporizer Support).
3. In Vivo Imaging System, IVIS Spectrum systems (PerkinElmer).
4. Fz (5 μg/mouse in 100 μL PBS) [14].

3 Methods

3.1 Preparation of PalmReNL-EV Producer Cells and Conditioned Media

1. Co-transfect 4 T1 cells at a high density (80–90% confluence) with pKT/CAGXSP/PalmReNL and pCMV(CAT)T7-SB100 at a mass ratio of 2:1 using a lipofection reagent, and culture the cells overnight (*see* Fig. 1).
2. Trypsinize the cells the next day and culture them for 1–2 weeks, adding puromycin (2 μg/mL) at 48–72 h to establish cells constitutively expressing PalmReNL.
3. Select PalmReNL-EV producer cells that maintain a high level of PalmReNL expression by fluorescence-activated cell sorting (FACS) or limiting dilution and fluorescence microscopy for tdTomato signals (*see* **Note 2**).
4. Seed the required number of PalmReNL-4 T1 cells into 10 cell culture dishes (1.5×10^6 cells per 100 mm dish with 10 mL medium). Replace the medium with EV-depleted medium the next day, and culture the cells for another 48–72 h, followed by collecting the conditioned medium (100 mL) (*see* **Note 3**).
5. Centrifuge the conditioned medium at $600 \times g$ for 10 min to remove cells and debris, and then $2000 \times g$ for 20 min to remove apoptotic bodies.
6. Supernatants may continue to be processed by immediate EV isolation or be kept in a − 80 °C freezer.

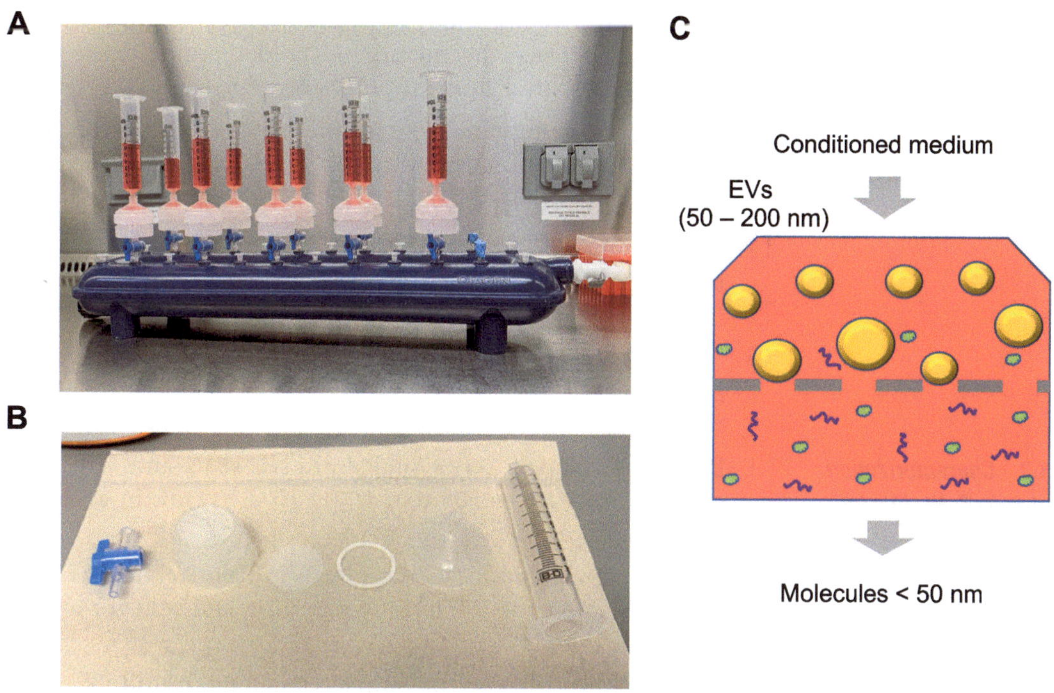

Fig. 2 Enrichment of sEVs by size-based EV isolation. (**a**) An EV isolation system using 50 nm membrane filters with holders, connected to a vacuum manifold with valves. (**b**) The assembly of a 50 nm membrane filter (valve, bottom holder, membrane, rubber o-ring, top holder, syringe). (**c**) After isolating m/lEVs, conditioned medium is processed through 50 nm membrane filters to enrich sEVs on the membranes

3.2 Enrichment of PalmReNL-EVs from Conditioned Media

1. Pellet medium/large EV (m/lEV)-enriched fraction by centrifugation at 20,000 × *g* for 60 min at 4 °C.
2. Remove the supernatant and keep it on ice for the following sEV isolation.
3. Resuspend the pellet with PBS and centrifuge at 20,000 × *g* for 60 min 4 °C to wash the m/lEV-enriched fraction.
4. Filter the supernatant from Step 2 through 0.22 μm membrane filters with pressure to remove large EVs.
5. Assemble a 50 nm membrane filter with a holder and a vacuum manifold (*see* Fig. 2a, b).
6. Wash membrane filters with 10 mL of PBS by applying vacuum suction.
7. Apply the remaining supernatant from Step 3 to the filter to trap the small EVs (sEVs) on the membrane (10 mL of conditioned medium per membrane is recommended to avoid clogging) (*see* Fig. 2c and **Note 4**).
8. Add 5 mL of PBS to the syringe for washing the enriched sEVs when ~500 μL of the supernatant remains in the membrane holder.

9. When 500–1000 μL of the sample remains, turn off the vacuum and place the membrane holder upside down on a new 15 mL conical sterile tube (*see* **Note 5**).
10. Loosen the membrane holder and remove the bottom half. Remove the membrane letting the sample drop into the tube.
11. Carefully collect the sEV-enriched suspension by using a pipet.
12. Measure the EV protein concentration by Bradford assay and the EV particle concentration by nanoparticle tracking analysis (NTA).
13. Aliquot and store sEV- and m/lEV-enriched fractions in a −80 °C freezer.

3.3 Characterization of PalmReNL-EVs by Bioluminescence Measurement and Fluorescence Microscopy

1. Prepare a dilution series of PalmReNL-sEVs and -m/lEVs ranging from 1.0×10^8 to 1.0×10^{10} EVs/mL in a 96-well or 384-well plate.
2. Measure the bioluminescence signals in PalmReNL-EVs using 25 μM Fz after mixing well by pipetting and wait for at least 5 min (*see* Fig. 3a).
3. Make a 10-μL droplet of buffer containing PalmReNL-sEVs and -m/lEVs ranging from 1.0×10^9 to 1.0×10^{10} EVs/mL in a glass-bottom chamber slide.
4. Fluorescence microscopy: Focus on the droplet rim to visualize individual single or aggregated PalmReNL-EVs (*see* Fig. 3b).

3.4 Retro-Orbital (RO) or Intraperitoneal (IP) Injection of PalmReNL-EVs in Mice

1. Thaw frozen PalmReNL-sEVs or -m/lEVs ($1.0 \times$ *g* 10^{10} EVs/mL) and fill a 29 G insulin syringe with 100 μL reporter EV suspension.
2. Anesthetize 8-week-old BALB/c female mice with isoflurane vapor anesthesia.

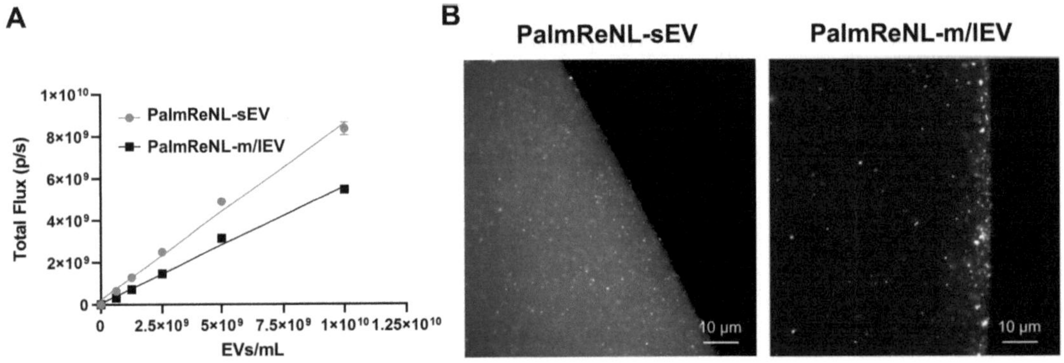

Fig. 3 Characterization of PalmReNL-sEVs and -m/lEVs. (**a**) Bioluminescence analysis of reporter EVs. (**b**) Fluorescence microscopy images of a droplet of buffer containing reporter EVs. Scale bars, 10 μm. (Adapted from Perez, Broadbent, Zarea et al. [15] with permission from Advanced Genetics)

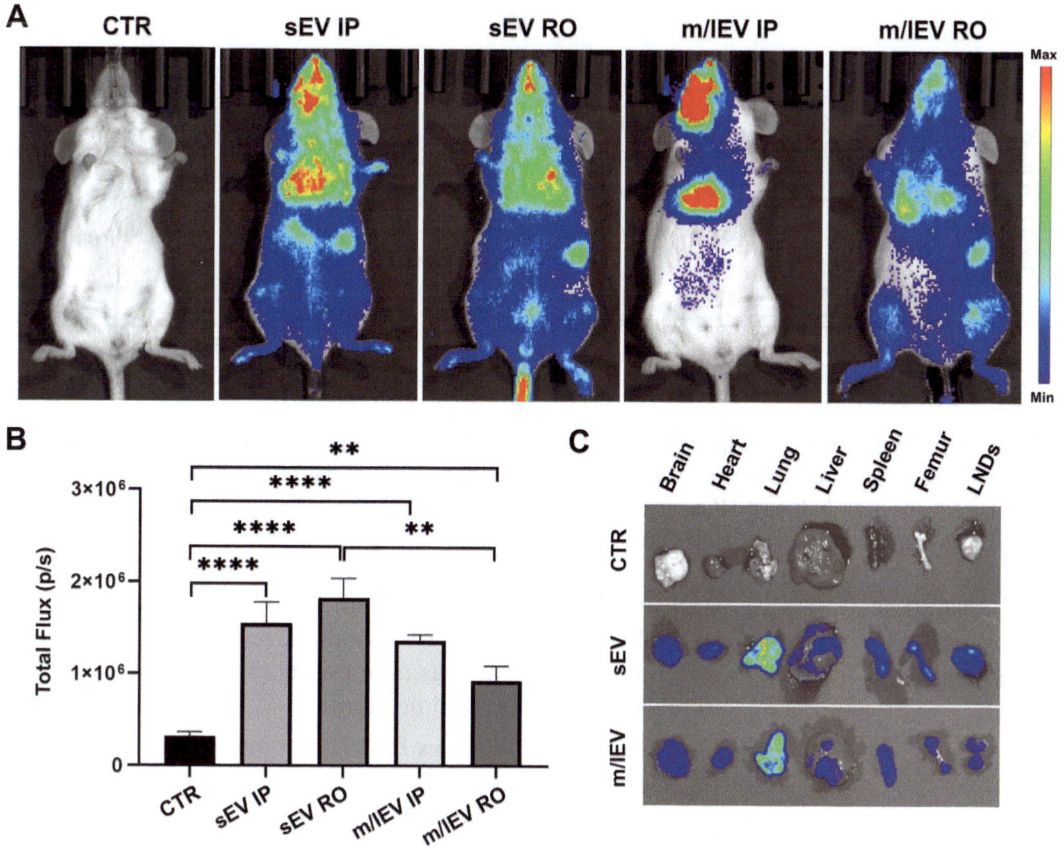

Fig. 4 Biodistribution of reporter EVs derived from metastatic 4 T1 mammary carcinoma cells. (**a**) PalmReNL-sEVs or -m/lEVs (1.0 × 10^9 EVs/100 μL) were administered by RO or IP injection. Furimazine was injected RO. Photons emitted from each mouse were captured by the IVIS, and the resulting cooled chip signals were processed using Living Image software (Perkin Elmer) and overlaid onto a black-and-white image of each mouse using a rainbow scale. (**b**) Analysis of the bioluminescence data in A. Plotted are total flux values (photons/second) for mice injected with reporter EVs. CTR = control (no PalmReNL-EVs injected) (control [$n = 10$]; sEV IP [$n = 3$]; sEV RO [$n = 18$]; m/lEV IP [$n = 3$]; m/lEV RO [$n = 15$]). Error bars, SEM. Statistical comparison by Student t-test: **$p < 0.01$; ****$p < 0.0001$. (**c**) Bioluminescence signals were measured in excised organs. (Adapted from Perez, Broadbent, Zarea et al. [15] with permission from Advanced Genetics)

3. Inject the suspension of PalmReNL-sEVs or -m/lEVs (1.0 × 10^9 EVs/mouse; via RO or IP) into the mice (*see* **Note 6**).
4. Inject Fz (0.25 mg/kg; RO) 5 min after RO EV injection or 2 h after IP EV injection (*see* **Note 7**).
5. Visualize the EV biodistribution overtime by in vivo bioluminescence imaging (BLI) (*see* Fig. 4a, b).
6. Euthanize the mice and harvest major organs for ex vivo BLI (*see* Fig. 4c).

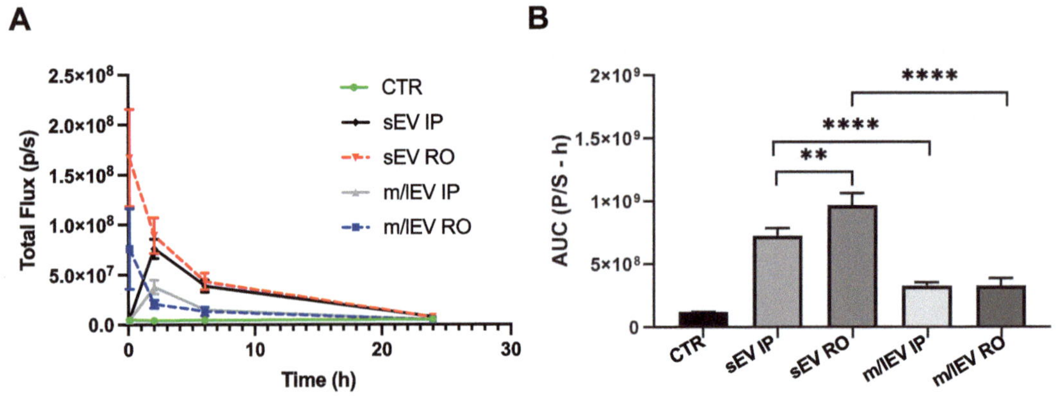

Fig. 5 Bioluminescence analysis of circulating reporter EVs in plasma. (**a**) Time course analysis of circulating PalmReNL-sEVs and -m/lEVs administered RO or IP. (**b**) AUC analysis of the bioluminescence signals in the plasma samples over 24 h post-EV injection. Error bars, SD ($n = 15$), $^{**}p < 0.01$; $^{****}p < 0.0001$. (Adapted from Perez, Broadbent, Zarea et al. [15] with permission from Advanced Genetics)

3.5 Analysis of PalmReNL-EVs Circulating in the Blood

1. Collect a drop of blood (20–50 μL) from the mouse eye using a sterile heparinized capillary blood collection with conical bottom inner tube at various time points (5 min, 2, 6, and 24 h) following RO or IP EV injection.
2. Centrifuge blood samples for ~5 min at ~1300 × *g* to collect the supernatants.
3. After collecting samples from all time points, mix ~10 μL supernatant with Fz (final conc. = 25 μM) and capture bioluminescence signals using the IVIS (*see* Fig. 5a).
4. Using Living Image software quantify the bioluminescence signals obtained from reporter EVs in plasma, plot these values over time (Fig. 5a), and calculate the total area under the curve (AUC) for PalmReNL-sEVs and -m/lEVs RO or IP injected (*see* Fig. 5b).

4 Notes

1. All animal procedures in this article were reviewed and approved by the Institutional Animal Care and Use Committee of Michigan State University.
2. By directly exciting tdTomato in PalmReNL, cells expressing PalmReNL can be identified by fluorescence microscopy or sorted by standard flow cytometry.
3. Since the doubling time is cell type-dependent, determine the duration to reach confluence within 2–3 days when different EV producer cells are used.

4. To overcome clogging problems and increase output, an ultra-fast isolation system called EXODUS has been developed [17]. The device uses negative-pressure oscillation and double-coupled harmonic oscillator-enabled membrane vibrations.
5. Since the total number of released EVs is cell type-dependent, the final volume of the enriched EV suspension needs to be optimized when EVs are isolated from different EV producer cells.
6. IP injection of EVs may be performed without anesthesia, whereas RO injection requires anesthesia for the procedure.
7. EVs injected RO vs. IP show different blood circulation kinetics in mice despite their similar biodistribution [15], so imaging time points need to be optimized.

Acknowledgments

We thank the Institute for Quantitative Health Science and Engineering (IQ) for providing facilities and resources for executing this work. This work was supported by start-up funds from MSU to MK.

References

1. Willms E et al (2018) Extracellular vesicle heterogeneity: subpopulations, isolation techniques, and diverse functions in cancer progression. Front Immunol 9:738
2. Thery C et al (2018) Minimal information for studies of extracellular vesicles 2018 (MISEV2018): a position statement of the International Society for Extracellular Vesicles and update of the MISEV2014 guidelines. J Extracell Vesicles 7(1):1535750
3. Gupta D et al (2020) Quantification of extracellular vesicles in vitro and in vivo using sensitive bioluminescence imaging. J Extracell Vesicles 9(1):1800222
4. Lai CP et al (2014) Dynamic biodistribution of extracellular vesicles in vivo using a multimodal imaging reporter. ACS Nano 8(1):483–494
5. Lazaro-Ibanez E et al (2021) Selection of fluorescent, bioluminescent, and radioactive tracers to accurately reflect extracellular vesicle biodistribution in vivo. ACS Nano 15(2):3212–3227
6. Takahashi Y et al (2013) Visualization and in vivo tracking of the exosomes of murine melanoma B16-BL6 cells in mice after intravenous injection. J Biotechnol 165(2):77–84
7. Contag CH, Bachmann MH (2002) Advances in in vivo bioluminescence imaging of gene expression. Annu Rev Biomed Eng 4:235–260
8. Cox E, Kanada M (2020) Advances in imaging of extracellular vesicles i/n vivo. Membrane 45(2):73–80
9. Wu AY et al (2020) Multiresolution imaging using bioluminescence resonance energy transfer identifies distinct biodistribution profiles of extracellular vesicles and exomeres with redirected tropism. Adv Sci (Weinh) 7(19): 2001467
10. Lai CP et al (2015) Visualization and tracking of tumour extracellular vesicle delivery and RNA translation using multiplexed reporters. Nat Commun 6:7029
11. Schaub FX et al (2015) Fluorophore-NanoLuc BRET reporters enable sensitive in vivo optical imaging and flow cytometry for monitoring tumorigenesis. Cancer Res 75(23):5023–5033
12. Suzuki K et al (2016) Five colour variants of bright luminescent protein for real-time multicolour bioimaging. Nat Commun 7:13718
13. Multhaup M et al (2010) Cytotoxicity associated with Artemis overexpression after

lentiviral vector-mediated gene transfer. Hum Gene Ther 21(7):865–875

14. Mates L et al (2009) Molecular evolution of a novel hyperactive sleeping beauty transposase enables robust stable gene transfer in vertebrates. Nat Genet 41(6):753–761
15. Perez GI et al (2022) In vitro and in vivo analysis of extracellular vesicle-mediated metastasis using a bright, red-shifted bioluminescent reporter protein. Adv Genet 3:2100055
16. Shelke GV et al (2014) Importance of exosome depletion protocols to eliminate functional and RNA-containing extracellular vesicles from fetal bovine serum. J Extracell Vesicles 30:3
17. Chen Y et al (2021) Exosome detection via the ultrafast-isolation system: EXODUS. Nat Methods 18(2):212–218

Chapter 4

Characterization of Extracellular Vesicles by Transmission Electron Microscopy and Immunolabeling Electron Microscopy

Maribel Lara Corona, Ilse Hurbain, Graça Raposo, and Guillaume van Niel

Abstract

Transmission electron microscopy (TEM) is currently the only method that enables the observation of extracellular vesicles (EVs) at a nanometer scale. Direct visualization of the whole content of EV preparation provides not only crucial insights on the morphology of EVs but also an objective evaluation of the content and purity of the preparation. Coupled to immunogold labeling, TEM allows the detection and association of proteins at the surface of EVs. In these techniques, EVs are deposited on grids and are chemically immobilized and contrasted to withstand a high-voltage electron beam. Under high vacuum, the electron beam hits the sample and the electrons that scatter forward are collected to form an image. Here, we describe the steps needed to observe EVs by classical TEM and the extra steps required to label proteins through immunolabeling electron microscopy (IEM).

Key words Transmission electron microscopy (TEM), Immunolabeling electron microscopy (IEM), Immunogold labeling, Extracellular vesicles (EVs), Antibodies (Ab)

1 Introduction

Transmission electron microscopy (TEM) witnessed the birth of the extracellular vesicle (EV) field providing the first observation of these entities [&, 1–3] and seminal knowledge about their heterogeneity, composition, and biogenesis mechanisms [4, 5]. While complementary methods of characterization of these vesicles are constantly completing the tool set, TEM remains an essential technique to analyze these vesicles and extend our knowledge on their features.

The working principle of TEM and immunolabeling electron microscopy (IEM) relies on an electron beam in high vacuum that hits the sample, which scatters forward some electrons to form an image [6]. Considering that increasing the velocity of an electron will decrease its wavelength, the TEM microscope accelerates

Seppo Vainio (ed.), *Cell-Secreted Vesicles: Methods and Protocols*, Methods in Molecular Biology, vol. 2668,
https://doi.org/10.1007/978-1-0716-3203-1_4,

electrons in order to decrease their wavelength to the nano or even atomic scale [7]. As a result, the resolution is greatly increased and the observation of even small EVs can be achieved with fine detail [8]. Low-atomic number atoms such as C, H, O, N, P, and S, which are abundant in biological samples, scatter few electrons. For this reason, the sample is coated with heavy metals to produce electron dense areas that will increase the contrast [9], this is often referred to as negative staining.

Oftentimes, the classification of EVs seems difficult since EV composition heavily depends on the cell type and the physiological state of the producing cell [10]. In spite of their heterogenicity, common EV features comprise their ability to reach target cells [11] and their capacity to modify recipient cell functions through various mechanisms [12]. In the last decades, EVs have gained wide recognition since they have been described for most biofluids [13], rendering them quite promising as non-invasive biomarkers [14, 15]. In addition, the extensive work done on EVs regarding chemical and genetic modifications highlights their potential as nano-vehicles for therapeutical molecules [16].

Vesicle size ranges from 30 nm to several μm of diameter [17, 18], and often, EV isolation protocols will result in solutions containing EVs of different sizes. TEM and IEM are techniques used to distinguish between various vesicle populations. In addition, these techniques are helpful to discriminate between EVs, particles of similar size or contaminants that might have been co-isolated with classical EV isolation protocols such as differential ultracentrifugation (UC), size exclusion chromatography (SEC), or density gradient [19, 20]. In this regard, TEM and IEM can be used to determine the purity of the EV preparation for downstream applications, notably in vivo [21].

To date, there is no gold standard for the complete characterization of EVs. However, it is undeniable that electron microscopy is still the one of the most robust and reliable techniques to visualize and label EVs. While various protocols have been employed to visualize and label EVs by TEM, we present here two standard protocols that allow the user to observe and label EVs in TEM and IEM, respectively.

2 Materials

Prepare all solutions with ultra-pure autoclaved water (prepared by purifying water to reach a resistivity of 18.2 MΩ × cm at 25 °C).

Follow all security and waste disposal rules displayed in the product information sheet and institutional procedures.

All materials are prepared with analytical grade reagents.

2.1 EV Deposition

1. Prepare or order TEM Grids 200 Mesh Cu/Pd. Formvar/ carbon supported grids (*see* **Note 1**).

2.2 Fixation

1. Phosphate buffer (PB): 200 mM, pH 7.4. Weigh 27.59 g of $NaH_2PO_4.H_2O$. Add water to a volume of 1 L. Weigh 35.60 g of $Na_2HPO_4.2H_2O$. Add water to a volume of 1 L. Mix 1 volume of 200 mMNaH_2PO_4 and 4 volumes of −200 mM Na_2HPO_4. Adjust pH to 7.4 if necessary. Store at room temperature (*see* **Note 2**).
2. Fixative A: 2% paraformaldehyde (PFA) (*see* **Note 3**), 1% glutaraldehyde in 100 mM PB, pH 7.4. To prepare 10 mL of fixative A, mix 5 mL of 200 mM phosphate buffer pH 7.4 with 0.4 mL of 25% glutaraldehyde, then add 1.25 mL of 16% formaldehyde. Make up to 10 mL with 3.35 mL of water. Use the fixative fresh.
3. Fixative B: 2% PFA in 100 mM PB, pH 7.4. To prepare 10 mL of fixative B, mix 5 mL 200 mM PB pH 7.4 with 1.25 mL 16% of PFA/H2O. Make up to 10 mL with 3.75 mL of water. Use the fixative fresh.
4. Fixative C: 1% glutaraldehyde in 100 mM phosphate buffer pH 7.4. To prepare 10 mL of fixative C, mix 5 mL of 200 mM PB pH 7.4 with 0.4 mL of 25% glutaraldehyde/ H2O. Make up to 10 mL with 4.6 mL of water. Use the fixative fresh.

2.3 Quenching/ Blocking

1. Phosphate buffered saline (PBS): 0.017 M KH_2PO_4, 0.05 M Na_2HPO_4, 1.5 M NaCl, pH 7.4 (*see* **Note 4**).
2. 1% PBS/BSA: Weigh 100 mg of Bovine Serum Albumin (BSA) and add PBS to a final volume of 10 mL.
3. 0.1% PBS/BSA: Mix 1 volume of 1% PBS/BSA and 9 volumes of PBS.
4. 50 mM PBS/glycine: Weigh 375 mg of glycine. Add PBS to a final volume of 100 ml. Store at 4 °C.

2.4 Contrasting

1. 2% Methylcellulose: Weigh 4 g of methylcellulose (25 centipoises) and transfer it to a cylinder. Add water at 90 °C to a volume of 200 mL and stir the solution. While stirring, cool down the methyl cellulose solution on ice rapidly, then seal it and leave it overnight at 4 °C still mixing. The next morning, stop mixing and leave it for 2 days more at 4 °C for polymer maturation. Add water to a final volume of 200 mL and centrifuge at 100000 × *g* during 1.5 h at 4 °C. Divide the solution in aliquots of 15 mL and store at 4 °C.

2. 4% Uranyl acetate: Weigh 4 g of Uranyl acetate and transfer it to a cylinder (*see* **Note 5**). Add water to a volume of 100 mL. Let it mix on a rotor (120 rpm) until it dissolves. Store it protected from light at 4 °C
3. Contrasting solution: Mix gently with a micropipette 900 μL of 2% methylcellulose and 100 μL of 4% uranyl acetate. Keep the solution in the dark at 4 °C. Prepare fresh before use.

2.5 Immunolabeling

1. Primary antibody: Primary Antibody of choice, diluted according to the manufacturer's recommendation (approximatively 1–20 μg/mL). Of note, the antibody should be directed against a luminal epitope exposed at the surface of the vesicle. We cannot guarantee that cytosolic protein are reachable pour the proposed protocol of IEM.
2. Protein A Gold (PAG): EM-Grade 10 nm, diluted according to the manufacturer's recommendations.
3. Bridging antibody: if the primary Antibody of choice has low or no affinity for Protein A (mouse IgG1 for instance), use a rabbit antibody raised against the isotype of the primary antibody (anti mouse IgG for instance) as a bridge and diluted according to the manufacturer's recommendation (approximatively 1–20 μg/mL).

3 Method

3.1 Conventional Transmission Electron Microscopy (TEM)

The following steps are performed at room temperature, unless indicated otherwise.

1. Unroll parafilm on the bench for EV deposition, fixation, washing and contrasting steps (*see* **Note 6**).
2. Deposit 4 μL of EVs at a concentration of 10^9 to 10^{11} particles/mL on formvar coated, carbonated grids for 20 min (*see* **Note 7**). Cover the grids with a Petri dish lid to protect them from air impurities and dehydration during all the incubation steps (*see* Fig. 1a).
3. Prepare on a parafilm firmly stretched on the bench, the drops of fixative A and water. One line of drops per EV samples as grids are not distinguishable (*see* Fig. 1b).
4. Recover the grids with tweezers and place them upside down on a drop of fixative. Fix the grids in fixative A for 20 min at room temperature (*see* **Note 8**).
5. Wash the grids with water 6 × 1 min at room temperature (*see* **Note 9**).

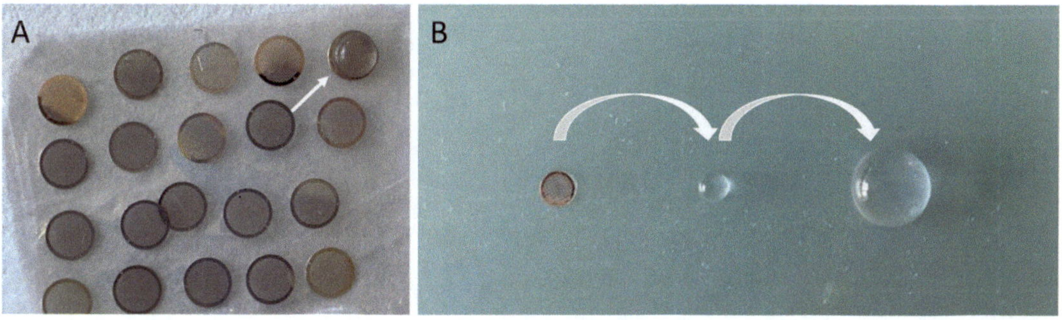

Fig. 1 Main steps of negative contrast and immunogold labeling. Steps showing the drop of sample deposited on the grid (white arrow) (**a**), the grid floating on the top of a droplet followed by the droplets for subsequent steps (**b**)

6. Work on ice. Put a parafilm in a Petri dish and deposit 50–100 μl drops of the contrasting solution of Methylcellulose/Uranyl Acetate. Incubate the grids upside down on the solution for 10–12 min. Protect the sample from light with aluminum foil (*see* **Note 10**).
7. Remove the grids from the drop with loops that have a slightly larger diameter than the grid.
8. Soak the excess of fluid by gently pushing the loop sideways (with an inclination of 45°) on a Whatman filter paper No. 1 placed on a parafilm (to avoid blotting uranyl acetate on the bench) so that a thin film is left covering the section side of the grid (*see* **Note 11**). Discard the parafilm and the filter paper which were in contact with UA as indicated in institutional procedures (*see* Fig. 2a, b).
9. Air-dry the grid for at least 10 min up to 1 h. After drying, the film should show a dark gold to blue interference color (*see* Fig. 2c).
10. Remove the grids from each loop with fine tweezers. Store the grids in an EM grid box at room temperature.
11. Use an electron microscope to take pictures of EVs (*see* Fig. 3a,b).

3.1.1 Immunolabeling

The following steps are performed at room temperature, unless indicated otherwise.

1. Unroll parafilm on the bench for EV deposition, fixation, immunolabeling, washing, and contrasting steps (see **Note 6**).
2. Deposit 4 μL of EVs at a concentration of 10^9 to 10^{11} particles/mL on formvar coated, carbonated grids for 20 min (see **Note 7**). Cover the grids with a Petri dish lid to protect them from air impurities and dehydration during all the incubation steps (*see* Fig. 1a).

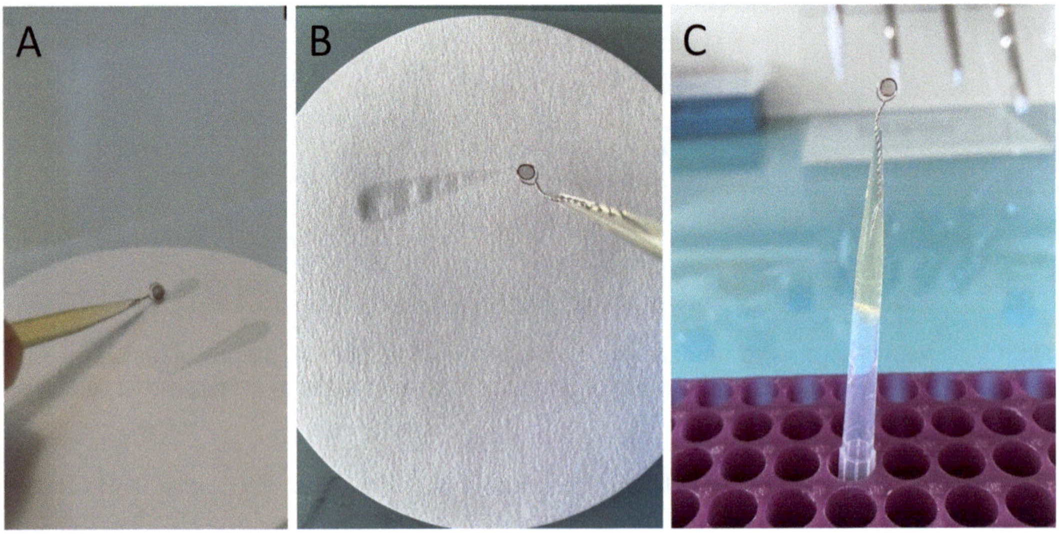

Fig. 2 Blotting step to remove the uranyl acetate and methylcellulose or uranyless. The grid has been recovered on a loop and blotted with an inclination of 45° on Whatman paper (**a**). Blotting should be carefully performed to control the thickness of uranyl and methylcellulose film (**b**). Air-dried sample on the grid will be removed from the loop to be observed by electron microscopy (**c**)

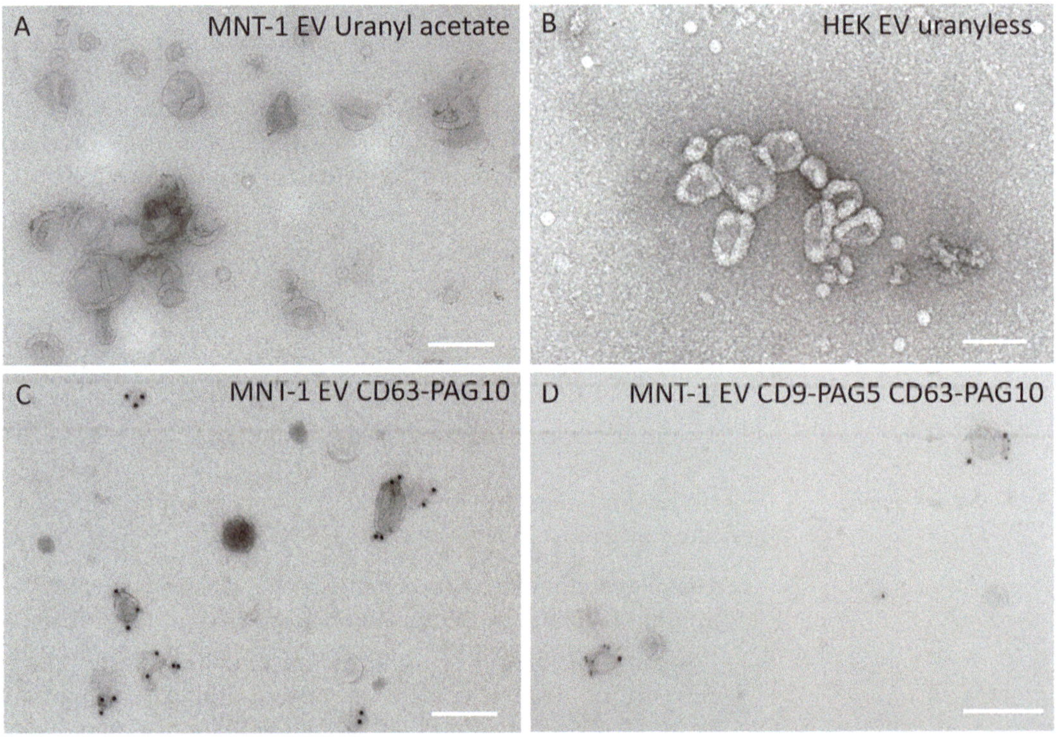

Fig. 3 (**a**) MNT1-derived EVs observed by TEM after negative contrast with uranyl/methylcellulose contrasting. (**b**) HEK-derived EVs observed by TEM after negative contrast with uranyless. (**c**) MNT1 derived EVs immunogold-labelled for CD63 and revealed with Protein A-gold particle of 10 nm diameter. (**c**) MNT1-derived EVs double immunogold-labelled for CD63 and revealed with Protein A-gold particle of 10 nm diameter and CD9 and revealed with Protein A-gold particle of 5 nm diameter. Scale bars = 200 nm (Credit Roberta Palmulli)

3. Prepare on a parafilm firmly stretched on the bench, the drops of fixative B and water. One line of drops per EV samples as grids is not distinguishable (*see* Fig. 1b).
4. Recover the grids with tweezers and place them upside down on a drop of fixative. Fix the grids in fixative B for 20 min (*see* **Note 8**).
5. Wash the grids with water 6 × 1 min at room temperature (*see* **Note 9**).
6. Incubate the grids in 50 mM PBS/Glycine 4 × 2 min (*see* **Note 12**).
7. Incubate the grids in 1% PBS/BSA for 5 min (*see* **Note 13**).
8. Prepare the antibody solution in 0.1% PBS/BSA and incubate the grids in this solution for 30 min (a 10–20 μL drop for one grid) (*see* **Note 14** for antibody dilution).

   For a primary antibody raised in rabbit or if it is a mouse IgG2a or IgG2b, go directly to step 10. If the primary raised in rat or is a mouse IgG1, perform the bridging steps 7 and 8 and continue with step 9 (*see* **Note 15**).
9. Incubate the grids in 0.1% BSA/PBS 5 × 2 min.
10. Prepare the bridging antibody in 0.1% BSA/PBS and incubate the grids for 20 min (a 10–20 μl drop for one grid).
11. Wash the grids in 0.1% BSA/PBS 5 x 2 min.
12. Incubate the grids with Protein-A-Gold (PAG) in BSA/PBS 0.1% for 20 min (*see* **Notes 16** and **17**) (a 10–20 μl drop for one grid).
13. Wash with PBS 5 × 2 min.
14. Fix in Fixative C for 5 min.

    For a single protein labeling, go directly to the step 14. For multiple labeling, go back to step 4 and repeat the instructions (*see* **Notes 18** and **19**).
15. Wash with water 6 × 1 min (*see* **Note 9**).
16. Work on ice. Put a parafilm in a Petri dish and deposit 50–100 μl drops of the contrasting solution of Methylcellulose/Uranyl Acetate. Incubate the grids upside down on the solution for 10–12 min. Protect the sample from light with aluminum foil (*see* **Note 10**).
17. Remove the grids from the drop with loops that have a slightly larger diameter than the grid.
18. Soak the excess of fluid by gently pushing the loop sideways (with an inclination of 45°) on a Whatman filter paper No. 1 placed on a parafilm (to avoid blotting uranyl acetate on the bench) so that a thin film is left covering the section side of the grid (*see* **Note 11**; *see* Fig. 2a,b.

19. Air-dry the grid for at least 10 min up to 1 h. After drying, the film should show a dark gold to blue interference color (*see* Fig. 2c).
20. Remove the grids from each loop with fine tweezers. Store the grids in a grid box at room temperature.
21. Use an electron microscope to take pictures of EVs (*see* Fig. 3c,d.

4 Notes

1. Put the EV solution on the formvar/carbon coated side. Formvar is a transparent, insulating film that allows the deposition and observation of EVs on the mesh. On the other hand, carbon is highly conductive, but hydrophobic. The glow discharge used to deposit carbon on the grids helps to render the surface grid more hydrophilic. If you carbon-coat the grids yourself, then coat the grids with around 2–3 nm of carbon. A very fine coat will quickly damage the sample and a very thick coat will prevent the electrons to be scattered forward.
2. Filter the PB solution with a 0.22 μm pore size filter before use.
3. Handle paraformaldehyde with gloves in a fume hood.
4. We use a 10X PBS solution diluted in water. Filter the PBS solution with a 0.22 μm pore size filter before use.
5. Uranyl acetate is a radioactive and carcinogenic substance. Use gloves, glasses, and a fume hood when handling this reagent. Follow all institutional rules and the product information sheet to dispose of the materials safely.
6. This procedure is performed by placing the grids (EVs facing the buffer) on droplets successive placed on the bench. It is important to keep the grids wet on the section side during all steps, but dry at the opposite side. Each rinsing step is performed on drops of 200–500 μL, while small droplets of 5–15 μL of immunoreagents are sufficient (this prevents the use of large amounts of antibodies and probes that are generally precious).
7. Do not touch the formvar with the tip while depositing the sample and use fine tweezers to handle the grids paying attention to only touch the perimeter of the grid to avoid damaging the area of observation.
8. After the EV incubation time, one 200 μl droplet of fixative can be used to hold five grids at the same time, making sure that they do not touch each other or overlap. The combined use of paraformaldehyde and glutaraldehyde allows a quick fixation protocol that maintains EV morphology and structural integrity.

9. The washing steps can be simply done by placing the grids on different drops of water. As previously mentioned, several grids of the same condition can be washed on the same water droplet at the same time. This washing step is extremely important to remove traces of phosphate that would precipitate uranyl acetate.
10. As uranyl acetate is a radioactive and carcinogenic substance, it may be prohibited in some institutions. Alternative contrasting agents are developed such as Uranyless®. The protocol remains the same than with the uranyl acetate/methylcellulose mix except that Uranyless can be used pure.
11. During blotting, a quick removal of the uranyl acetate/methylcellulose (UA/MC) will result in a thicker film of UA/MC, whereas a slower removal will result in a thinner film. This step is critical: if the methylcellulose film is too thin, EVs can dry out and will not be stable enough under the electron beam. If the methylcellulose film is too thick, structural details on the-EVs will hardly be visible and will not show any contrast.
12. Glycine is used to quench unreacted aldehydes after the fixation step.
13. To prevent non-specific binding between the antibody or antibodies and the sample, we use BSA as a blocking agent. Another alternative to reduce the background is the Cold Water Fish Skin Gelatin.
14. As for other applications, the optimal antibody dilution may vastly differ from one antibody to another. Try using the manufacturer's recommended dilution for immunohistochemistry (IHC) as a starting point if you have not previously worked with the antibody.
15. If you want to perform double labeling, you should do it for antibodies that do not require bridging or for maximum one antibody that requires bridging (using the bridge during the first labeling step).
16. For conventional IEM, labeling can only be performed for proteins found at the surface of EVs.
17. Protein-A Gold strongly binds to several mammal IgG (for example, rabbit) and this interaction facilitates the labeling with antibodies used for other applications like IHC, immunofluorescence (IF), or Western Blot (WB). For multiple labeling, use different sizes of gold particles. It is recommended to prepare the first antibody with the largest gold particle (for example, 10 nm) and continue with a smaller one (for example, 5 nm). Take in consideration that having several bead sizes in the same sample increases the probability of cross-reactions.

18. Before performing any double labeling, perform first a single labeling with each antibody separately to get acquainted with the labeling pattern. It is possible to perform labeling with two polyclonal antibodies in combination with protein A-Gold or one polyclonal and one monoclonal antibody (from different species). Labeling with two monoclonal antibodies is not recommended even if they are of different isotypes, because the bridging steps will most likely cause false positives.
19. For multiple labeling, it is advised to start with the less abundant antigen, and then repeat the procedure for a more abundant antigen.

Acknowledgments

We would like to thank Roberta Palmulli to provide electron microscopy pictures and Maryse Romao for the pictures of grids handling. MLC is financed by ANR ZENITH (ANR-20-CE18-0026_01). GVN is supported by the fondation ARC and Institut National du Cancer-projets libres "Biologie et Sciences du Cancer"' (INCa-PLBIO2019-125).

References

1. Wolf P (1967) The nature and significance of platelet products in human plasma. Br J Haematol 13(3):269–288). Wiley. https://doi.org/10.1111/j.1365-2141.1967.tb08741.x
2. Harding C, Stahl P (1983) Transferrin recycling in reticulocytes: pH and iron are important determinants of ligand binding and processing. Biochem Biophys Res Commun 113(2):650–658). Elsevier BV. https://doi.org/10.1016/0006-291x(83)91776-x
3. Pan B-T, Johnstone RM (1983) Fate of the transferrin receptor during maturation of sheep reticulocytes in vitro: selective externalization of the receptor. Cell 33(3):967–978). Elsevier BV. https://doi.org/10.1016/0092-8674(83)90040-5
4. Colombo M, Moita C, van Niel G, Kowal J, Vigneron J, Benaroch P, Manel N, Moita LF, Théry C, Raposo G (2013) Analysis of ESCRT functions in exosome biogenesis, composition and secretion highlights the heterogeneity of extracellular vesicles. J Cell Sci Comp Biolog. https://doi.org/10.1242/jcs.128868
5. Jeppesen DK, Fenix AM, Franklin JL, Higginbotham JN, Zhang Q, Zimmerman LJ, Liebler DC, Ping J, Liu Q, Evans R, Fissell WH, Patton JG, Rome LH, Burnette DT, Coffey RJ (2019) Reassessment of exosome composition. Cell 177(2):428–445). Elsevier BV. https://doi.org/10.1016/j.cell.2019.02.029
6. Knoll M, Ruska E (1932) Das Elektronenmikroskop. Zeitschrift für Physik 78(5–6): 318–339). Springer Science and Business Media LLC. https://doi.org/10.1007/bf01342199
7. Winey M, Meehl JB, O'Toole ET, Giddings TH Jr (2014) Conventional transmission electron microscopy. In: Drubin DG (ed) Molecular biology of the cell, vol 25, pp 319–323). American Society for Cell Biology (ASCB). https://doi.org/10.1091/mbc.e12-12-0863
8. Chuo ST-Y, Chien JC-Y, Lai CP-K (2018) Imaging extracellular vesicles: current and emerging methods. J Biomed Sci 25(1):91. Springer Science And Business Media LLC. https://doi.org/10.1186/s12929-018-0494-5
9. Harris JR (2015) Transmission electron microscopy in molecular structural biology: a historical survey. Arch Biochem Biophys 581: 3–18). Elsevier BV. https://doi.org/10.1016/j.abb.2014.11.011
10. Yáñez-Mó M, Siljander PR-M, Andreu Z, Bedina Zavec A, Borràs FE, Buzas EI, Buzas K, Casal E, Cappello F, Carvalho J,

Colás E, Cordeiro-da Silva A, Fais S, Falcon-Perez JM, Ghobrial IM, Giebel B, Gimona M, Graner M, Gursel I et al (2015) Biological properties of extracellular vesicles and their physiological functions. J Extracell Vesicles 4(1):27066). Wiley. https://doi.org/10.3402/jev.v4.27066
11. Raposo G, Nijman HW, Stoorvogel W, Liejendekker R, Harding CV, Melief CJ, Geuze HJ (1996) B lymphocytes secrete antigen-presenting vesicles. J Exp Med 183(3):1161–1172). Rockefeller University Press. https://doi.org/10.1084/jem.183.3.1161
12. van Niel G, D'Angelo G, Raposo G (2018) Shedding light on the cell biology of extracellular vesicles. Nat Rev Molec Cell Biol 19(4): 213–228). Springer Science and Business Media LLC. https://doi.org/10.1038/nrm.2017.125
13. Bano R, Ahmad F, Mohsin M (2021) A perspective on the isolation and characterization of extracellular vesicles from different biofluids. RSC Adv 11(32):19598–19615). Royal Society of Chemistry (RSC). https://doi.org/10.1039/d1ra01576a
14. Lane RE, Korbie D, Hill MM, Trau M (2018) Extracellular vesicles as circulating cancer biomarkers: opportunities and challenges. Clin Transl Med 7(1). Wiley. https://doi.org/10.1186/s40169-018-0192-7
15. Urabe F, Kosaka N, Ito K, Kimura T, Egawa S, Ochiya T (2020) Extracellular vesicles as biomarkers and therapeutic targets for cancer. Am J Physiol Cell Physiol 318(1):C29–C39). American Physiological Society. https://doi.org/10.1152/ajpcell.00280.2019
16. Mentkowski KI, Snitzer JD, Rusnak S, Lang JK (2018) Therapeutic potential of engineered extracellular vesicles. AAPS J 20(3):50. Springer science and business media LLC. https://doi.org/10.1208/s12248-018-0211-z
17. Doyle L, Wang M (2019) Overview of extracellular vesicles, their origin, composition, purpose, and methods for exosome isolation and analysis. Cells 8(7):727). MDPI AG. https://doi.org/10.3390/cells8070727
18. Raposo G, Stoorvogel W (2013) Extracellular vesicles: exosomes, microvesicles, and friends. J Cell Biol 200(4):373–383). Rockefeller University Press. https://doi.org/10.1083/jcb.201211138
19. Stam J, Bartel S, Bischoff R, Wolters JC (2021) Isolation of extracellular vesicles with combined enrichment methods. J Chromatogr B 1169:122604). Elsevier BV. https://doi.org/10.1016/j.jchromb.2021.122604
20. Willms E, Cabañas C, Mäger I, Wood MJA, Vader P (2018) Extracellular vesicle heterogeneity: subpopulations, isolation techniques, and diverse functions in cancer progression. Front Immunol 9:738. Frontiers Media SA. https://doi.org/10.3389/fimmu.2018.00738
21. Rikkert LG, Nieuwland R, Terstappen LWMM, Coumans FAW (2019) Quality of extracellular vesicle images by transmission electron microscopy is operator and protocol dependent. J Extracell Vesicles 8(1):1555419). Wiley. https://doi.org/10.1080/20013078.2018.1555419

Chapter 5

Extracellular Vesicle Isolation by a Tangential-Flow Filtration-Based Large-Scale Purification Method

Runjie Yuan, Yijun Zhou, Gabriel F. Arias, and Dirk P. Dittmer

Abstract

Extracellular vesicle (EV) isolation from conditioned cell culture medium has been a challenging topic. It is particularly difficult to obtain pure and intact EVs at a large scale. The commonly used methods such as differential centrifugation, ultracentrifugation, size exclusion chromatography, polyethylene glycol (PEG) precipitation, filtration, and affinity-based purification each have their advantages and limitations. Here, we present a tangential-flow filtration (TFF) based, multi-step purification protocol that combines filtration, PEG precipitation, and Capto Core 700 multimodal chromatography (MMC) to isolate EVs at high purity from large volumes of cell culture conditioned medium. Inserting the TFF step before PEG precipitation removes proteins, which may aggregate in subsequent steps and co-purify with EVs.

Key words Extracellular vesicles (EVs), Purification, Tangential-flow filtration (TFF), Capto Core 700 multimodal chromatography (MMC)

1 Introduction

Extracellular vesicles (EVs) are lipid-bound membrane particles. They are secreted by all mammalian and plant cells, as well as fungi and bacteria. Their roles in cell-to-cell communication, cargo transport, and disease biomarkers have gained increasing attention [1]. Thus, developing an efficient purification protocol is essential to isolate EVs from conditioned medium or biofluids.

Differential ultracentrifugation (UC) has been widely used as the benchmark EV isolation method [2]. In UC, larger impurities are removed by low-speed differential centrifugation, and EVs are pelleted and harvested after high-speed UC [3]. However, the yield and quality of EVs isolated by UC are low [4, 5]. Aside from UC, several other techniques have been proved to isolate EVs. Polyethylene glycol (PEG)-based precipitation, for instance, uses soluble polymer solutions to precipitate EVs by low-speed centrifugation [6, 7]. The yield of EVs purified by PEG precipitation is high,

Seppo Vainio (ed.), *Cell-Secreted Vesicles: Methods and Protocols*, Methods in Molecular Biology, vol. 2668,
https://doi.org/10.1007/978-1-0716-3203-1_5,

but contaminants such as non-exosome-associated proteins, immunoglobulins, and viral particles are prone to co-precipitate with the EVs [4, 6, 8]. Size-exclusion chromatography (SEC) separates EVs based on their size. In SEC, larger molecules are restricted from entering the pores on the resins and only travel between the interstitial space and thus elute early. Smaller molecules spend more time in the pores and elute later [9]. EVs isolated by SEC are of good integrity and purity [4, 10], but an additional step is needed to concentrate the diluted EVs. Ultrafiltration (UF) can be used to both concentrate and purify EVs and has been shown with better yield and quality than UC [4, 11]. However, the purification efficiency is limited by the membrane's molecular weight cut-off (MWCO). Other isolation techniques, such as bead-based immunoprecipitation, likewise have their specific advantages and drawbacks. These are discussed elsewhere [12]. We believe that a combination of several isolating techniques carefully selected to preserve EV integrity and improve yield, and purity is needed to circumvent these problems. Here, we present a tangential-flow filtration (TFF)-based, multi-step purification protocol that combines filtration, PEG precipitation, and Capto Core 700 multimodal chromatography (MMC). In addition, this protocol includes the enzymatic digestion of DNA and RNA that is not encased within the EVs. Our method is an easy-to-perform and scalable pipeline.

2 Materials

2.1 Medium Conditioning and Initial Clarification

1. Cell Culture Medium
   - For U2OS cells: Dulbecco's Modified Eagle Medium (DMEM) (Gibco, Cat: 11995–065) supplemented with 10% EV-depleted fetal bovine serum (FBS), 50 U/ mL Penicillin/Streptomycin (Gibco, Cat:15140-122), and 20 mM L-Glutamine (Gibco, Cat:25030-081).
   - For BCBL-1 cells: Roswell Park Memorial Institute (RPMI) 1640 Medium (Gibco, Cat: 11875–093) supplemented with 10% EV-depleted FBS, 50 U/ mL Penicillin/streptomycin (Gibco, Cat:15140-122, and 20 mM L-Glutamine (Gibco, Cat:25030-081).

     To make EV-depleted FBS, add PEG-8000 to FBS to a final concentration of PEG at 4%, mix thoroughly, and rotate overnight at 4 °C. Centrifuge FBS-PEG mix at 1000 × *g* for 60 min at 4 °C to pellet and remove serum EVs. Keep the supernatant and store it at −20 °C.

     Cell lines are from the American Type Culture Collection (ATCC).

2. 1× Dulbecco's Phosphate Buffered Saline (DPBS) (Gibco, Cat:14190-144): Stored at 4 °C
3. T175 Flasks
4. 0.05% Trypsin-EDTA (Gibco, Cat:25300-054): Stored at 4 °C
5. 0.22 μm 1 L Vacuum-Driven Filter System (Genesee, Cat. # 25–229)
6. 0.45 μm 1 L Vacuum-Driven Filter System (Genesee, Cat. # 25–230)
7. 0.22 μm Syringe Filters
8. Syringes
9. ZetaView BASIC NTA-Nanoparticle Tracking Video Microscope PMX-120

2.2 Tangential Flow Filtration

1. ÄKTA Flux Tangential Flow Filtration System (Cytiva ÄKTA Flux s, P: 29038437).
2. Double-distilled water (ddH_2O).
3. Cleaning-in-place buffer: 0.5 M NaOH.
4. Neutralizing buffer: 20 mM, Tris-HCl pH 7.0. Dissolve 121.1 g Tris Base into 1 L H_2O. Adjust pH with HCl. Filter through a 0.22 μm pressure-driven filter system in a biosafety cabinet to ensure sterility.
5. MidGee Hoop UF hollow fiber cartridges (Cytiva, Model #: UFP-750-C-H24LA) with MWCO of 750 kDa (*see* **Note 1**).
6. pH testing papers (Baxter, Cat.#P1119-6A).
7. 1× DPBS: Stored at 4 °C.
8. Storage buffer: 0.1 M NaOH.

2.3 PEG Precipitation and PBS Resuspension

1. DNA-free RNase (ThermoFisher, Cat# EN0531): Stored at −20 °C.
2. RNA-free DNase (Promega, Cat#: M6101): Stored at −20 °C.
3. 1× DPBS: Stored at 4 °C.
4. 40% w/v Polyethylene Glycol 8000 (PEG-8000, Fisher BioReagents, Cat.#BP233–1): dissolve 160 g PEG-8000 in 0.4 L ddH_2O. Filter through a 0.22 μm pressure-driven filter system in a biosafety cabinet to ensure sterility.

2.4 Capto Core 700 Chromatography

1. ÄKTA Start Chromatography System (Cytiva, Cat: 29022094).
2. Capto Core 700 multimodal chromatography resin (Cytiva, Cat: 17548102).

3. 1 mL HiTrap Capto Core 700 columns (Cytiva, P: 17–5481-51).
4. Column running buffer: 1x Ca^{++}/Mg^{++}-free phosphate-buffered saline (CMF-DPBS). To make 1 L stock solution, add 8 g NaCl (137 mM), 200 mg KCl (2.7 mM), 2.16 g $Na_2HPO_4 \cdot 7H_2O$ (8.1 mM), 0.2 g KH_2PO_4 (1.1 mM), and add ddH_2O to 1 L. Filter through a 0.45 μm filter to remove debris.
5. Column regeneration buffer: 0.5 M NaOH in 30% v/v isopropanol.
6. Column storage buffer: 20% v/v Ethanol.

3 Methods

3.1 Medium Conditioning and Initial Clarification

Procedures described in this method are based on BCBL-1 (suspension) and U2OS (adherent) cell lines but can be adapted to other cell lines. Cells were kept in an incubator at 37 °C with 5% CO_2. All media and solutions should be warmed up before use. Sterile techniques should be followed. This step corresponds to Fig. 1a-c.

1. Cell culture.
    - U2OS (ATCC HTB-96): U2OS is an adherent cell line. Cells are seeded at 10–20% confluency in a T175 flask in DMEM medium with all supplements and grow to confluency, followed by 48 h of medium conditioning.
    - BCBL-1: BCBL-1 is a suspension cell line. Cells are seeded at 2×10^5 cells/mL in RPMI medium with all supplements and allowed to grow to 1×10^6 cells/mL, followed by 24 h of medium conditioning.
2. The conditioned medium should be sampled before medium harvesting. Take 1–5 mL conditioned medium and centrifuge at 1000 × *g* for 5 min at 4 °C to pellet cells and large debris. Filter the supernatant through 0.2 μm syringe filters. Measure particle concentrations by Nanoparticle Tracking Analysis (NTA) with ZetaView. Conditioned medium is harvested when particle concentration reaches 5×10^9 to 1×10^{10} particles/mL.
3. Centrifuge the conditioned media at 1000 × *g* for 5 min at 4 °C to remove cells and large debris. Then, filter the supernatant through 0.45 and 0.22 μm vacuum-driven pressure filters. Filtered media can be stored at 4 °C for 24 h if not used immediately.

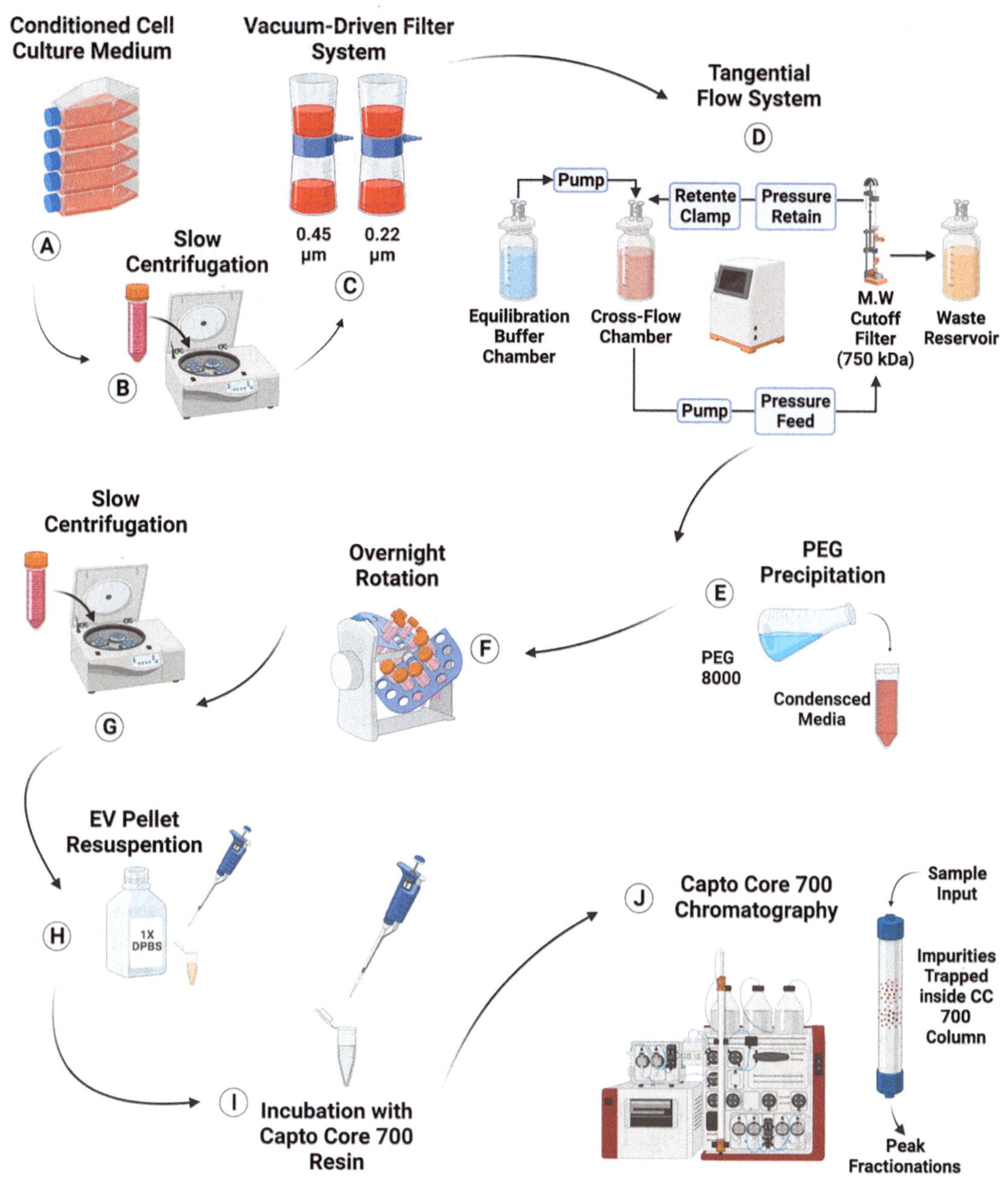

Fig. 1 Step-by-step illustration of the EV purification pipeline. (**a**) Cell culture medium conditioning. (**b**) Slow centrifugation to remove cells and large debris. (**c**) Initial clarification with 0.45 and 0.22 μm vacuum-driven filters. (**d**) Media concentration and further clarification by tangential flow filtration (TFF) with 750 kDa MWCO. (**e–g**) PEG precipitation of EVs. (**h**) Resuspension of EV pellet in 1× sterile DPBS. (**i**) EVs incubation with Capto Core 700 resin removes protein and lipid aggregates. (**j**) Capto Core 700 column chromatography further purifies EVs

3.2 Tangential Flow Filtration

1. Follow the manufacturer's recommendations and install the UF hollow fiber cartridge (750 kDa MWCO) to the ÄKTA Flux Tangential Flow Filtration System (*see* Fig. 1d). Make sure all connections are tightly closed to prevent leaking (*see* **Note 1**).
2. Prime the system with 1× DPBS. Rinse both cross-flow and equilibration buffer chamber and the cartridge, tubing, and outlets. Fully drain the DPBS in the cross-flow chamber after priming. Avoid introducing bubbles into the cartridge.
3. Add conditioned medium into the sample reservoir. Open the permeate line and connect it to a collection bottle. Maintain flow rate at 30–35 mL/min. Gradually and carefully apply pressure by adjusting the retentate pressure (Pr) control clamp till feed pressure (Pf) stabilizes at 30–35 psi and trans-membrane pressure (TMP) is at 28–33 psi. Make sure Pf is lower than 40 psi at all times (*see* **Note 2**).
4. Concentrate the medium to 10% of its original volume. Take both the sample volume in the cross-flow chamber and the system volume into account when calculating the final sample volume (*see* **Note 3**). Release the Pr control valve and close the permeate outlet to stop concentration. Collect samples through the upper drain valve.
5. To clean the system, wash the system thoroughly with 0.5 M NaOH and then ddH_2O. If used immediately after, neutralize the system with 20 mM Tris-HCl pH 6.8 solution ddH_2O and then 1× DPBS. Flush all outlets with 20 mM Tris-HCl pH 6.8 and test the pH of the buffer from the outlets to ensure the system is adequately neutralized. If the instrument is not used immediately after, store the system in 0.1 M NaOH. Rinse both cross-flow and equilibration buffer chambers, as well as the cartridge, tubing, and drain valves in each wash.

3.3 PEG Precipitation and EV Resuspension

1. Add sterile PEG-8000 to a final PEG-8000 concentration of 4% w/v. Mix thoroughly and seal the tube with parafilm to prevent leaking. Incubate at 4 °C with rotation overnight (*see* Fig. 1e).
2. Centrifuge the mixture at 1000 × *g* for 1 h at 4 °C to pellet EVs (*see* Fig. 1f, g).
3. Discard the supernatant. Carefully resuspend the EV pellet in 1× DPBS. Gently pipette up and down until the EV pellet fully resuspends. Try to avoid bubbles when mixing (*see* Fig. 1h).
4. Add RNase and DNase at 1:100 dilution, 100 μg/mL, and 10 U/mL, respectively, and incubate the mixtures at 4 °C overnight (*see* **Note 4**).

3.4 Capto Core 700 Chromatography

1. To prepare Capto Core 700 resin for pre-incubation (*see* **Note 5**), take 200 μL resin slurry, centrifuge to remove the storage buffer and resin fines, add 1 mL of 1× DPBS, and incubate for 5 min. Centrifuge at 500 × *g* for 5 min. Discard the supernatant. Repeat this step three times (*see* Fig. 1i).
2. Mix EV samples with the prepared resins and incubate for 5 min. Centrifuge at 500 × *g* for 5 min. Transfer the supernatant to a sterile tube. This is the EV sample pretreated with Capto Core 700 resin.
3. Turn on ÄKTA Start Chromatography System and prime the system with 1× CMF-DPBS. Connect the 1 mL HiTrap Capto Core 700 column and prime the column with 1× CMF-DPBS at a flow rate of 1 mL/min. Wash with at least three times column volumes of buffer. Install sterile tubes to the fraction collector (*see* Fig. 1j).
4. Inject the Capto Core 700 resin pretreated EV sample into the column. Start the system with a flow rate of 0.5 mL/min. EVs will remain in the flow-through while impurities bind to the column. Collect and save peak flow-through fractions (*see* **Note 6**).
5. The Capto Core 700 column can be regenerated by 0.5 M NaOH in 30% isopropanol followed by neutralization with 1× CMF-DPBS and stored in 20% ethanol. The column should be washed with ddH_2O before and after washes with organic solvents to prevent salt precipitation. Each wash should be at least five times column volumes.

4 Notes

1. Different filter cartridges may be used here with careful consideration of the operation parameters based on the cartridge design. Detailed information on the cartridge used in this method is listed in Table 1 for reference. Consult manufacturer's instructions for proper filter preparation and cleaning procedures. A routine water flux test should be performed to determine filter performance.
2. This protocol generates an estimated shear rate of 8000 s^{-1} under the aforementioned conditions. Samples prepared under 2000 s^{-1} to 4000 s^{-1} and 8000 s^{-1} at a constant TMP of 15–17 psi showed little to no impact on EV size distribution (*see* Fig. 2a), EV marker protein level (*see* Fig. 2b), average EV size (*see* Fig. 2c), and purity (*see* Fig. 2d and e). However, skipping the TFF step led to a significantly reduced level of EV marker proteins (*see* Fig. 2b) and purity (*see* Fig. 2d and e).

Table 1
Cartridge details

Supplier and model number	Pore size (NMWC)	Fiber i.d. (mm)	Membrane area (cm^2)	Nominal flow path number (cm)
Cytiva UFP-750-C-H24LA	750,000	0.5	42	60

Volumetric flow rate (mL/min)	8.5	17	33
Shear rate (s^{-1})	2000	4000	8000

3. The system volume, or hold-up volume, is the total volume of liquid in the tubing and filter cartridge outside the cross-flow chamber that would contribute to the final product volume. The system volume varies between different TFF units. The system volume of the instrument used in this method is 12 mL.
4. The minimum incubation time with RNase and DNase is 2 h. Lipid dyes or drugs can also be added in this step to incubate with EVs for labeling or drug loading purposes [13–15].
5. The EV solution can be very viscous and may cause high pressure during sample injection. To avoid pressure jumps and column damage, samples are pretreated with 1/3 volumes of primed Capto Core 700 resin slurry and centrifuged at 500 × *g* in a 5424R Eppendorf Centrifuge for 5 min at 4 °C (*see* Fig. 1i).
6. Three independent experiments with U2OS conditioned medium at a starting volume of 500 mL were performed to validate the consistency of the method. Samples were collected from the conditioned medium after 0.22 μm filtration (M), post TFF (CM), after PBS resuspension of the PEG precipitated pellet (PR), and after Capto Core 700 (FT). EV sizes were consistent throughout the purification method (*see* Fig. 3a). Samples were concentrated 1000-fold post-PBS resuspension of the PEG precipitated pellet, and the final concentration of EVs after Capto Core 700 is 100-fold higher than in the starting material (*see* Fig. 3c). EV sizes (*see* Fig. 3b) and quality (*see* Fig. 3d) are consistent from all three independent experiments. The particle recovery rate is listed in Table 2.

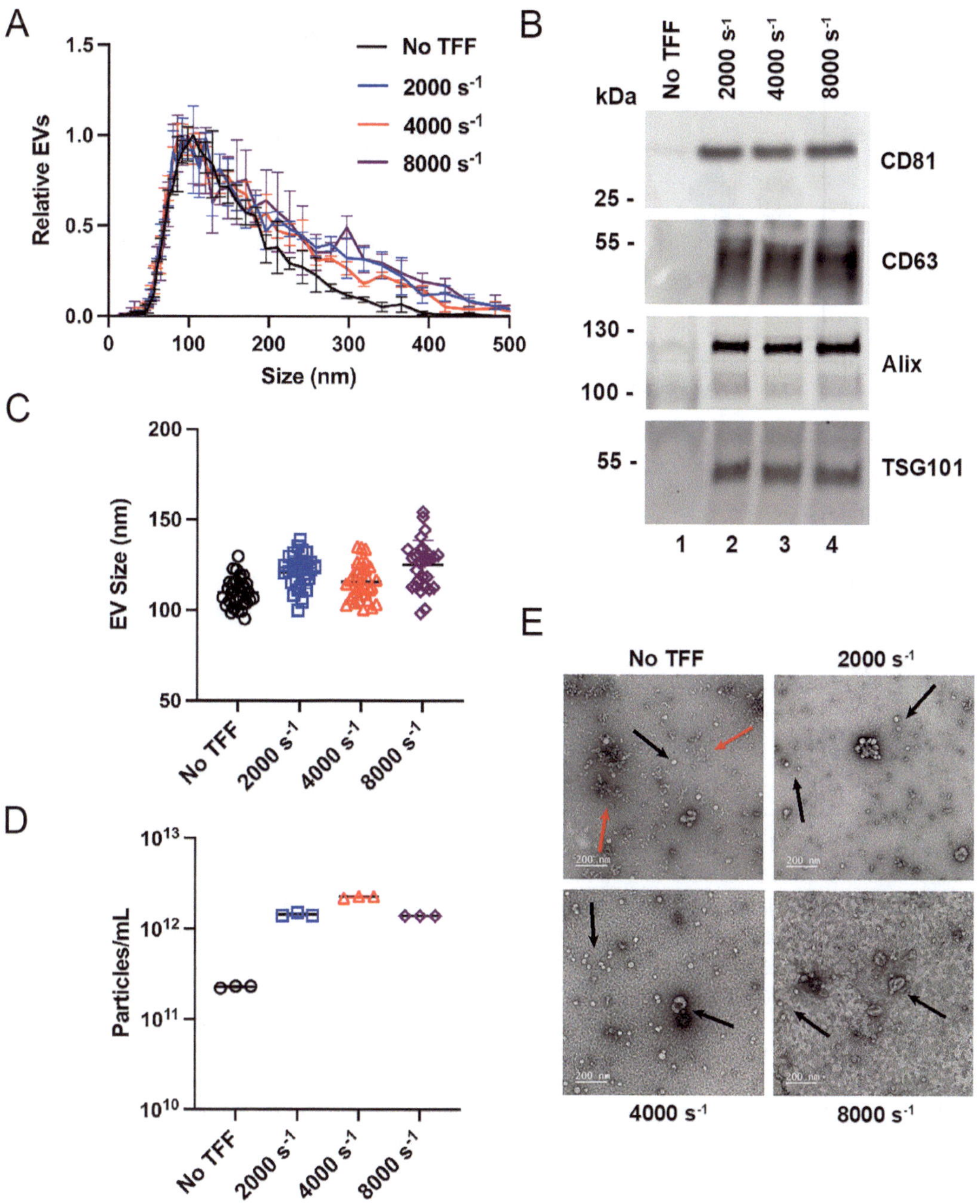

Fig. 2 TFF significantly improves sample purity by removing protein aggregates while little to no differences were observed between 2000 s^{-1}, 4000 s^{-1}, and 8000 s^{-1} samples. Data in A, C, and D were analyzed by Nanoparticle Tracking Analysis (NTA). (**a**) Size distribution. Black, no TFF; blue, shear rate = 2000 s^{-1}; red, shear rate = 4000 s^{-1}; purple, shear rate = 8000 s^{-1}. (**b**) Western blot. Lane 1, no TFF; lane 2, shear rate = 2000 s^{-1}; lane 3, shear rate = 4000 s^{-1}; lane 4, shear rate = 8000 s^{-1}. (**c**) Average EV size. Black circle, no TFF; blue rectangle, shear rate = 2000 s^{-1}; red triangle, shear rate = 4000 s^{-1}; purple rhombus, shear rate = 8000 s^{-1}. (**d**) Purity index. The purity index was calculated by dividing particle concentration with protein concentration (ANOVA, $N = 3$). (**e**) Transmission electron microscope images. Peak fractions from

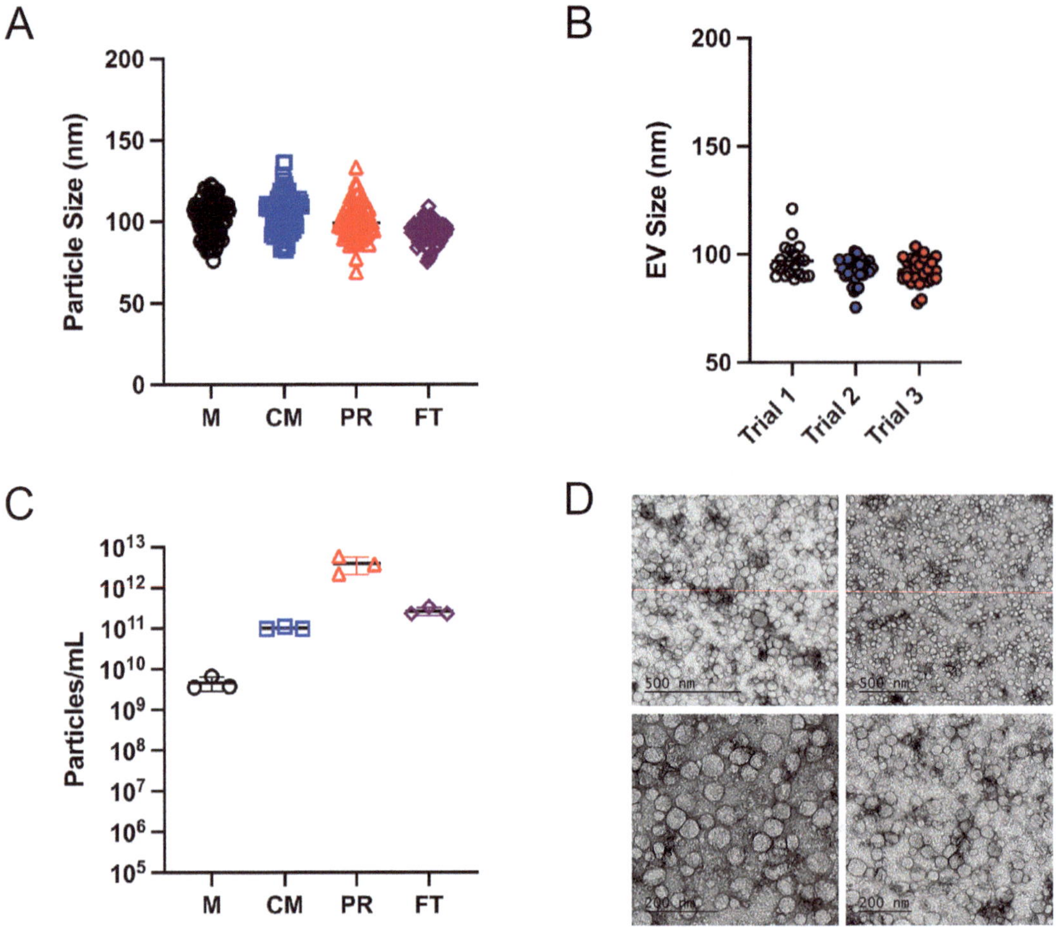

Fig. 3 The TFF-based, large-scale EV purification protocol consistently generated EVs at high yield and quality. (**a**) Particle size distribution throughout the EV purification protocol. M (black circle): conditioned media; CM (blue rectangle): condensed media; PR (red triangle): PBS resuspension; FT (purple rhombus): pooled CC 700 chromatography flow-through peak fractions. (**b**) Final EV size from pooled CC 700 chromatography flow-through peak fractions. Data from three individual experiments were reported with mean and SD. (**c**) Average particle concentration throughout the EV purification protocol. (**d**) Transmission electron microscope of EVs post-Capto Core C700 (at 8000 s^{-1} shear rate). Peak fractions from CC700 were absorbed on glow discharged carbon-coated 400 mesh copper grids. The grids were then stained with 2% w/v uranyl acetate in water. Samples were visualized on an FEI Tecnai 12 TEM at 80 kV. EM micrographs were captured using a Gatan Orius CCD camera and Gatan Digital Micrograph software

Fig. 2 (continued) CC700 were absorbed on glow discharged carbon-coated 400 mesh copper grids. The grids were then stained with 2% w/v uranyl acetate in water. Samples were visualized on an FEI Tecnai 12 TEM at 80 kV. EM micrographs were captured using a Gatan Orius CCD camera and Gatan Digital Micrograph software. Black arrows, EVs; red arrows, protein or lipid aggregates

Table 2
Particle recovery rate throughout the purification procedure

	Volume (mL)	Particle concentration (Particles/mL)	Total particle count (Particles)	Recovery (%)
M	500	4.6 ± 1.7 E+09	2.30E+12	100%
CM	50	8.5 ± 1.53 E+10	4.25E+12	100%
PR	0.5	5.1 ± 3.3 E+12	2.55E+12	100%
FT	2.0	2.8 ± 1.32 E+11	5.53E+11	24%

M: conditioned media, *CM:* concentrated media, *PR:* PBS resuspension, *FT:* pooled CC 700 chromatography flow-through fractions. The total particle count was calculated by multiplying particle concentrations with volume. The recovery rate was calculated by dividing particle count throughout each step by that of the conditioned medium (M)

Acknowledgments

This work was supported by public health service grant R01-CA228172 to D.P.D.

References

1. Kalluri R, LeBleu VS (2020) The biology, function, and biomedical applications of exosomes. Science 367(6478):eaau6977
2. Théry C, Witwer KW, Aikawa E et al (2018) Minimal information for studies of extracellular vesicles 2018 (MISEV2018): a position statement of the International Society for Extracellular Vesicles and update of the MISEV2014 guidelines. J Extracell Vesicles 7(1):1535750
3. Théry C, Amigorena S, Raposo G et al (2006. **Chapter 3**) Isolation and characterization of exosomes from cell culture supernatants and biological fluids. Curr Protoc Cell Biol 3:22
4. Lobb RJ, Becker M, Wen SW et al (2015) Optimized exosome isolation protocol for cell culture supernatant and human plasma. J Extracell Vesicles 4:27031
5. McNamara RP, Caro-Vegas CP, Costantini LM et al (2018) Large-scale, cross-flow based isolation of highly pure and endocytosis-competent extracellular vesicles. J Extracell Vesicles 7(1): 1541396
6. Konoshenko MY, Lekchnov EA, Vlassov AV et al (2018) Isolation of extracellular vesicles: general methodologies and latest trends. Biomed Res Int 2018:8545347
7. Rider MA, Hurwitz SN, Meckes DG (2016) ExtraPEG: a polyethylene glycol-based method for enrichment of extracellular vesicles. Sci Rep 6(1):23978
8. Sidhom K, Obi PO, Saleem A (2020) A review of Exosomal isolation methods: is size exclusion chromatography the best option? Int J Mol Sci 21(18):6466
9. Burgess RR (2018) A brief practical review of size exclusion chromatography: rules of thumb, limitations, and troubleshooting. Protein Expr Purif 150:81–85
10. Gámez-Valero A, Monguió-Tortajada M, Carreras-Planella L et al (2016) Size-exclusion chromatography-based isolation minimally alters extracellular Vesicles' characteristics compared to precipitating agents. Sci Rep 6(1): 33641
11. Benedikter BJ, Bouwman FG, Vajen T et al (2017) Ultrafiltration combined with size exclusion chromatography efficiently isolates extracellular vesicles from cell culture media for compositional and functional studies. Sci Rep 7(1):15297
12. McNamara RP, Dittmer DP (2020) Modern techniques for the isolation of extracellular vesicles and viruses. J Neuroimmune Pharmacol 15(3):459–472
13. McNamara RP, Eason AB, Zhou Y et al (2022) Exosome-encased nucleic acid scaffold chemotherapeutic agents for superior anti-tumor and anti-angiogenesis activity. ACS Bio Med Chem Au, vol 2, p 140
14. McNamara RP, Zhou Y, Eason AB et al (2022) Imaging of surface microdomains on individual extracellular vesicles in 3-D. J Extracell Vesicles 11(3):e12191–e12191
15. Chambers MG, McNamara RP, Dittmer DP (2021) Direct stochastic optical reconstruction microscopy of extracellular vesicles in three dimensions. J Vis Exp 174

Chapter 6

Metabolomics Analysis of Urinary Extracellular Vesicles by Nuclear Magnetic Resonance and Liquid Chromatography–Mass Spectrometry

Marta Martin-Lorenzo, Dolores Molero, and Gloria Alvarez-Llamas

Abstract

Extracellular vesicle (EV) release and their content are influenced by diverse clinical conditions. EVs participate in inter-cellular communication and have been postulated as reflectors of the pathophysiology of the cells, tissues, organs or the whole system with which they are in contact. Urinary EVs have been proved to reflect pathophysiology not only of renal system related diseases constituting an additional source of potential biomarkers easily accessible in a non-invasive way. The interest in EVs cargo has been mostly focused on proteins and nucleic acids and more recently it has been extended to metabolites. Metabolites represent the downstream changes in the genome, transcriptome, and proteome as a reflection of processes occurring in living organisms. For their study, nuclear magnetic resonance (NMR) and mass spectrometry in tandem (LC-MS/MS) are widely used. NMR is a reproducible and non-destructive technique and we show here methodological protocols for the metabolomics analysis of urinary EVs by NMR. Additionally, we also describe the workflow for a targeted LC-MS/MS analysis that is extensible to untargeted studies.

Key words Extracellular vesicles, Metabolites, Metabolomics, Nuclear magnetic resonance, Mass spectrometry, LC–MS

1 Introduction

Extracellular vesicles (EVs) were initially considered as a mere way of cellular disposal but they have been proved to be inter-cellular communicators, effective vehicles for material transfer and a rich source of potential biomarkers in various diseases. Depending on the clinical context, the significantly increased (or decreased) number of released EVs may be directly considered a pathological indicator. Beyond that, it has also been shown that the EV cargo is influenced by different clinical conditions. Most studies focus on abundance variation in nucleic acids and proteins between compared groups. As a result of its formation process within the cell, EVs may contain both intracellular and transmembrane

Seppo Vainio (ed.), *Cell-Secreted Vesicles: Methods and Protocols*, Methods in Molecular Biology, vol. 2668,
https://doi.org/10.1007/978-1-0716-3203-1_6,

components so being a potential reflector of what is happening directly in the cell, the tissue, the organ, and the whole system with which they are in contact. EVs even act as mediators of disease or directly participate in reparation mechanisms [1–4].

In cancer, tumor-derived EVs appear to be valuable non-invasive surrogates for biopsies [5]. When targeting renal diseases, mainly urogenital system cancers, diabetic nephropathy, renal fibrosis, or kidney allograft function post-transplantation, urine is the biological fluid of choice [6–8]. The composition of this biological fluid is the result of the blood filtration in the glomeruli, minus what it is reabsorbed in the renal tubule plus the active secretion into the tubule. Cells that comprise the nephron (podocytes, endothelial, mesangial, parietal, and tubular cells) secrete EVs along the urinary tract [9] and once isolated from urine they can be investigated for pathophysiological clues, making particular sense as a complementary source to direct urinalysis [10]. Apart from this logical utility in the study of renal diseases, urinary EVs have been also investigated in the study of cardiovascular pathology, as diabetes mellitus, hypertension, or atherosclerosis development [11–15] and in neurologic disease [16, 17].

More recently, the interest in EVs cargo has been extended to their metabolic content [18] mainly in cancer studies [19–22], but also in cardiovascular disease [11], acute pancreatitis [23], gestational hypertension [24], and infection by SARS-CoV-2 [25]. The number of studies is very limited but sufficiently to envisage their enormous potential individually or in combination with information provided by proteins and nucleic acids.

Here we show methodological protocols for the metabolomics analysis of urinary EVs isolated by ultracentrifugation as previously published by our group (*see* Fig. 1) [11, 26, 27]. We include analyses by nuclear magnetic resonance (NMR and high-resolution magic angle spinning, HR-MAS) and by liquid chromatography–mass spectrometry in selected reaction monitoring mode (SRM-LC–MS/MS) (targeted analysis). Protocols for 2D-NMR acquisition and recommendations for subsequent metabolites identification are also included.

2 Materials

2.1 Urine Collection

1. Protease inhibitor cocktail: 104 mM 4-(2-aminoethyl) benzenesulfonyl fluoride hydrochloride (AEBSF), 80 μL aprotinin, 2 mM leupeptin, 1.5 mM pepstatin, and 1.4 mM E-64 (P8340, Sigma).
2. Sterile urine container.

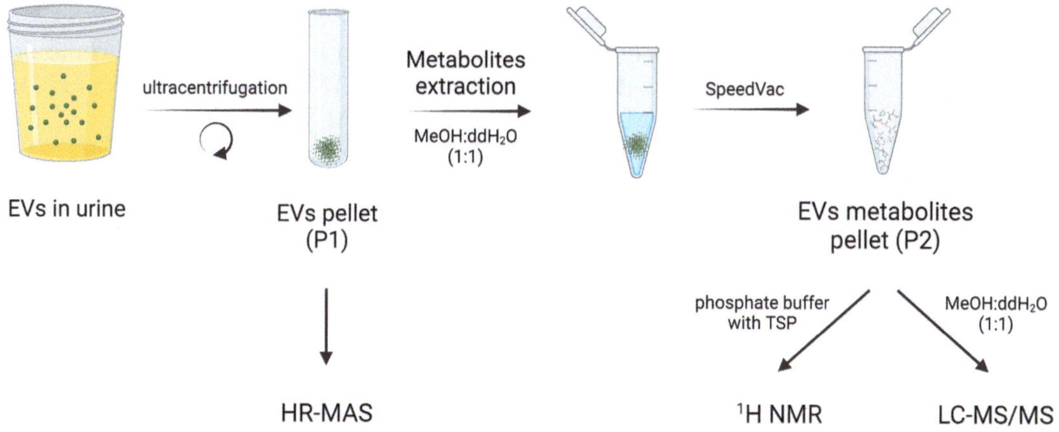

Fig. 1 The urinary EVs are isolated by ultracentrifugation and the resulting pellet can be directly analyzed by HR-MAS or metabolites can be extracted in organic solution for ^{1}H NMR and LC-MS/MS

2.2 Urinary EVs Isolation

1. Isolation solution: 10 mM triethanolamine and 250 mM sucrose in double-distilled water (ddH_2O), pH to 7.6.
2. Phosphate-buffered saline (PBS) (10X): 11.5 mM NaH_2PO_4, 12.5 mM Na_2HPO_4 and 150 mM NaCl in ddH_2O, pH 7.2.
3. Dithiothreitol (DTT).
4. Polypropylene centrifuge tubes (50 mL, ref.: 357005 Beckman).
5. Polycarbonate Aluminum Bottle with Cap Assembly (26.3 mL ref.: 355618 Beckman).
6. JA 20 Rotor-fixed angle (Beckman).
7. 70 Ti Rotor-fixed angle (Beckman).

2.3 EVs Characterization by Electron Microscopy

1. 4% Paraformaldehyde in PBS.
2. 2% Uranyl acetate in ddH_2O.
3. Parafilm.
4. Formvar–carbon-coated EM grids (Ted Pella Inc.).
5. JEM 1010 transmission microscope (JEOL Peabody).

2.4 Metabolites Extraction

1. Cold methanol (MeOH): ddH_2O (1:1).

2.5 Untargeted Analysis by Nuclear Magnetic Resonance (NMR)

1. Bruker AVIII500 spectrometer 11.7 T operating at 500.13 MHz.
2. Bruker 700-MHz AVANCE III instrument operating at 700.17 MHz equipped with a 5 mm triple resonance cryo-probe with a z-axis gradient.
3. Acquisition & Processing software: TOPSPIN (Bruker).

4. Deuterated water (D_2O) (99.9%).
5. Buffered phosphate solution with 3-(Trimethylsilyl)propionic-2,2,3,3-d4 acid sodium salt (TSP): 200 mM Na_2HPO_4, 200 mM NaH_2PO_4 and 0.01 mM TSP in D_2O.
6. NMR 5 mm tubes (for ^{1}H NMR analysis).
7. 50-μL zirconium oxide rotor with cylindrical insert (for HR-MAS analysis).

2.6 Targeted Analysis by Liquid Chromatography and Mass Spectrometry in Tandem (LC–MS/MS)

1. 6460 Triple Quadrupole LC-MS/MS (1200 Series, Agilent Technologies).
2. Acquisition software: Mass Hunter software (Agilent Technologies).
3. Atlantis T3 column (Waters).
4. Mobile phase A: 0.1% formic acid in ddH_2O.
5. Mobile phase B: 0.1% formic acid in acetonitrile (LC/MS grade).
6. 0.20 μm PTFE membrane from aqueous and organic solutions syringe filters.
7. 12 × 32 mm glass screw neck vial.

3 Methods

3.1 Urine Collection for EVs Analysis

1. Collect urine in a sterile container (*see* **Notes 1** and **2**).
2. Add 50 μL protease inhibitor cocktail per 100 mL urine volume and vortex vigorously (*see* **Note 3**).
3. Process immediately or store at −80 °C.

3.2 EVs Isolation by Ultracentrifugation

1. In case of frozen urine, thaw the sample first at room temperature for 10 min and then at 37 °C in a water bath. Vigorous vortex for 2 min.
2. Low-speed centrifugation (17,000 × *g*) for 10 min at 4 °C (*see* **Note 4**). Collect the supernatant (S1) by decanting and keep it (*see* **Note 5**).
3. Treat the pellet from the low-speed centrifugation with 500 μL PBS and 0.1 g DTT. Vortex. Heat at 37 °C for 10 min (*see* **Note 6**).
4. Add 7–10 mL PBS to the pellet treated with DTT and centrifuge again at 17,000 × *g* for 10 min at 4 °C. Discard the resulting pellet and keep the supernatant (S2).
5. Mix S1 and S2, and distribute the mixture in the polypropylene centrifuge tubes. Use a precision balance to weight them and

equalize the weighs 2 to 2 by adding PBS if needed. Face the tubes with the same weight and ultracentrifuge at 175,000 × *g* for 70 min at 4 °C (*see* **Note 7**).

6. Discard the supernatant and keep the EVs containing pellet (P1) (*see* **Note 8**).
7. Store the resultant EVs pellet (P1) at −80 °C or go directly to characterization or metabolite extraction (*see* **Note 9**).

3.3 EVs Characterization by Electron Microscopy

1. Resuspend the EVs pellet (P1) in 500 μL 4% paraformaldehyde in PBS and spot 15 μL on parafilm (*see* **Note 10**).
2. Float a Formvar-carbon-coated grid on the sample droplet for 3 min.
3. Use a filter paper to eliminate the excess fluid from the grid.
4. Float the grid on a drop of ddH_2O for 1 min.
5. Negative stain by floating the grid on a drop of 2% uranyl acetate in ddH_20 for 30 s and remove the excess fluid.
6. Visualize using a transmission microscope (*see* **Note 11**).

3.4 Metabolites Extraction from EVs

1. Solubilize the EVs pellet (P1) in cold methanol: ddH_20 (1:1) and vortex vigorously (*see* **Notes 12** and **13**).
2. Sonicate for 30 min at 4 °C.
3. Centrifuge at 10,000 × *g* for 10 min at 4 °C.
4. Discard the pellet and keep the supernatant (S3). Perform SpeedVac up to dryness on S3 and keep the EVs metabolites pellet (P2) at −80 °C until analysis (*see* **Note 14**).

3.5 Metabolomic Analysis by NMR

NMR spectroscopy is a robust, reproducible, and non-destructive technique widely used in the study of metabolites present in different biological matrices. It has a limited sensitivity overcome thanks to high-field equipments. Differential clinical studies are mostly performed using (1D)-^{1}H NMR experiments. The one-dimensional version of the NOESY (nuclear Overhauser enhancement spectroscopy) is the most commonly used. The capacity of monodimensional NMR is limited for the unequivocal identification of metabolites because of the potential signal overlapping particularly in complex samples (e.g., biological matrices). Thus, two-dimensional (2D) experiments are recommended including H-H correlation experiments COSY, total H-H correlation TOCSY, and heteronuclear H-C correlation HSQC for metabolite assignment confirmation. Acquisition and processing data here presented were performed using Bruker instruments and TOPSPIN software (Bruker Biospin Ltd).

3.5.1 Metabolomics Analysis by ^{1}H NMR

1. Resuspend the EVs metabolites pellet (P2) in 300 μL ddH_20 and 300 μL of buffered phosphate solution with TSP. Vortex vigorously.
2. Sonicate for 20 min at 4 °C.
3. Centrifuge at 10,000 × *g* for 30 s and discard the pellet.
4. Transfer the supernatant to a NMR cryotube of 5 mm diameter.
5. Perform acquisition and processing of data with TOPSPIN. Set the temperature to 278 K and insert the sample into the magnet. Wait for 5–10 min to allow the sample reaching the optimal temperature.
6. Select the solvent and lock. Adjust tune and match. Shimming using Bruker "topshim" command in topspin.
7. Perform a standard proton experiment with 90° pulse of 2 μs. Check that the line shape is good enough to ensure a good resolution and measure the chemical shift of the water signal to set the value of the offset for water suppression.
8. Calculate the 90° pulse width using Bruker "pulsecal" command.
9. Carefully adjust the power level for water suppression without suppressing any metabolite signal.
10. NMR acquisition: Use the 1H one-dimensional version of nuclear Overhauser effect spectroscopy NOESY 1D (noesypr1d; Bruker Biospin) with pre-saturation during relaxation time and mixing time. Parameters: spectral width 16 ppm; time domain size 32 K; mixing time 150 ms; relaxation time 2 s; number of scans 512.
11. Process the spectrum with a 1-Hz exponential multiplication line broadening, prior to the Fourier transformation; and once in the frequency domain, adjust the phase of the peaks and the baseline. Internally reference the chemical shift to the TSP signal at δ: 0.00 ppm. (*see* **Note 15**) (*see* Fig. 2a).
12. For differential analysis, load and perform alignment of the spectra into AMIX (Analysis of MIXtures, Bruker Rheinstetten, Germany). Reduce ^{1}H NMR spectra by subdivision into integral regions of a fixed or customized ppm range (typical values are 0.02–0.04 ppm bucket size) between δ 0.4 and 9 ppm (excluding the water region). Normalize the buckets and perform the desired statistical analysis (customize the normalization and statistical analysis parameters for the particular comparisons).
13. Perform signals assignment by Chenomx NMR Suite (Chenomx NMR Suit, Edmonton, Canada) or other software and the database HMBD.

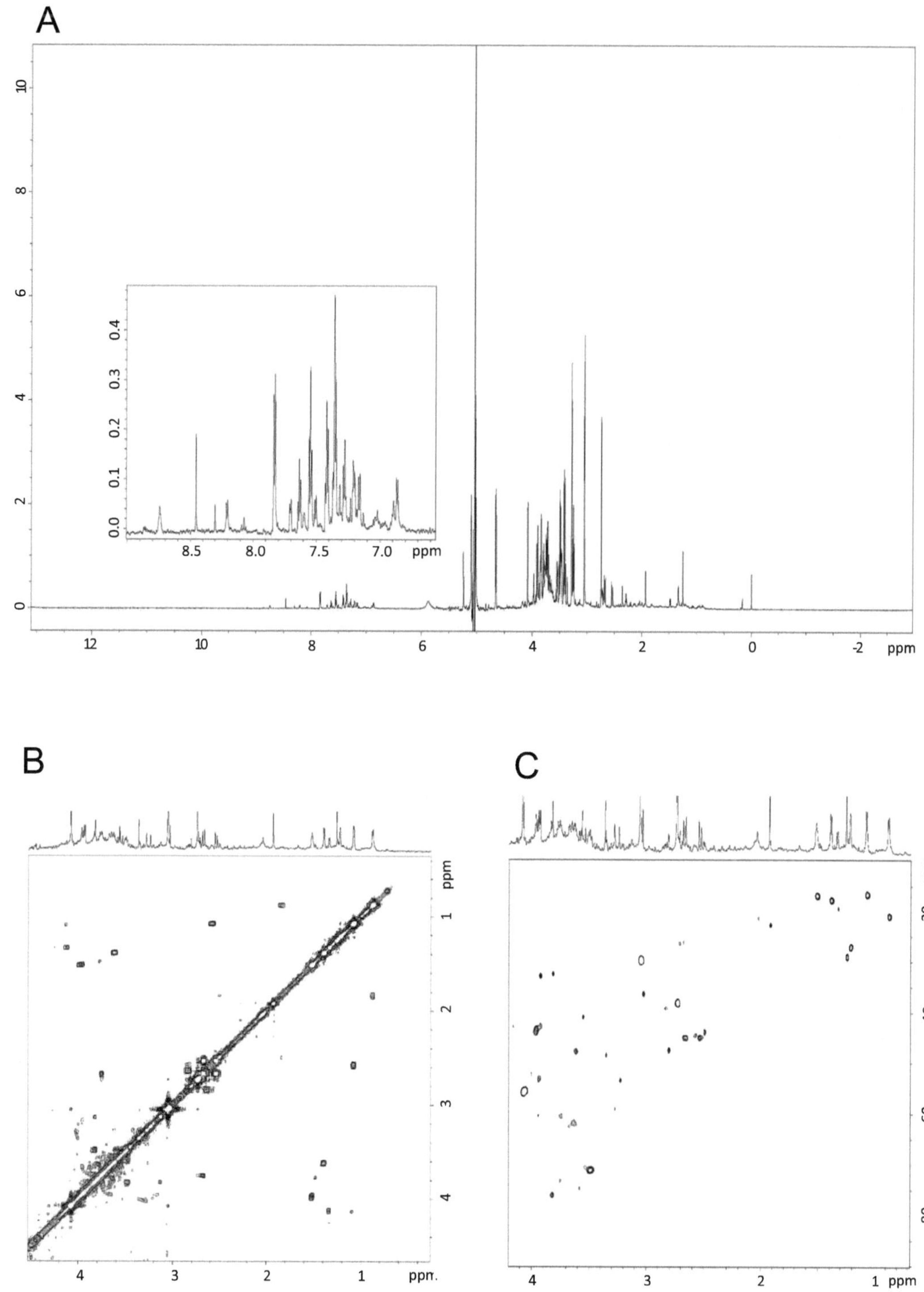

Fig. 2 Representative 700 MHz ^{1}H NMR (**a**), H-H COSY (**b**) and H-C HSQC (**c**) spectra of urinary EVs sample

14. H-H COSY (2D homonuclear shift correlation; for 2D-NMR assignment confirmation). Use the sequence cosygpprqf, Bruker Biospin, with pre-saturation during relaxation delay and using gradient pulses for selection. Parameters: spectral width 12.4844 ppm in both dimensions; domain size 2048 in f2 and 256 in f1; d1 2s number of scans192 (*see* Fig. 2b).
15. H-C HSQC (2D heteronuclear shift correlation; for 2D-NMR assignment confirmation). Use the sequence hsqcetgpprsisp2.2, Bruker Biospin, with pre-saturation during relaxation delay and decoupling during acquisition. Parameters: spectral width 15.9 ppm in f2 and 249.9 ppm in f1; domain size 2048 in f2 and 256 in f1; number of scans 192 (*see* Fig. 2c).

3.5.2 Metabolomic Analysis by HR-MAS

1. In order to avoid degradation perform the NMR analysis at 4°C and minimize shimming and NMR set up times at maximum
2. Transfer the EVs pellet (P1) to a 50 μL zirconium oxide rotor with cylindrical insert and spun at 5000 Hz spinning rate to remove the effects of spinning side bands from the spectra acquired.
3. Acquire the spectrum using a sequence based on the first increment of the NOESY pulse sequence and the following parameters: 512 scans, a spectral width of 6009 Hz and 32k data points. Perform water suppression through irradiation of water signal during 2 s.
4. To suppress broad signals from macromolecules a Carr-Purcell-Meibom-Gill (CPMG) pulse sequence is recommended as a T2 filter.
5. Phase the spectra, correct the baseline and reference to δ 0 ppm the TSP singlet using the same software of the acquisition, TOPSPIN software.
6. For differential analysis and metabolite identification follow the **steps 12** to **15** of subheading 3.5.1: Metabolomic analysis by ^{1}H NMR.

3.6 Metabolomic Analysis by Targeted Mass Spectrometry

There are many analytical procedures to carry out metabolomics analyses by liquid chromatography and mass spectrometry in tandem (LC–MS/MS). Two main approaches can be distinguished, untargeted and targeted analysis both of them requiring parameters set-up depending on the available instrumentation. In untargeted LC–MS/MS, no preselection of metabolites is made beforehand and it is applied for discovery purposes. When there is a special interest in a particular metabolite, group of metabolites or in a specific metabolic pathway, a targeted metabolic analysis is performed. We describe here an LC–MS/MS protocol in Selected

Reaction Monitoring mode that can be used as complementary analysis to NMR.

1. Dissolve the EVs metabolites pellet (P2) in 100 μL cold methanol: ddH_20 (1:1).
2. Vortex and sonicate for 10 min (*see* **Note 16**).
3. Centrifuge at 15000 × *g* for 10 min and discard the undissolved material.
4. Filter the solution through a 0.20 μm syringe filter directly to the MS vial.
5. Inject 10 μL of the filtered extract onto an Atlantis T3 column (Waters) thermostatically controlled at 40 °C.
6. Analyze by LC–MS/MS at 0.4 mL/min in an acetonitrile gradient: (1) at 0 min 0% B, (2) at 1 min 0% B, (3) at 2.5 min 95% B, (4) at 2.51 min 0% B, and (5) at 5 min 0% B. Dwell time = 50 ms, delta EMV = 400 V (positive mode) or 600 V (negative mode) (*see* **Note 17**).

4 Notes

1. Specific clinical cases may deeply condition the urine dilution factor; e.g., diabetic patients may provide unusually diluted urine samples. In general, a minimum volume of 50 mL per individual sample ensures enough material for EVs characterization, metabolomic untargeted analysis, and targeted confirmation.
2. Second morning urine is suitable if not dietary interference is expected to influence result. Mid-stream urine is recommended.
3. If urine must be immediately frozen, without the possibility of adding the protease inhibitor cocktail, it should be added during the thawing process as soon as possible.
4. Cellular debris is removed in the low-speed centrifugation.
5. Separate the supernatant immediately afterwards to avoid redissolution of the pellet.
6. By treatment with DTT the Tamm–Horsfall protein (THP) network is disaggregated and the unspecifically entrapped EVs can be released. THP is a major glycoprotein component of urine thus inherent to the samples.
7. It is important to activate the centrifuge brake to stop the centrifugation smoothly but fast in order to avoid pellet redissolution.

8. To avoid urine matrix contamination, carefully wipe out with paper any possible rest of supernatant without disturbing the EVs pellet.
9. In case of biomarker analysis, weight the pellet (P1) before freezing for normalization purposes.
10. Whenever possible perform the analysis without freezing the samples; they can be stored overnight at 4 °C.
11. EVs can also be characterized by Western blot, nano tracking analysis, confocal microscopy, ExoView® platform, among others.
12. Solubilization in 400–600 μL organic solution works for NMR and LC–MS/MS analyses. Adapted from [28].
13. In case of HR-MAS analysis, skip this step and go directly to subheading 3.5.2.
14. The metabolites extract (S3) can be also kept at 4 °C for 24 h or −80 °C in case of long-term storage before SpeedVac up to dryness.
15. Phase adjustment and correct 0 ppm setting are key for spectra comparison.
16. Prolong the sonication time up to complete dissolution of the pellet.
17. A standard analytical workflow is shown for targeted metabolomics analysis by LC–MS/MS in Selected Reaction Monitoring (SRM) mode. Analytical conditions in terms of collision energy and fragmentor potential should be optimized for every metabolite by using commercial standards. Conditions for different instruments should be optimized.

References

1. Azevedo CAB, da Cunha RS, Junho CVC, da Silva JV, Moreno-Amaral AN, de Moraes TP et al (2021) Extracellular vesicles and their relationship with the heart-kidney Axis, uremia and peritoneal dialysis. Toxins (Basel) 13(11). https://doi.org/10.3390/toxins13110778
2. Sharma M, Ravichandran R, Bansal S, Bremner RM, Smith MA, Mohanakumar T (2018) Tissue-associated self-antigens containing exosomes: role in allograft rejection. Hum Immunol 79(9):653–658. https://doi.org/10.1016/j.humimm.2018.06.005
3. Weng Q, Wang Y, Xie Y, Yu X, Zhang S, Ge J et al (2022) Extracellular vesicles-associated tRNA-derived fragments (tRFs): biogenesis, biological functions, and their role as potential biomarkers in human diseases. J Mol Med (Berl) 100:679. https://doi.org/10.1007/s00109-022-02189-0
4. Zhang Y, Ma KL, Gong YX, Wang GH, Hu ZB, Liu L et al (2018) Platelet microparticles mediate glomerular endothelial injury in early diabetic nephropathy. J Am Soc Nephrol 29(11):2671–2695. https://doi.org/10.1681/asn.2018040368
5. Tang D, Liu S, Shen H, Deng G, Zeng S (2022) Extracellular vesicles promote the formation of pre-metastasis niche in gastric cancer. Front Immunol 13:813015. https://doi.org/10.3389/fimmu.2022.813015
6. Huang W, Zhu XY, Lerman A, Lerman LO (2022) Extracellular vesicles as Theranostic tools in kidney disease. Clin J Am Soc Nephrol 17:1418. https://doi.org/10.2215/cjn.16751221

7. Lu Y, Liu D, Feng Q, Liu Z (2020) Diabetic nephropathy: perspective on extracellular vesicles. Front Immunol 11:943. https://doi.org/10.3389/fimmu.2020.00943
8. Zhang PL, Liu ML (2021) Extracellular vesicles mediate cellular interactions in renal diseases-novel views of intercellular communications in the kidney. J Cell Physiol 236(8):5482–5494. https://doi.org/10.1002/jcp.30268
9. Martinez-Arroyo O, Ortega A, Redon J, Cortes R (2021) Therapeutic potential of extracellular vesicles in hypertension-associated kidney disease. Hypertension 77(1):28–38. https://doi.org/10.1161/hypertensionaha.120.16064
10. Erdbrügger U, Blijdorp CJ, Bijnsdorp IV, Borràs FE, Burger D, Bussolati B et al (2021) Urinary extracellular vesicles: a position paper by the urine task force of the International Society for Extracellular Vesicles. J Extracell Vesicles 10(7):e12093. https://doi.org/10.1002/jev2.12093
11. Agudiez M, Martinez PJ, Martin-Lorenzo M, Heredero A, Santiago-Hernandez A, Molero D et al (2020) Analysis of urinary exosomal metabolites identifies cardiovascular risk signatures with added value to urine analysis. BMC Biol 18(1):192. https://doi.org/10.1186/s12915-020-00924-y
12. Bazzell BG, Rainey WE, Auchus RJ, Zocco D, Bruttini M, Hummel SL et al (2018) Human urinary mRNA as a biomarker of cardiovascular disease. Circ Genom Precis Med 11(9):e002213. https://doi.org/10.1161/circgen.118.002213
13. Gonzalez-Calero L, Martínez PJ, Martin-Lorenzo M, Baldan-Martin M, Ruiz-Hurtado-G, de la Cuesta F et al (2017) Urinary exosomes reveal protein signatures in hypertensive patients with albuminuria. Oncotarget 8(27):44217–44231. https://doi.org/10.18632/oncotarget.17787
14. Herrera-Van Oostdam AS, Toro-Ortíz JC, López JA, Noyola DE, García-López DA, Durán-Figueroa NV et al (2020) Placental exosomes isolated from urine of patients with gestational diabetes exhibit a differential profile expression of microRNAs across gestation. Int J Mol Med 46(2):546–560. https://doi.org/10.3892/ijmm.2020.4626
15. Perez-Hernandez J, Riffo-Campos AL, Ortega A, Martinez-Arroyo O, Perez-Gil D, Olivares D et al (2021) Urinary- and plasma-derived exosomes reveal a distinct MicroRNA signature associated with albuminuria in hypertension. Hypertension 77(3):960–971. https://doi.org/10.1161/hypertensionaha.120.16598
16. Sun R, Wang H, Shi Y, Gao D, Sun Z, Chen Z et al (2019) A pilot study of urinary exosomes in Alzheimer's disease. Neurodegener Dis 19(5–6):184–191. https://doi.org/10.1159/000505851
17. Wang S, Kojima K, Mobley JA, West AB (2019) Proteomic analysis of urinary extracellular vesicles reveal biomarkers for neurologic disease. EBioMedicine 45:351–361. https://doi.org/10.1016/j.ebiom.2019.06.021
18. Williams C, Palviainen M, Reichardt NC, Siljander PR, Falcón-Pérez JM (2019) Metabolomics applied to the study of extracellular vesicles. Meta 9(11). https://doi.org/10.3390/metabo9110276
19. Puhka M, Takatalo M, Nordberg ME, Valkonen S, Nandania J, Aatonen M et al (2017) Metabolomic profiling of extracellular vesicles and alternative normalization methods reveal enriched metabolites and strategies to study prostate cancer-related changes. Theranostics 7(16):3824–3841. https://doi.org/10.7150/thno.19890
20. Smolarz M, Kurczyk A, Jelonek K, Żyła J, Mielańczyk Ł, Sitkiewicz M et al (2021) The lipid composition of serum-derived small extracellular vesicles in participants of a lung cancer screening study. Cancers (Basel). 13(14). https://doi.org/10.3390/cancers13143414
21. de Bruyn DP, Beasley AB, Verdijk RM, van Poppelen NM, Paridaens D, de Keizer ROB et al (2022) Is tissue still the issue? The promise of liquid biopsy in uveal melanoma. Biomedicine 10(2). https://doi.org/10.3390/biomedicines10020506
22. Strybel U, Marczak L, Zeman M, Polanski K, Mielańczyk Ł, Klymenko O et al (2022) Molecular composition of serum exosomes could discriminate rectal cancer patients with different responses to neoadjuvant radiotherapy. Cancers (Basel) 14(4). https://doi.org/10.3390/cancers14040993
23. Lou D, Shi K, Li HP, Zhu Q, Hu L, Luo J et al (2022) Quantitative metabolic analysis of plasma extracellular vesicles for the diagnosis of severe acute pancreatitis. J Nanobiotechnol 20(1):52. https://doi.org/10.1186/s12951-022-01239-6
24. Weingrill RB, Paladino SL, Souza MLR, Pereira EM, Marques ALX, Silva ECO et al (2021) Exosome-enriched plasma analysis as a tool for the early detection of hypertensive gestations. Front Physiol 12:767112. https://doi.org/10.3389/fphys.2021.767112

25. Alzahrani FA, Shait Mohammed MR, Alkarim S, Azhar EI, El-Magd MA, Hawsawi Y et al (2021) Untargeted metabolic profiling of extracellular vesicles of SARS-CoV-2-infected patients shows presence of potent anti-inflammatory metabolites. Int J Mol Sci 22(19). https://doi.org/10.3390/ijms221910467
26. Zubiri I, Posada-Ayala M, Benito-Martin A, Maroto AS, Martin-Lorenzo M, Cannata-Ortiz P et al (2015) Kidney tissue proteomics reveals regucalcin downregulation in response to diabetic nephropathy with reflection in urinary exosomes. Transl Res 166(5):474–84.e4. https://doi.org/10.1016/j.trsl.2015.05.007
27. Zubiri I, Posada-Ayala M, Sanz-Maroto A, Calvo E, Martin-Lorenzo M, Gonzalez-Calero L et al (2014) Diabetic nephropathy induces changes in the proteome of human urinary exosomes as revealed by label-free comparative analysis. J Proteome 96:92–102. https://doi.org/10.1016/j.jprot.2013.10.037
28. Martin-Lorenzo M, Gonzalez-Calero L, Maroto AS, Martinez PJ, Zubiri I, de la Cuesta F et al (2016) Cytoskeleton deregulation and impairment in amino acids and energy metabolism in early atherosclerosis at aortic tissue with reflection in plasma. Biochim Biophys Acta 1862(4):725–732. https://doi.org/10.1016/j.bbadis.2015.12.006

Chapter 7

Methodologies for Scalable Production of High-Quality Purified Small Extracellular Vesicles from Conditioned Medium

Gabriella Andriolo, Elena Provasi, Andrea Brambilla, Stefano Panella, Sabrina Soncin, Viviana Lo Cicero, Marina Radrizzani, Lucia Turchetto, and Lucio Barile

Abstract

The development of an extracellular vesicles (EV)-based therapeutic product requires the implementation of reproducible and scalable, purification protocols for clinical-grade EV. Commonly used isolation methods including ultracentrifugation, density gradient centrifugation, size exclusion chromatography, and polymer-based precipitation, faced limitations such as yield efficiency, EV purity, and sample volume. We developed a GMP-compatible method for the scalable production, concentration, and isolation of EV through a strategy involving, tangential flow filtration (TFF). We applied this purification method for the isolation of EV from conditioned medium (CM) of cardiac stromal cells, namely cardiac progenitor cells (CPC) which has been shown to possess potential therapeutical application in heart failure. Conditioned medium collection and EV isolation using TFF demonstrated consistent particle recovery (~10^{13} particle/mL) enrichment of small/medium-EV subfraction (range size 120–140 nm). EV preparations achieved a 97% reduction of major protein-complex contaminant and showed unaltered biological activity. The protocol describes methods to assess EV identity and purity as well as procedures to perform downstream applications including functional potency assay and quality control tests. The large-scale manufacturing of GMP-grade EV represents a versatile protocol that can be easily applied to different cell sources for wide range of therapeutic areas.

Key words Cardiac Progenitors Cells (CPC), Master Cell Bank (MCB), Conditioned Medium (CM), Extracellular Vesicles (EV), Advanced Therapy Medicinal Products (ATMP), Good Manufacturing Practices (GMP), Production methods, Quality Control methods (QC) Tangential Flow Filtration (TFF)

Gabriella Andriolo and Elena Provasi contributed equally with all other contributors.

Seppo Vainio (ed.), *Cell-Secreted Vesicles: Methods and Protocols*, Methods in Molecular Biology, vol. 2668,
https://doi.org/10.1007/978-1-0716-3203-1_7,

1 Introduction

Extracellular vesicles (EV) are a heterogeneous group of secreted lipid bilayer-delimited particles that vary in size typically between 50 nm and 1 μm [1]. Studies of EV composition revealed that they carry proteins, lipids, and nucleic acids, and this content can vary widely depending on cell-source types and conditions [2]. Specific EV-enclosed cargoes act as signaling biomolecules in normal cell homeostatic and pathological processes by directly affecting the fate and function of neighboring cells as well as target cells at distance [3]. Such capability to deliver cargo in specific target cells, which have been demonstrated in numerous in vivo studies [4–7], is gaining huge interest envisioning EV from cell culture medium, as therapeutic delivery tool to treat disease [8]. In such context is crucial to purify homogeneous EV preparation for functional dissection, as cell secretome contains a soluble fraction rich in secreted proteins, RNA granules, and vesicles subtypes (including apoptotic bodies) [9]. Commonly used EV isolation methods include ultracentrifugation, density gradient centrifugation, size exclusion chromatography, and polymer-based precipitation. However, the challenge has now shifted from proof of concept to developing reliable and efficient large-scale protocol to manufacture consistently pure and potent EV-based drug for broad-based patients access [10]. Here, we outline a detailed protocol to obtain highly purified small EV from cell-conditioned medium (CM) in xeno-free conditions. This protocol employs a tangential flow filtration (TFF) as rapid and efficient method for separation and purification of EV as well as analytic approaches to characterize their protein composition. Although the protocol is applicable to different types of cells, it referred to specific cardiac explant-derived cells, also known as cardiac progenitor cells (CPC) that were used as cellular source for EV production. CPC represent promising candidates for the development of cell-based therapies, as novel and interesting treatment for cardioprotective strategy in heart failure. CPC cultured as "cardiosphere-derived cells" went through favorable Phase 1 and 2 studies demonstrating safety and possible efficacy. [11–13] We recently developed GMP-grade methods for production and testing of a CPC master cell bank (MCB), which may be used either as Advanced Therapy Medicinal Products (ATMP) or as source for the manufacturing of EV for clinical trials [14]. We also provide useful insights in various downstream applications including functional assay to test EV potency and quality control (QC) methods to assess the safety (sterility, endotoxin). Methods to assess identity/purity (EV-specific markers and/or absence of contaminants) in close compliance with the International Society for Extracellular Vesicle (ISEV) guidelines, are also described [15]. We believe a set of detailed instructions for a specific experimental method, is a valuable resource for academic and industries to implement the translation of EV into clinical testing.

2 Materials

2.1 EV Production

High-quality and properly certified reagents and materials are mandatory for the whole process of CPC expansion, Master Cell Bank (MCB) and EV production, in compliance with regulatory guidelines for ATMP and related products [16]. For cardiac tissue manipulation, MCB manufacturing and reagents refer to Andriolo et al. [14]

2.1.1 MCB (P2) Thawing

1. Cell culture medium (MSC-Brew GMP Medium, Miltenyi Biotec GmbH, www.miltenyibiotec.com) (*see* **Note 1**).
2. 15 mL tubes.
3. Centrifuge.
4. Instrument ThermoStat Plus (Eppendorf, www. eppendorf. com).
5. Single use sterile pipettes.

2.1.2 CPC Expansion

1. Different size of plastic supports for CPC culture expansion (*see* **Note 2**).
2. Cell culture medium (MSC-Brew GMP Medium, Miltenyi Biotec GmbH).
3. DPBS (Dulbecco's Phosphate-Buffered Saline without calcium and magnesium, Gibco/ Thermo Fisher Scientific, www. thermofisher.com).
4. TrypLE™ Select Enzyme (Gibco/Thermo Fisher Scientific) (*see* **Note 3**).
5. Single use sterile pipettes.

2.1.3 Production of Conditioned Medium

1. DPBS.
2. Dulbecco's Modified Eagle Medium (DMEM) 4.5 g/l glucose without phenol red (Gibco/Thermo Fisher Scientific).
3. Sterile bottles (Corning, www.corning.com).
4. Single use sterile pipettes.

2.1.4 CPC End of Production Cell (EPC) Freezing

1. Freezing medium: Cryostor® CS10 (BioLife Solutions, Stem-Cell Technologies, www.stemcell.com) (*see* **Note 4**).
2. 1.8 mL cryovials.

3. Controlled rate biological freezer (Biofreeze® BV45, Consarctic, www.consarctic.de).
4. Single use sterile pipettes.

2.1.5 Tangential Flow Filtration (TFF) and EV Enrichment

1. ÄKTA™ flux 6 system (GE/Cytiva, www.cytivalifesciences.com) (*see* **Note 5**).
2. Sterile bottles (Corning).
3. Sterile Transfer Cap (Corning).
4. Filters ULTA Pure Capsule HC 0.6 / 0.2 μm (GE/Cytiva).
5. 300 kDa cut-off hollow fiber cartridge (GE/Cytiva).
6. 0.22 μm capsule filter filtration.
7. Formulation buffer (*see* **Note 6**).

2.2 CPC-EV Quality Control Tests

High-quality and properly certified reagents and materials are mandatory for QC assays.

2.2.1 Identity and Potency

- *Qualitative Assessment of EV-Specific Luminal Antigens-WB.*
  1. RIPA Buffer 1X, (Sigma-Aldrich,www.sigmaaldrich.com).
  2. Protease inhibitor (Sigma-Aldrich).
  3. Pierce™ BCA Protein Assay Kit (ThermoFisher, www.thermofisher.com).
  4. Loading buffer for western blot (ThermoFisher).
  5. 4–20% Mini-PROTEAN® TGX™ Precast Protein Gels, (Bio-Rad, www.bio-rad.com).
  6. Protein marker (Bio-Rad).
  7. Trans-Blot Turbo Transfer System (Bio-Rad).
  8. Intercept (TBS) Blocking buffer (LI-COR Biotechnology, www.licor.com).
  9. TBS 1× (24 g Tris–HCl 88 g NaCl in 1 L H_2O, pH 7.4) (Bio-Rad).
  10. Antibody diluent buffer (Intercept blocking buffer/TBS 1: 1 with 0.2% Tween® 20) (Bio-Rad).
  11. TBS-TWEEN (1x TBS supplemented with 0.1% Tween® 20) (Bio-Rad).
  12. Primary antibody: TSG101 (cat. ab83), SYNTENIN (cat. ab133267), ALIX (cat. ab186429) (Abcam, www.abcam.com).
  13. Fluorescent Secondary antibodies (LI-COR).
  14. Single use pipettes and pipetus.
  15. Fluorescent imager system (i.e., Odyssey, Li-COR).
  16. Orbital shaker.

- *Quantitative* Assessment of EV-Specific Luminal Antigens.

  1. Human TSG101 ELISA Kit (Sandwich ELISA) - LS-F8581 (LSBio LifeSpan Biosciences Inc., www.lsbio.com) (*see* **Note 7**).
  2. Multichannel pipettes, micropipettes, and filter tips.
  3. Reservoirs.
  4. Single use pipettes and pipetus.
  5. 37 °C incubator.
  6. Plate reader equipped to read absorbance at 450 nm.

- *Analysis of EV-Specific Surface Antigens by MACsPlex Exosomes Kit.*

  1. MACSPlex Kit, human - 130-108-813 (Miltenyi Biotec GmbH).
  2. Micropipettes and filter tips.
  3. Single use pipettes and pipetus.
  4. 5 ml FACS tubes.
  5. Centrifuge.
  6. Orbital shaker.
  7. Cytometer with blue (e.g., 488 nm) and red (e.g., 635 nm) lasers capable of detecting FITC, PE, and APC fluorochromes.

- *Analysis of EV Size & Concentration (NTA).*

  1. NanoSight LM10 (Malvern Instruments, www.malvernpanalytical.com).
  2. DPBS (Gibco/Thermo Fisher Scientific).
  3. Micropipettes and filter tips.

- *Assessment of EV Functional Activity.*
  Antiapoptotic Activity of EV

  1. Human CPC.
  2. Micropipettes and sterile filter tips.
  3. Single use sterile pipettes and pipetus.
  4. 1.5 ml, 15 and 50 ml sterile tubes.
  5. Black 96-well plate with transparent flat bottom (Corning).
  6. Complete medium (StemMACS™ MSC expansion Media kit XF, Miltenyi Biotec GmbH).
  7. 2% gelatin (Sigma-Aldrich).

8. DPBS (Gibco/Thermo Fisher Scientific).
9. DMEM, no glucose, no glutamine (Biological Industries, www.bioind.com).
10. Staurosporine (Sigma-Aldrich).
11. Plasmalyte A (Baxter, www.baxter.com).
12. 37 °C, 5% CO_2 incubator.
13. Fetal Bovine Serum (FBS, Gibco/Thermo Fisher Scientific).
14. Propidium Iodide (PI, Thermo Fisher Scientific).
15. Calcein AM (Thermo Fisher Scientific).
16. Plate reader equipped to read fluorescence (calcein excitation ~485 nm - emission ~535 nm, PI excitation ~560 nm - emission ~630 nm).

Proangiogenic Activity of EV

1. Va2 Kit protocol (Cellworks, www.cellworks.life) (*see* **Note 8**).
2. Plasmalyte A (Baxter).
3. Micropipettes and sterile filter tips.
4. Single use sterile pipettes and pipetus.
5. 70% ethanol (Sigma-Aldrich/Merck).
6. SIGMAFAST™ p-Nitrophenyl phosphate Tablets (Sigma-Aldrich/Merck).
7. 96-well plate flat bottom (Corning).
8. NaOH (Sigma-Aldrich/Merck).
9. 37 °C, 5% CO_2 incubator.
10. Optical microscope with camera.
11. Plate reader equipped to read absorbance at 405 nm.

2.2.2 Purity

- *Total Protein Content.*

1. QuantiPro™ BCA Assay kit (Sigma-Aldrich/Merck).
2. Micropipettes and sterile filter tips.
3. Single use sterile pipettes and pipetus.
4. 1.5 ml tubes.
5. 37 °C incubator.
6. 96-well plate flat bottom (Corning).
7. Plate reader equipped to read absorbance at 560 nm.

- *Total DNA Content.*
  1. Quant-iT™ PicoGreen™ dsDNA Reagent and Kit (Thermo Fisher Scientific).
  2. Micropipettes and filter tips.
  3. Single use pipettes and pipetus.
  4. 96-well plate flat bottom (Corning).
  5. Plate reader equipped to read fluorescence (excitation ~480 nm, emission ~520 nm).
- *Analysis of Cellular-Derived Contaminants.*
  1. Human CANX/Calnexin ELISA Kit (Sandwich ELISA) - LS-F22571–1 (LSBio LifeSpan Biosciences Inc.).
  2. Multichannel, micropipettes, and filter tips.
  3. Reservoirs.
  4. Single use pipettes and pipetus.
  5. 37 °C incubator.
  6. Plate reader equipped to read absorbance at 450 nm.

2.2.3 Safety

- *Sterility.*
  1. BacT/ALERT FA PLUS (aerobic) (FA bottles) and BacT/ALERT FN PLUS (anaerobic) (FN bottles) (BioMérieux, www.biomerieux.com) (*see* **Note 9**).
  2. Sterile syringes and needles.
  3. Agar plates: Tryptic Soy Agar (TSA) or Columbia Blood Agar (CBA) and Sabouraud Dextrose Agar (SDA) (BioMérieux).
- *Endotoxin.*
  1. Endosafe® PTS™ cartridges - 0.05 EU/mL sensitivity (Charles River, www.criver.com).
  2. LAL Reagent Water (Charles River).
  3. Sterile and nonpyrogenic tubes and sterile and nonpyrogenic tips.

3 Methods

3.1 EV Production

The whole large-scale production process of CM (6–8 L) is performed under sterile conditions and if necessary in a GMP-certified classified area (Class A laminar flow hood in class B background). *See* Fig. 1 for scheme of production steps and QC tests. For cardiac

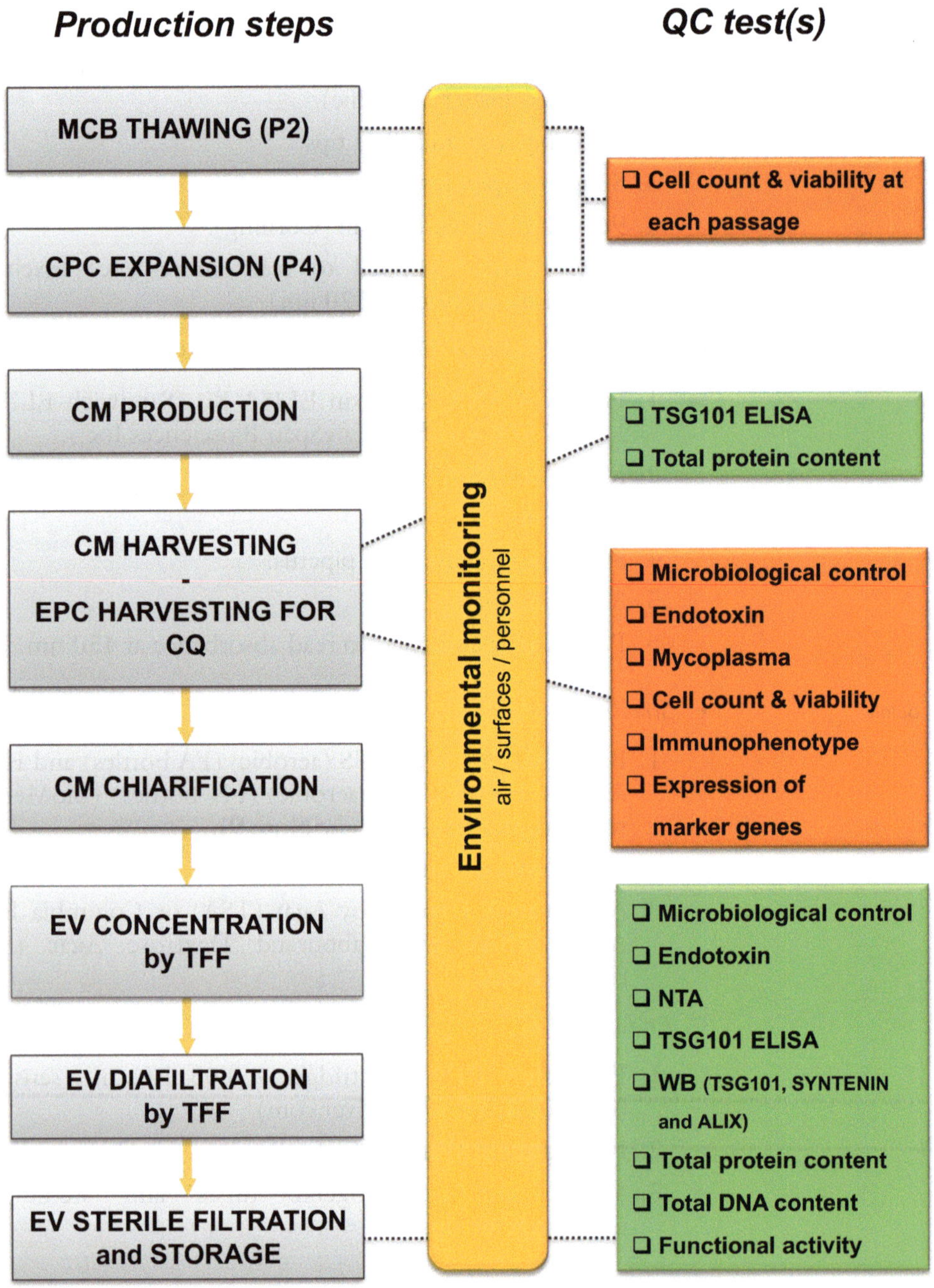

Fig. 1 Schematics depict of production steps with the corresponding QC activities
Acronyms: MCB (P2) master cell bank at passage 2, CPC cardiac progenitor cells, CM conditioned medium, EPC end of production cells, EV extracellular vesicles
Legend for colors: yellow = environmental monitoring; orange = QC activities on cells; green = QC activities on EV

tissue manipulation and MCB methods referred to Andriolo et al. [14]

The high-quality CPC-EV isolation from CM is carried out in a sterile closed system using TFF [17].

3.1.1 MCB (P2) Thawing

1. Preheat the complete medium for 30′ at 37 °C.
2. Withdraw the desired number of vials of frozen MCB at passage 2 (P2) from the vapor-phase nitrogen tank (*see* **Note 10**).
3. Transport the vials to the laboratory in an insulated container.
4. Transfer the vials at 37 °C in a thermostat.
5. When warmed, transfer the vials under a laminar flow box and disinfect them with isopropanol 70%.
6. Transfer the content of each vial into a 15 mL tube.
7. Gently dispense, dropwise, 8 mL of the prewarmed medium into each tube.
8. Wash the vial with 1 mL of medium and add it to the corresponding tube. Mix gently the cell suspension in the tube by three aspiration/dispensing cycles with the same pipette.
9. Centrifuge the cell suspension at 500 × g for 5 min at room temperature, discard the supernatant, and resuspend the cells in 1 mL of medium.
10. Count total and viable cells (*see* **Note 11**).

3.1.2 CPC Expansion

1. Seed CPC at 8-10 × 10^4 cells/cm^2 in appropriate culture vessels (T flasks, HYPERFlask® or HYPERStack®-12) in presence of culture medium (*see* **Note 1**) and incubate at 37 °C 5% CO_2.
2. At about 80% confluence, harvest CPC using TrypLE™ Select Enzyme (*see* **Note 3**).
3. Collect cell suspension in appropriate tubes and centrifuge at 500 × g for 5/10 min at room temperature.
4. Discard the supernatant and resuspend the cells in an appropriate volume of medium for the cell count (*see* **Note 11**).
5. Seed cells and proceed to expand until passage 4 (P4) as described above in this paragraph (*see* **Note 12**).

3.1.3 Production of Conditioned Medium

1. Remove from the vessels the culture medium (*see* **Note 13**).
2. Wash cells twice with DPBS.
3. Add to each vessel DMEM 4.5 g/l glucose without phenol red, thereafter indicated as CM (*see* **Note 14**).
4. Incubate cells at 37 °C with 5% CO_2.

Table 1
QC test for CM

Test	Method	Specification
TSG101	ELISA	FIO
Total proteins	BCA assay	FIO

5. Every 2–3 days, control cells at the microscope to ensure they are safe and without contaminations.
6. Collect in sterile bottles the CM (total 6–8 L) after 7–14 days of culture.
7. After 1–2 weeks, about 6–8 L of CM are collected and analyzed for TSG101 and total proteins content (Table 1).

3.1.4 End of Production Cell (EPC) Freezing

1. Harvest EPC (*see* **Note 15**) using TrypLE™ Select Enzyme (*see* **Note 3**).
2. Centrifuge cells and remove the culture medium.
3. Collect suspension containing cells in appropriate tubes and centrifuge at 500 × g for 5/10 min at room temperature.
4. Discard the supernatant and resuspend the cells an appropriate volume of medium for the cell count (*see* **Note 11**).
5. Suspend the cell pellet in freezing medium (*see* **Note 4**) at the concentration of about 5×10^6 cells/mL (optimal range: 2-6 × 10^6 cells/mL): gently add the solution (previous maintained at 4 °C) to the cells.
6. Transfer 1 mL aliquot in each cryovial, then place the vials into a controlled rate biological freezer and proceed according to a suitable freezing program (*see* **Note 16**).
7. Store vials in a vapor-phase nitrogen tank.

3.1.5 Tangential Flow Filtration (TFF) and EV Enrichment

The process of EV enrichment is performed through TFF using ÄKTA™ Flux 6 instrument (previously sanitized according to the manufacturer's instructions) [17, 18], in accordance with following process phases (*see* Fig. 1):

1. Clarification Phase (*see* **Note 17**):
   (a) Connect the sterile bottle containing the CM through a sterile transfer cap to the peristaltic pump.
   (b) Connect on-line a 0.6/0.22 μm filter capsule.
   (c) Proceed to the clarification of the CM through the capsule filter activating the peristaltic pump and filling directly in the instrument tank (process vessel).

2. Concentration Phase:
   (a) Activate the feed pump of the instrument to initiate the EV concentration by TFF through the hallow fiber cartridge.
   (b) Set-up the instrument parameters (flow rate, transmembrane pressure) to minimize the "shear stress" in order to preserve EV integrity (*see* **Note 18**).
   (c) Collect the permeate, containing components below the 300 kDa cut-off, in a waste container.
   (d) Recirculate the retentate (concentrated CM enriched in EV) to the process vessel.
   (e) Stop the concentration phase when the volume of retentate is about 300–500 ml in the process vessel.
3. Diafiltration Phase (*see* **Note 19**):
   (a) Dilute the concentrated CM contained in the process vessel with the formulation buffer.
   (b) Proceed to the concentration of the diluted CM enriched in EV following the above **step 2** Concentration Phase, until reaching a 200–300 ml volume in the process vessel.
4. Recovery Phase:

   The final EV-enriched solution contained into the process vessel is collected through the bottom sample port in a sterile postprocess bag.
5. Sterilizing Filtration Phase:
   (a) The postprocess bag is connected to the instrument circuit on-line with sterile filter unit 0.22 μm.
   (b) The instrument transfer pump is used and the sterile product (275–350 ml) is collected directly in a final product bag.
6. Filling Phase:

   The final EV enriched product is filled in 0.5 ml, 1 ml, and 3 ml aliquots in 1.8 ml and 4.5 ml cryovials, as appropriate, under sterile conditions.
7. Storage:

   The vials containing high-purified EV are frozen and stored at −80 °C.

The yield of the EV enrichment and purity grade at different process phases are determined by measuring the content of TSG101 and total proteins respectively.

The results of three lots are shown in Fig. 2.

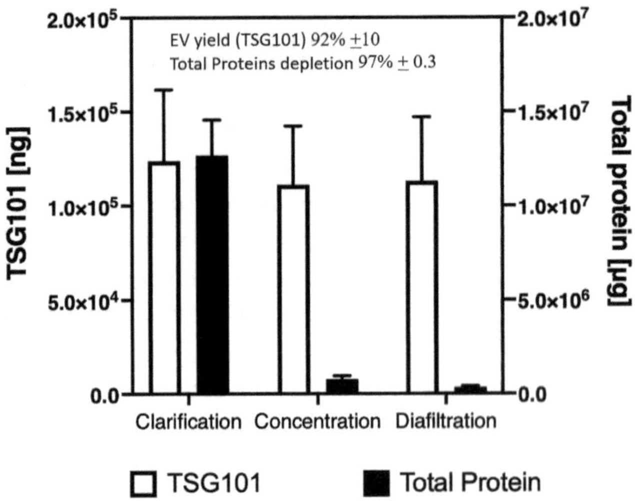

Fig. 2 TFF purification process. (**a**) Histograms plot of EV recovery in terms of quantitative analysis of TSG101 and depletion of total proteins content, during the TFF purification steps (Mean + SD with $n = 3$). Overall EV yield and overall total protein depletion are indicated (Mean + SD with $n = 3$)

3.2 EV Quality Control

QC test activities (Table 2) are performed in suitable GMP-certified labs and are intended to verify EV's:

- Identity: by the detection of luminal and surface EV-specific antigens.
- Potency: by functional test available on the market or developed and validated "in house".
- Purity: by the detection of most common process contaminants.
- Safety: by the detection of bacteria and endotoxins.
- Stability: to determine how long the product can be stored while maintaining its characteristics.

The tests are carried out on the final product aliquoted in cryovials and stored at −80 °C.

Results of three lots are shown in Figs. 3 and 4.

3.2.1 Identity and Potency

- *Qualitative Assessment of EV-Specific Luminal Antigens-WB.*

  The identity is evaluated by the qualitative detection of specific luminal EV marker: TSG101, SYNTENIN, and ALIX.

  1. Lyse samples using 1x RIPA buffer supplemented with Protease inhibitor for 30 min in ice.
  2. Quantify the protein extract (*see* **Note 20**).
  3. Prepare three different samples: for each sample add 30 µg of total protein, then add ddH_2O to a final volume of 30 µl.
  4. Add 6 µl of Loading buffer containing SDS and β-mercaptoethanol.

Table 2
CQ test for EV

Test	Method	Specification
Microbiological control of cellular products	EP 2.6.27	No growth of microorganisms
Endotoxin assay	EP 2.6.14	< 5.00 EU/mL
EV size and concentration	NTA	Size: 50–150 nm Concentration: $\geq 10^{10}$/mL[a]
TSG101	ELISA	≥ 200 ng/mL[a]
Total proteins	BCA assay	≥ 90% removal respect to CM[a]
Total DNA	Quant-iT™ PicoGreen™ dsDNA	≥ 90% removal respect to CM[a]
Cell-derived contaminants	ELISA	FIO
Functional activity	Cytotoxicity protection Angiogenetic activity	Protection % ≥ 20%[a] FIO

[a]These specifications were determined on the three GMP runs of 6–8 liters of CM

5. Incubate 10 min at 95 °C.
6. Prepare the 4–20% Mini-PROTEAN® TGX™ Precast Protein Gels in the cassette as the manufacturer suggests (*see* **Note 21**).
7. Quickly spin samples at 2000 rpm for 1 min.
8. Load samples in the wells: alternate protein marker and sample in order to have three couples. Let the gel run until the marker shows a nice separation.
9. Transfer the gel-containing protein to a PVDF membrane using Trans-Blot Turbo Transfer System (Biorad) following manufacturer instructions.
10. Remove the membrane from the cassette, use tweezers to avoid membrane contamination.
11. Cut the membrane to obtain three pieces, each with a lane of marker and a lane of sample.
12. Block the membranes with 10 ml Intercept blocking buffer for 1 h at room temperature with gentle shaking.
13. Dilute primary antibodies in diluent buffer to obtain a final concentration of 1 μg/ml or using the one suggested by the vendor.
14. Carefully remove blocking buffer, then incubate the three membranes in different primary antibody solution for 1–3 h

A

Parameter	Method	Specification	Results		
			EV lot 1	EV lot 2	EV lot 3
Sterility	Microbiological control for cellular products	Absence of growth	Absence of growth	Absence of growth	Absence of growth
Bacterial Endotoxin (EU/ml)	Quantitative LAL test	< 5.0	< 5.0	< 5.0	< 5.0
TSG101 (ng/ml)	ELISA	FIO	285	272	203
Total Proteins (µg/ml)	BCA assay	FIO	1008	840	1094
Particle concentration (p/ml)	Nanoparticle Tracking Analysis	FIO	8.26×10^{10}	1.12×10^{11}	1.02×10^{11}
Particle size (nm)		50-150	121	143	131
CD9 expression	Flow cytometry	Positive	Positive	Positive	Positive
CD63 expression		Positive	Positive	Positive	Positive
CD81 expression		Positive	Positive	Positive	Positive

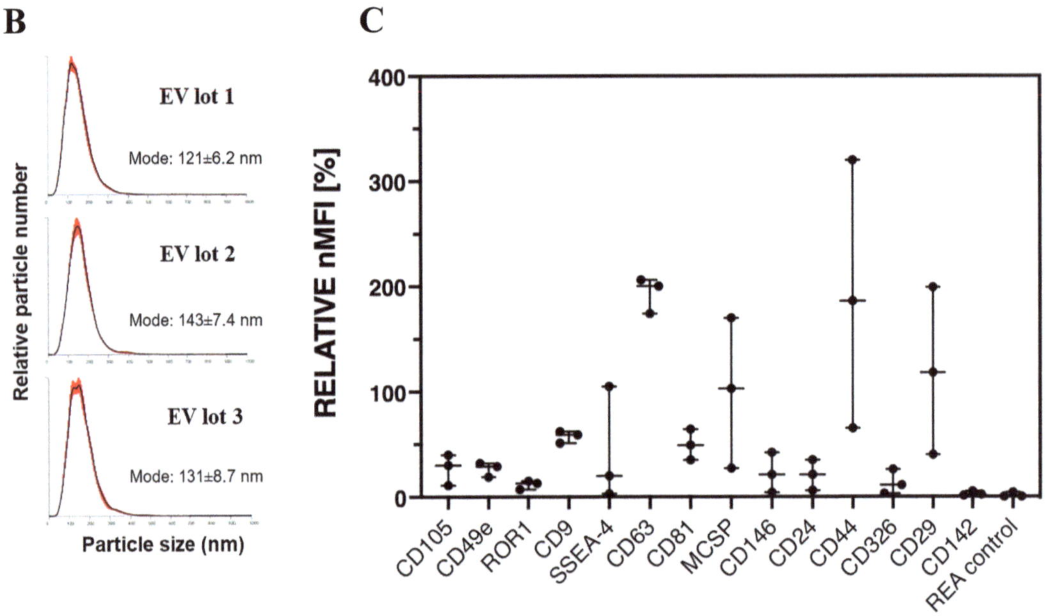

Fig. 3 Characterization of EV: (**a**) Results of QC test carried on three different lots of large-scale GMP-grade EV production. (**b**) Nanoparticle tracking analysis profiles. (**c**) Flow cytometry analysis of superficial markers on three EV lots. Expression levels for each EV surface antigen was reported after subtraction for the respective fluorescence values of blank control and normalization for mean MFI for CD9/CD63/CD81 (normalized MFI, nMFI; expressed as percentage, %)

at room temperature or overnight at 4 °C with gentle shaking.

15. Wash each blot with a 1× TBS-TWEEN for 5 min with gentle shaking.
16. Repeat the washing step three times.

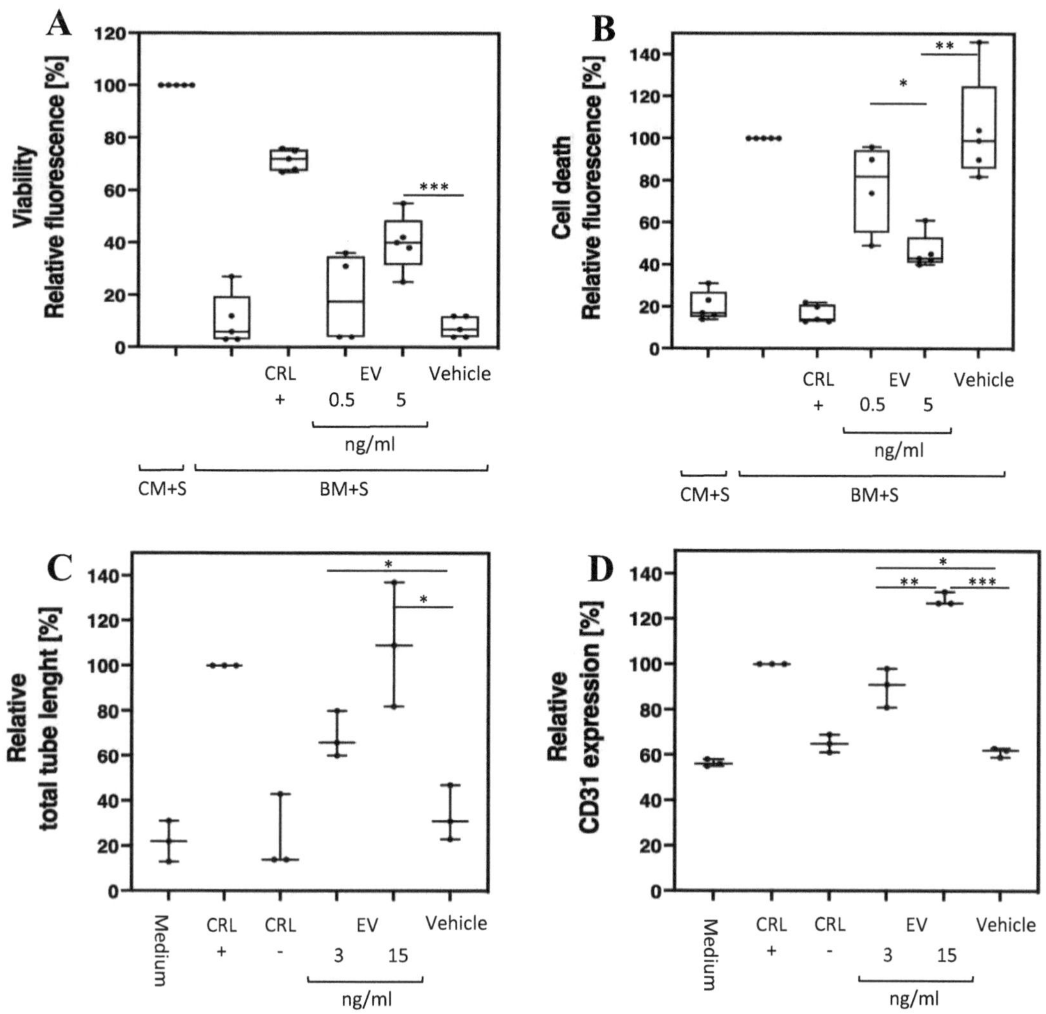

Fig. 4 Antiapoptotic activity of EV: (**a**) Representative images of cells (10×) after treatments. (**a**, **b**) Results from calcein and PI staining, respectively ($n = 5$). Complete medium + staurosporine (CM+S) is the 100% survival reference for calcein, while basal medium + staurosporine (BM + S) is the 100% apoptosis reference for PI. CRL+ = 10% FBS
Proangiogenic activity of EV: (**c**, **d**) Results ($n = 3$). CRL + = VEGF; CRL - = Suramin; Vehicle = Plasma-Lyte A®. (**c**) Total tube length (AngioSys 2.0 Image Analysis Software). (**d**) CD31 expression detected by ELISA

17. Dilute secondary antibody in diluent buffer. The antibody final concentration depends by the manufacturer instructions and by the detection system used.
18. Carefully remove washing buffer, then incubate the three membranes in secondary antibody solution for 1 h at room temperature.
19. Wash each blot with a 1x TBS-TWEEN for 5 min with gentle shaking.
20. Repeat the washing step three times.

21. Scan the membranes with an appropriate imaging system and check for the presence of the specific band.

- *Quantitative Assessment of EV-Specific Luminal Antigens.*

  The amount of EV in the different steps of production and purification is evaluated by the quantitative detection of TSG101 (*see* **Note 7**). The "Human TSG101 ELISA Kit" is used following manufacturer instructions.

1. Prepare reagents, standard, and samples following User Manual instructions.
2. Add 100 μl of Standard, Blank, or Sample per well, cover with a plate sealer, and incubate for 1 h at 37 °C.
3. Aspirate the liquid of each well, do not wash.
4. Add 100 μl of Detection Reagent A working solution to each well, cover with a plate sealer, and incubate for 1 h at 37 °C.
5. Aspirate the liquid from each well and wash three times. Wash by adding approximately 350 μl of 1x Wash Buffer using a squirt bottle, multichannel pipette, manifold dispenser, or automated washer. Allow each wash to sit for 1–2 min before completely aspirating. After the last wash, invert the plate and tap against clean absorbent paper.
6. Add 100 μl of Detection Reagent B working solution to each well, cover with a new plate sealer, and incubate for 30 min at 37 °C.
7. Aspirate the liquid from each well and wash five times as outlined in **step 5**.
8. Add 90 μl of TMB Substrate solution to each well, cover with a new plate sealer, and incubate for 10–20 min at 37 ° C. Protect from light and monitor periodically until optimal color development has been achieved.
9. Add 50 μl of Stop Solution to each well. The blue color will change to yellow immediately. If color change does not appear uniform, gently tap the plate to ensure thorough mixing. The Stop Solution should be added to wells in the same order and timing as the TMB Substrate solution.
10. Determine the optical density (OD value) of each well immediately using a microplate reader set to 450 nm.
11. Data analysis: average the duplicate readings for each standard, control, and sample and subtract the average zero standard optical density. Construct a standard curve by plotting the log of the target antigen concentrations versus the log of the OD and determine by regression analysis the best fit line. Use the fit line with the ODs from the unknown samples to infer the TSG101 concentration.

- *Analysis of EV-Specific Surface Antigens by MACsPlex Kit.*

  The MACSPlex Kit allows detection of 37 EV surface epitopes plus two isotype controls. This kit comprises a cocktail of various fluorescently labeled bead populations, each coated with a specific antibody binding the respective surface epitope. The 39 bead populations can be distinguished by different fluorescence intensities by flow cytometry.

  Hereby we report on "Short protocol" for the use of 1.5 mL tubes; however, other type of vessels and instruments can be used with slight changes respect to the original protocol depending on the starting material and the instruments available in the laboratory (*see* **Note 22**).

  Be sure to determine the EV concentration by quantifying the protein concentration (*see* **Note 23**).

1. Label reagent tubes for the blank control and samples.
2. Add to each 1.5 mL tube either isolated EV (4–20 μg protein) from each sample diluted to 120 μL using the MACSPlex Buffer.
3. Resuspend MACSPlex Capture Beads by vortexing for at least 30 s and transfer 15 μL of Capture Beads to each tube.
4. Add 5 μL of MACSPlex Detection Reagent CD9, CD63, or CD81 or 15 μL of detection cocktail to each tube and mix by pipetting up and down.
5. Incubate tubes for 1 h at room temperature protected from light using an orbital shaker (450 rpm).
6. Add 500 μL of MACSPlex Buffer to each tube.
7. Centrifuge at room temperature at 3000 × g for 5 min.
8. Aspirate 500 μL of the supernatant carefully, leaving about 150 μL in the tube.
9. Add 500 μL of MACSPlex Buffer to each tube.
10. Incubate tubes for 15 min at room temperature protected from light using an orbital shaker (450 rpm).
11. Centrifuge at room temperature at 3000 × g for 5 min.
12. Aspirate 500 μL of the supernatant carefully, leaving about 150 μL in the tube.
13. Resuspend sample by pipetting up and down and transfer the samples to a 5 ml FACS tube.
14. Sample acquisition:
    (a) Calibrate the MACSQuant® Analyzer using MACSQuant Calibration Beads (*see* **Note 24**).

(b) Run the MACSPlex Setup Beads to ensure proper recognition of all bead populations.

(c) Run EV samples.

15. Data analysis (*see* **Note 25**):

(a) Subtract the median signal intensity of each bead obtained from your control sample (buffer only) from the signal intensities of the respective beads incubated with your sample.

(b) Repeat **step 1** for all samples to be analyzed.

(c) For each sample, calculate the median signal intensity of the signals detected for the MACSPlex Capture Beads CD9, CD63, and CD81.

(d) Expression levels for each EV surface antigen are reported as normalized for mean MFI of CD9/CD63/CD81 (normalized MFI, nMFI; expressed as percentage, %).

- *Analysis of EV Size & Concentration (NTA).*

  EV particles were evaluated by Nanoparticle Tracking Analysis (NTA), with the instrument NanoSight LM10 (Malvern Instruments, UK).

1. Dilute EV samples to obtain a final nanoparticle concentration in the linear range of the instrument (from $2x10^8$ to $2x10^9$ /ml).
2. Acquire at least 4–5 videos from different portions of each sample with the following parameters: Screen gain = 10, Camera level = 13.
3. Analyze the videos by the software NTA (version 2.3, build 0034) with the minimal expected particle size, minimum track length, and blur setting, all set to automatic. The expected optimal Detection threshold (after the manual adjustments suggested by the Malvern manual) is 3. Average values with the standard errors are automatically generated by the software.

- *Assessment of EV Functional Activity.*

  Antiapoptotic Activity of EV.

  The antiapoptotic activity of EV is tested by an in vitro apoptosis/viability assay developed in house, based on the use of cardiac progenitor cells (CPC) and staurosporine as apoptosis inducer.

1. Plate human CPC were plated (10^4 cells/well) in black 96-well plate with transparent flat bottom in complete medium upon coating with 0.2% gelatin for 30 min at 37 ° C. Leave wells without cells to be used as blanks.

2. The day after, treat cells with different conditions:
   (a) Complete medium.
   (b) Complete medium with 1 μM staurosporine.
   (c) Basal medium (DMEM, no glucose, no glutamine) with 1 μM staurosporine.
   (d) Basal medium with 1 μM staurosporine with EV at 5 ng/ml of TSG101 protein.
   (e) Basal medium with 1 μM staurosporine with equal amount of vehicle (Plasmalyte A).
   (f) Basal medium with 1 μM staurosporine with 10% FBS as positive control for protection.
3. Incubate overnight at 37 °C, 5% CO_2.
4. Add 100 μl of 2.5 μM PI for 30 min at 37 °C and quantify cell death by fluorescence detection with the plate reader.
5. Wash cells with DPBS then stain with 100 μl of 3 μM calcein AM for 30 min at 37 °C and quantify viability by fluorescence detection with the plate reader.
6. Analyze data by subtracting the fluorescence of blank wells from the fluorescence obtained from respective sample wells.
7. Normalize viability data on 1 μM staurosporine complete medium condition and cell death data on 1 μM staurosporine basal medium condition.

Proangiogenic Activity of EV.

The proangiogenic activity of EVs was tested by the V2a Kit (*see* **Note 8**), an in vitro test designed to determine angiogenic activity of compounds or conditioned media. We tested EV at different concentrations (3 and 15 ng of TSG101/ml) and used same volumes of vehicle as control. The angiogenic stimulator VEGF and the inhibitor suramin were included as positive and negative controls, respectively. Complete medium was also used as negative control. Media were changed every 48–72 h and after 12–14 days, cells were fixed in cold ethanol 70% and subsequently stained by a tubule staining kit containing anti-CD31 primary antibody (PECAM-1) and a secondary antibody conjugated with alkaline phosphatase (AP). The level of angiogenesis was detected by two means:

1. CD31 ELISA Soluble Substrate: SIGMAFAST™ p-Nitrophenyl phosphate Tablets were dissolved in water according to manufacturer's instructions and added to stained cells. p-Nitrophenyl phosphate (p-NPP) is hydrolyzed by AP in ρ-Nitrophenol, a chromogenic product that absorbs maximally at 405 nm. After 20 min at 37 °C, two

100 μl aliquots were taken from each well and plated in a 96-well plate containing 25 μl of 3 M NaOH. Blank samples were also prepared by adding p-NPP directly to NaOH. Absorbance at 405 nm of sample wells was measured against the blank wells by the Infinite F-200 PRO plate reader. This read out not included in the V2a Kit was implemented to have a more quantitative and standardized measure of angiogenetic activity of EV.

2. CD31 Tubule Staining Insoluble Substrate: cells were washed three times with 0.5 ml of deionized water and stained with BCIP/NBT according to manufacturer's instructions (V2a Kit, Cellworks). Tubule formation was visualized by 4x magnification optical microscope and analyzed by the AngioSys 2.0 Image Analysis Software (Cellworks). For each treatment condition, mean of total tube length from three wells was plotted.

3.2.2 Purity

- *Total Protein Content.*

  The QuantiPro™ BCA Assay kit was used for total protein quantification. A protocol has been set up following manufacturer's instruction.

1. Prepare the reagent mix by assembling buffer QA (1 volume) + QB (1 volume) + QC (0.02 volumes).
2. Prepare standard samples according to the volumes in the following table and plate in duplicate 150 μl/well:

Final protein conc. (μg/ml)	Standard (μl)	H_2O (μl)
80	80	920
40	40	960
30	30	970
20	20	980
10	10	990
5	5	995
1	1	999
Blank	0	1000

3. Dispense the sample in duplicate, 150 μl/well.
4. Add 150 μl/well of the reagent mix to each well.
5. Seal the plate and incubate for 2 h at 37 °C.
6. Read the 560 nm absorbance by a plate reader.

7. Data analysis:
   (a) Subtract blanks and calculate means of samples and standards.
   (b) Calculate linear regression curve relating standards quantity to OD.
   (c) Infer samples concentration.

- *Total DNA Content.*

  DNA content is measured by the Quant-iT™ PicoGreen™ dsDNA Reagent and Kit following manufacturer's instructions.

  1. Prepare the assay buffer (TE) by diluting the concentrated buffer 20-fold with sterile, distilled, DNase-free water.
  2. Allow the Quant-iT™ PicoGreen™ dsDNA Reagent to warm to room temperature before opening the vial, then prepare the "working reagent" by diluting the concentrated DMSO solution 200-fold in TE.
  3. Prepare DNA standard curve by first diluting the lambda DNA standard (provided at 100 μg/ml in the kit) 50-fold in TE to make a 2 μg/ml working solution. Then prepare the standard curve by serially diluting the working solution. The suggested range is 500, 200, 80, 32, 12.8, 5.12, 2.05 ng/ml.
  4. Prepare samples by dilution in TE to a final volume of 100 μl.
  5. Add 100 μL of the "working reagent" to each well. Mix well and incubate for 2–5 min at room temperature, protected from light.
  6. Measure the sample fluorescence using a microplate reader set to read excitation ~480 nm and emission ~520 nm.
  7. Data analysis:
     (a) Subtract the blank value from all samples and standards.
     (b) Use corrected data to generate a standard curve of fluorescence versus DNA concentration.
     (c) Infer the DNA concentration of each sample from the standard curve.

- *Analysis of Cell-Derived Contaminants.*

  Purity controls include the check for absence of proteins commonly co-isolated within EV preparations. Depending on cell sources, expected contaminants, it can be useful to verify the removal of serum albumin, and proteins from the endoplasmic reticulum or plasma membrane (e.g., Calnexin, GRP94) [19, 20].

Calnexin is one of the contaminants evaluated in CPC-EV by ELISA assay (*see* **Note 26**).

1. Prepare reagents, standard, and samples following User Manual instructions (ref. Human CANX/Calnexin ELISA Kit (Sandwich ELISA) User Manual Catalog No. LS-F9150).
2. Follow the same instructions described for the TSG101 kit (*see* Subheading 2.2.1) with the following changes:
   (a) **Step 2**: increase the incubation time to 90 min.
   (b) **Step 8**: incubate for a fixed time of 15 min.

3.2.3 Safety

- *Sterility.*

  This assay is performed to detect fungal and/or bacterial contamination in the final products. It is carried out using an automated microbial detection system, the BacT/ALERT® 3D (BioMérieux, www.biomerieux.com) (*see* **Note 26**).

  The sample inoculum is performed in a clean room area (Class A in B background): microbial and particle contamination should be monitored according to GMP Annex 1 [21]. The subsequent steps (cultures, subcultures, and bacteria identification) are carried out in a standard QC microbiological laboratory.

1. Remove the FA and FN bottle caps and disinfect the rubber-septum.
2. Thaw the desired sample vials (*see* **Note 27**).
3. With a syringe inoculate the sample in FN PLUS and FA PLUS media bottles (*see* **Note 28**), taking care to avoid air injection in the bottle.
4. Mix the bottles.
5. Record all data on the dedicated forms or labels (*see* **Note 29**).
6. Transfer all samples outside the clean room area and load them into the Bact/Alert (bioMérieux) following the supplier's instructions.
7. Control daily the instrument for any growth alarm during the incubation time (*see* **Note 30**).
8. Evaluate the results: a sample is <u>negative</u> if no growth is detected (stable colorimetric signal), <u>positive</u> if growth is detected (increase in signal level). In case of positive sample, the microorganism must be isolated (*see* **Note 31**) and its genus and species must be identified (*see* **Note 32**).

- *Endotoxin.*

  This assay is performed to detect bacterial endotoxins in the final product. It is designed as a final product release test; it is thus performed on frozen EV suspension. The method is performed using the Endosafe®-PTS™ system (Charles River, www.criver.com) in combination with FDA-licensed disposable test cartridges.

  It is performed in standard QC laboratory, and sample manipulation is performed under a laminar flow box.

  1. Insert the PTS endotoxin cartridge into the slot of the PTS reader following the manual instructions (*see* **Notes 33** and **34**)
  2. Prepare at least 100 μl of the test sample diluted with LAL Reagent Water (*see* **Note 35**). Vortex the sample to prevent the endotoxin from sticking to the wall.
  3. Load 25 μl of sample (*see* **Note 36**) into each of the four sample reservoirs (*see* **Note 37**).
  4. Read and print the results. The assay detection limit is expressed as the sensitivity of the method (e.g., λ 0.05) multiplied by the dilution factor (e.g., 100) of the validated sample (e.g., 5.00EU/mL). The test results for samples below the detection limit are expressed as < detection limit (e.g., <5.00EU/mL).

  Acceptance criteria:

- The spike recovery is between 50% and 200% (*see* **Notes 38** and **39**).
- The percent Coefficient of Variation (CV%) for sample and Positive Control is less than 25%.

3.2.4 EV Stability

EV are stored in cryovials at −80 °C and their stability have been assessed through the maintenance of antiapoptotic activity on cardiac cells (as in "Antiapoptotic activity of EV" Subheading of 3.2.1).

Stability at −80 °C						
6 months	1 year	2 years	4 years	6 years[a]	8 years[a]	10 years[a]

[a]data for these time points are not yet available (*see* Fig. 5)

For each time point, results of cells viability in presence of EV at 5 ng/ml of TSG101 were normalized on the complete medium condition plus staurosporine.

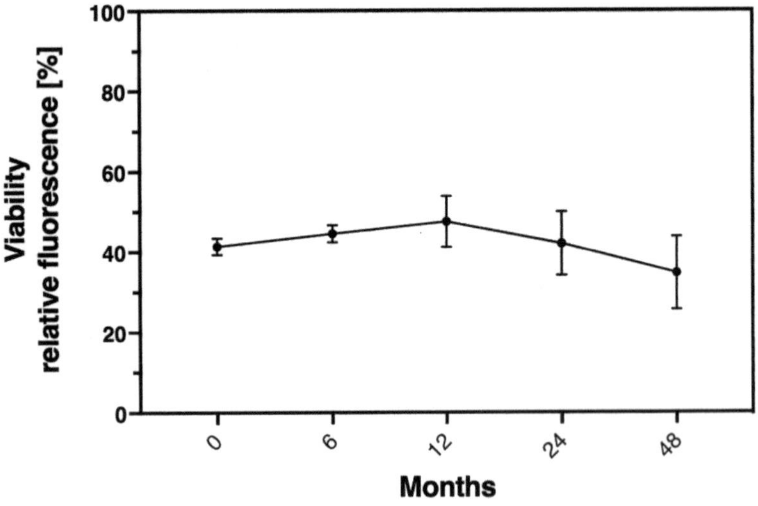

Fig. 5 EV stability at −80 °C storage: EV functional activity (as protection of cardiac cells viability during treatment with staurosporine) has been evaluated at different time points (as listed in Subheading 3.2.4)

4 Notes

1. This culture medium was selected among several commercially available serum-free xeno-free alternatives, on the basis of its superior performance in terms of CPC growth; moreover, it has provided of suitable quality and properly certified for GMP applications.
2. CPC are cultured on following plastic supports: tissue-culture flasks 115 cm^2 with reclosable lids (TPP, www.tpp.ch), T-flasks (75–150 cm^2), CellBIND® HYPERFlask® (1720 cm^2) or HYPERStack®-12 (6000 cm^2) culture vessels (Corning, www.corning.com). Flasks should be coated with CELLstart™ (Gibco/Thermo Fisher Scientific, www.thermofisher.com) or other adhesion substrates, provided that they are of suitable quality and properly certified.
3. Alternative enzymatic solutions may be used, provided that they are of suitable quality, nonanimal derived, and properly certified. The procedure for the use of TrypLE™ Select Enzyme follows the manufacturer instructions.
4. Cryostor® CS10 is a serum-free, animal component-free, and defined cryopreservation medium containing 10% dimethyl sulfoxide (DMSO) as cryoprotective agent; other freezing media may be used, provided that they are of suitable quality, nonanimal derived, and properly certified.
5. Efforts were made to develop an integrated closed circuit [17], encompassing the full downstream process (clarification, concentration, diafiltration, and final sterilizing filtration) in a

single work station, the ÄKTA™ flux 6 system (GE/Cytiva). The system is equipped with a 300 kDa (or 100-500 kDa) cut-off hollow fiber cartridge.

6. As formulation buffer is intended, solutions compatible with re-infusion in humans (such as Plasma-Lyte A® solution, Baxter, www.baxter.com) in which stability of EV studies has previously been performed.
7. Among several commercially available ELISA kits for EV proteins, the TSG101 ELISA used in this study was selected for its sensitivity, allowing for the detection of the low EV concentration present in the conditioned medium. Nevertheless, other EV-specific proteins can be used for EV quantification [19].
8. This kit is no longer produced by the manufacturer, so we do not report the detailed protocol but only the general principles that can be used to choose an alternative commercial kit or to develop an in-house assay. For further details on the kit principles and protocol, please refer to Va2 Kit protocol from Cellworks UK.
9. These media were chosen because of the presence of an antibiotic-inhibitor in their composition. The medium fertility is therefore guaranteed even for test samples potentially containing antibiotics, such as patient-derived tissues.
10. For each CPC-EV manufacturing run, frozen MCB P2 was thawed and cultured (8-10x10^4 cells/cm^2 initial seed) in culture medium, to obtain at passage 4 P4 about 16 CellBIND®-Hyperflask® culture vessels.
11. CPC are counted by an automated system based on standard trypan blue technique (Automated cell counter EVE, NanoEnTek, www.nanoentek.com). This instrument provides the number of total, viable and dead cells, and % of viability (Table 3).
12. Thanks to their negatively charged, highly hydrophilic CellBIND® surface (Corning), designed to facilitate cell attachment and spreading, HYPERFlasks® and HYPERStacks® do not require any coating. At about 80% confluence, harvest CPC using TrypLE™ Select Enzyme and count them as described above.

Table 3
QC test for MCB (P2). These tests are performed at the thawing of the MCB aliquot (P2) and after expansion (P4) before EV production

Test	Method	Specification
Cell count and viability	EP 2.7.29	≥ 80% total cells recovery ≥70% cell viability

Table 4
QC test for EPC

Test	Method	Specification
Microbiological control of cellular products	EP 2.6.27	No growth of microorganisms
Endotoxin assay	EP 2.6.14	<5.00 EU/mL
Mycoplasma assay by NAT	EP 2.6.7	Negative
Expression study of marker genes	RT-PCR	Expression of GATA4, TBX18,TBX5, and MESP1
Cell count and viability	EP 2.7.29	≥ 80% total cells recovery ≥70% cell viability
Immunophenotype(identity assay)	EP 2.7.24	≥ 90% stained cells positivefor CD90, CD105, and CD73 ≤ 10% stained cells positive forCD45/CD34/ CD14/CD20/HLA-DR

13. In view of large-scale manufacturing CPC-EV from about 6–8 L of CM, CPC are expanded up to culture passage 4 (P4) in about 16 HYPERFlasks®.
14. In case of HYPERFlasks® vessels, the EV production medium added is about 530 ml/vessel. Vessels are incubated at 37 °C with 5% CO_2 for 7–14 days without any change of medium.
15. The EPC harvested after CM production were frozen for QC purposes (Table 4) referred to Andriolo et al. [14] The average cell yield of EPC at P4 ($n = 3$ large-scale lots) was 6.5×10^8 cells (range 5.9×10^8–7.1×10^8).
16. Controlled-rate freezing is recommended to ensure maximum cell viability for a wide variety of cells, being aware that the profile should be optimized for each specific application, as different cell types and freezing media may have different requirements.
17. During the clarification step of CM, cellular debris and eventual microbial contamination are eliminated through the passage of a double filter capsule of 0.6 and 0.2 μm.
18. The concentration of EV is performed through the TFF of CM using a 300 kDa hollow fiber cartridge. The set-up of instrument parameters (flow rate, transmembrane pressure) is performed according to manufacturer's instructions.
19. The concentrated CM contained in the *process vessel* is diluted in formulation buffer (Plasma-Lyte A® solution, total 5 volumes in five diafiltration cycles), with the aim to obtain a

replacement of the initial production medium greater than or equal to 95%. The diluted solution is concentrated through the same hollow fiber cartridge used for the previous phase.

20. Several kit can be used for this purpose; we here describe the QuantiPro™ BCA Assay kit (*see* Subheading 3.2.2).
21. Alternatively it is possible to prepare and cast a manual gel using an acrylamide percentage that allows the separation of the specific band of the protein of interests (TSG101 = 47 kDa; SYNTENIN = 32 kDa; ALIX = 96 kDa).
22. Short or Overnight protocol can be chosen using the MACS-Plex Filter Plate or 1.5 ml tubes. For details, please refer to the MACSPlex Exosome Kit datasheet available at www.miltenyibiotec.com.
23. The MACSPlex Exosome Kit can be performed on precleared cell culture supernatant, ascites, or urine directly. For plasma, serum, or cell culture supernatant of cells scarcely secreting EV, it is recommended to isolate EV beforehand.
24. Data acquisition can be performed also with different cytofluorimeters. The analysis of MACSPlex Exosome Kit requires flow cytometers with blue (e.g., 488 nm) and red (e.g., 635 nm) lasers, which are capable of detecting FITC, PE, and APC. For setting up these instruments, the MACSPlex Exosome Setup Beads are included in the kit. For further information, refer to the application note "General instructions for MACSPlex Assay data acquisition and analysis" in the tab "Resources - Resource finder – Application notes and scientific reports" at www.miltenyibiotec.com.
25. For further details, please refer to the MACSPlex Exosome Kit datasheet available at www.miltenyibiotec.com.
26. Calnexin was under limit of detection for all tested lots (stated sensitivity of the kit ≥0.375 ng/ml), in both the starting CM and the concentrated EV.
27. Alternatively, other commercially available systems may be used, such as Bactec (Becton Dickinson, www.bd.com), according to the manufacturer's recommendations and upon proper validation.
28. The minimum amount to be tested depends on the total volume of the final cell product (as in EP 2.6.27):

Total product volume (V)	Inoculum volume / bottle (FA or FN medium)	Total volume to be tested (FA + FN medium)
V ≥ 10 mL	1% of total volume	2% of total volume
1 mL ≤ V < 10 mL	100 μl	200 μl
V < 1 mL	Not applicable	Not applicable

For a product such as frozen EV, consisting of multiple aliquots (frozen vials), the minimum volume to be tested determines the minimum number of aliquots to be tested.

Aliquot(s) should be representative of the whole batch: they must be homogenous (the EV suspension should be well re-suspended during the prefreezing phase) and, in case of several aliquots dedicated to microbiological assays, they have to be sampled among the first, middle, and final prepared aliquots.

The maximum product volume to be inoculated in every bottle must be determined during the method's validation.

29. Alternatively, the volume to be tested in both bottles may be drawn in a single syringe. In this case, completely remove the air, first transfer the inoculum to the FN bottle (half of the volume) and then to the FA bottle (all the remaining volume and air). This reduces air influx in the anaerobic bottle.

30. Forms or labels (in the latter case they can be stuck directly on the lab record book) must be sterile. Autoclaved paper or labels must be used and brought into the clean room in triple-sterilized bags.

31. The incubation period at 36 ± 1 °C is usually of 7 days. This time can be changed during the method's validation for each specific product.

32. Use TSA or CBA plates for bacteria, incubating them under aerobic (from FA bottles) or anaerobic (from FN bottles) conditions. Use SDA plates for yeast and molds.

    Plate around 1 mL of the positive sample to obtain single colonies after 3 days at 32.5 °C.

33. A possible identification technique is based on the use of the VITEK® 2 system (bioMérieux). Briefly, a pure colony must be sub-cultured on the appropriate plate and condition (see note above); a Gram staining with cell morphology identification is then performed and, according to the result, the sample is loaded on an appropriate VITEK card for microorganism classification.

34. Allow the cartridge to reach room temperature in the bag before use. During cartridge manipulation, pay attention not to touch the reservoirs: the cartridge has a rounded edge purposely designed to be touched.

35. The calibration code and the lot number required by the instrument are found in the cartridge's certificate of analysis.

36. Method set-up and validation must be performed to define the optimal dilution of the final product. Different sample's dilutions can be tested with the inhibition/enhancement cartridge (Charles River).

37. Keep the pipette tilted and do not touch the well bottom with the pipette tip. Pipette slowly in order to avoid bubbles formation and squirts: they can cause nonhomogenous samples, interfere with light transmission, and in the end produce false results.
38. Two out of four wells contain endotoxin spikes (0.5 EU/mL): they serve as positive control to assess for the presence of interfering substances in the sample.
39. The spike recovery (%) is calculated as: spike value/archived spike concentration * 100. The archived spike concentration is specific for each cartridge lot and it is defined in the certificate of analysis.

References

1. van Niel G, D'Angelo G, Raposo G (2018) Shedding light on the cell biology of extracellular vesicles. Nat Rev Mol Cell Biol 19:213–228
2. Lo Cicero A, Stahl PD, Raposo G (2015) Extracellular vesicles shuffling intercellular messages: for good or for bad. Curr Opin Cell Biol 35:69–77
3. Lionetti V, Barile L (2022) Fndc5/irisin-enriched extracellular vesicles: a new hormonal relay in the regular race against vascular ageing. Eur Heart J 43:4596. https://doi.org/10.1093/eurheartj/ehac517
4. Barile L et al (2018) Cardioprotection by cardiac progenitor cell-secreted exosomes: role of pregnancy-associated plasma protein-a. Cardiovasc Res 114:992–1005
5. Gangadaran P et al (2022) Interleukin-4 receptor targeting peptide decorated extracellular vesicles as a platform for in vivo drug delivery to thyroid cancer. Biomedicine 10:1978
6. Bellio MA et al (2022) Systemic delivery of large-scale manufactured Wharton's jelly mesenchymal stem cell-derived extracellular vesicles improves cardiac function after myocardial infarction. J Cardiovasc Aging 2:9
7. Herman S, Fishel I, Offen D (2021) Intranasal delivery of mesenchymal stem cells-derived extracellular vesicles for the treatment of neurological diseases. Stem Cells 39:1589–1600
8. Barile L, Vassalli G (2017) Exosomes: therapy delivery tools and biomarkers of diseases. Pharmacol Ther 174:63–78
9. Barile L, Guzik TJ (2020) The swan song of dying cells. Cardiovasc Res 116:e90–e92
10. Estes S, Konstantinov K, Young JD (2022) Manufactured extracellular vesicles as human therapeutics: challenges, advances, and opportunities. Curr Opin Biotechnol 77:102776
11. Makkar RR et al (2012) Intracoronary cardiosphere-derived cells for heart regeneration after myocardial infarction (CADUCEUS): a prospective, randomised phase 1 trial. Lancet 379:895–904
12. McDonald CM et al (2022) Repeated intravenous cardiosphere-derived cell therapy in late-stage Duchenne muscular dystrophy (HOPE-2): a multicentre, randomised, double-blind, placebo-controlled, phase 2 trial. Lancet 399:1049–1058
13. Ishigami S et al (2015) Intracoronary autologous cardiac progenitor cell transfer in patients with hypoplastic left heart syndrome: the TICAP prospective phase 1 controlled trial. Circ Res 116:653–664
14. Andriolo G et al (2021) GMP-grade methods for cardiac progenitor cells: cell Bank production and quality control. Methods Mol Biol 2286:131–166
15. Lötvall J et al (2014) Minimal experimental requirements for definition of extracellular vesicles and their functions: a position statement from the International Society for Extracellular Vesicles. J Extracell Vesicles 3:26913
16. Solomon J et al (2016) Current perspectives on the use of ancillary materials for the manufacture of cellular therapies. Cytotherapy 18:1–12
17. Andriolo G et al (2018) Exosomes from human cardiac progenitor cells for therapeutic applications: development of a GMP-grade manufacturing method. Front Physiol 9:1169
18. Andriolo, G. *et al.* Device and method for producing and purifying exosomes. (2018)
19. Théry C et al (2018) Minimal information for studies of extracellular vesicles 2018

(MISEV2018): a position statement of the International Society for Extracellular Vesicles and update of the MISEV2014 guidelines. J Extracell Vesicles 7:1535750

20. Davidson SM et al (2022) Methods for the identification and characterization of extracellular vesicles in cardiovascular studies - from exosomes to microvesicles. Cardiovasc Res cvac031. https://doi.org/10.1093/cvr/cvac031

21. Eudralex Volume 4 - EU Guidelines for Good Manufacturing Practice for Medicinal Products for Human and Veterinary Use. https://health.ec.europa.eu/latest-updates/eudralex-volume-4-eu-guidelines-good-manufacturing-practice-medicinal-products-human-and-veterinary-2022-02-21_en

Chapter 8

Automated On-Line Isolation and Fractionation Method for Subpopulations of Extracellular Vesicles

Susanne K. Wiedmer, Evgen Multia, Thanaporn Liangsupree, and Marja-Liisa Riekkola

Abstract

Immunoaffinity chromatography (IAC) with selective antibodies immobilized on polymeric monolithic disk columns enables selective isolation of biomacromolecules from human plasma, while asymmetrical flow field-flow fractionation (AsFlFFF or AF4) can be used for further fractionation of relevant subpopulations of biomacromolecules (e.g., small dense low-density lipoproteins, exomeres, and exosomes) from the isolates. Here we describe how the isolation and fractionation of subpopulations of extracellular vesicles can be achieved without the presence of lipoproteins using on-line coupled IAC-AsFlFFF. With the developed methodology, it is possible to have fast, reliable, and reproducible automated isolation and fractionation of challenging biomacromolecules from human plasma with a high purity and high yields of subpopulations.

Key words antibodies, asymmetrical flow field-flow fractionation, extracellular vesicles, immunoaffinity chromatography, isolation, on-line coupling, tetraspanins

1 Introduction

The extracellular vesicles (EV) membrane is enriched with a variety of biomolecules relevant for intercellular communication, including tetraspanins, major histocompatibility complex, transmembrane proteins, and adhesion molecules. Recently, we have shown that Raman spectroscopy in combination with comprehensive two-dimensional gas chromatography can be utilized for successful label-free characterization of plasma-derived EV subpopulations, followed by their isolation from plasma [1]. Antibodies targeting at proteins expressed on EV surfaces, such as tetraspanins, *e.g.*, CD9, CD63, and CD81, are among the most commonly used immunoaffinity ligands for isolation and characterization of EVs from endocytic origins, *i.e.*, exosomes [2–6]. Other surface expression markers can also be used for the isolation and detection of

Seppo Vainio (ed.), *Cell-Secreted Vesicles: Methods and Protocols*, Methods in Molecular Biology, vol. 2668,
https://doi.org/10.1007/978-1-0716-3203-1_8,

highly specific EV subpopulations. For instance, integrin CD61 is a specific marker for platelet-derived EVs, the most abundant EVs found in blood plasma, and antibodies targeting at CD61 have been used for specific capture and immunoaffinity isolation of platelet-derived EVs [7].

The mostly used physical characterization methods for EVs rely on dynamic light scattering, including nanoparticle tracking analysis, tunable resistive pulse sensing, and vesicle flow cytometry [8, 9]. Recently the use of capillary electrophoresis using laser-induced fluorescence [10] or ultraviolet detection [11] has also been demonstrated.

Fast and selective isolation of biomacromolecules can be achieved with affinity-based separation techniques such as affinity chromatography. In affinity chromatography, the binding ligand is immobilized on a stationary support and the ligand utilizes biological interactions to selectively capture its counterpart (*e.g.*, enzyme captures substrate, antibody captures antigen in immunoaffinity, or vice versa). Most commonly, the bound biomolecules can be released with nonspecific elution by changing the pH of the mobile phase. The most applied form of bioaffinity chromatography is immunoaffinity chromatography (IAC). Here, we describe the development of immunoaffinity chromatographic methods to capture EVs of different cellular origins from human plasma using anti-CD9, anti-CD63, and anti-CD81 monoclonal antibodies, while anti-CD61 monoclonal antibody was exploited to capture platelet-derived EVs [7].

AsFlFFF has been extensively utilized for studying biomacromolecules and nanoparticles [12–16]. In FlFFF, the cross flow, perpendicular to the main flow, creates a hydrodynamic field that results in the separation of macromolecules based on the differences in their diffusion coefficients. The channel in AsFlFFF consists of two plates separated by a spacer. A permeable membrane at the bottom plate is called the accumulation wall. Macromolecules are carried by a parabolic flow profile from the channel to detectors and biomacromolecules are separated by the perpendicular cross flow. The separation including sample injection, sample focusing, and sample elution is shown in Fig. 1.

Size-based separation of biomacromolecules by AsFlFFF is more flexible compared to that carried out, *e.g.*, by size exclusion chromatography (SEC) due to the cross flow that can be easily modified and optimized. Small biomacromolecules (high diffusion coefficient) remain further from the bottom of the channel, while larger biomacromolecules stay closer to the accumulation wall. Thus, in normal mode, smaller biomacromolecules arrive first to the detector followed by larger ones, opposite to the order in SEC. The main advantage of AsFlFFF for the separation of biomacromolecules is its gentle fractionation due to lack of shear forces from the stationary phase (present in SEC) that easily deteriorate the

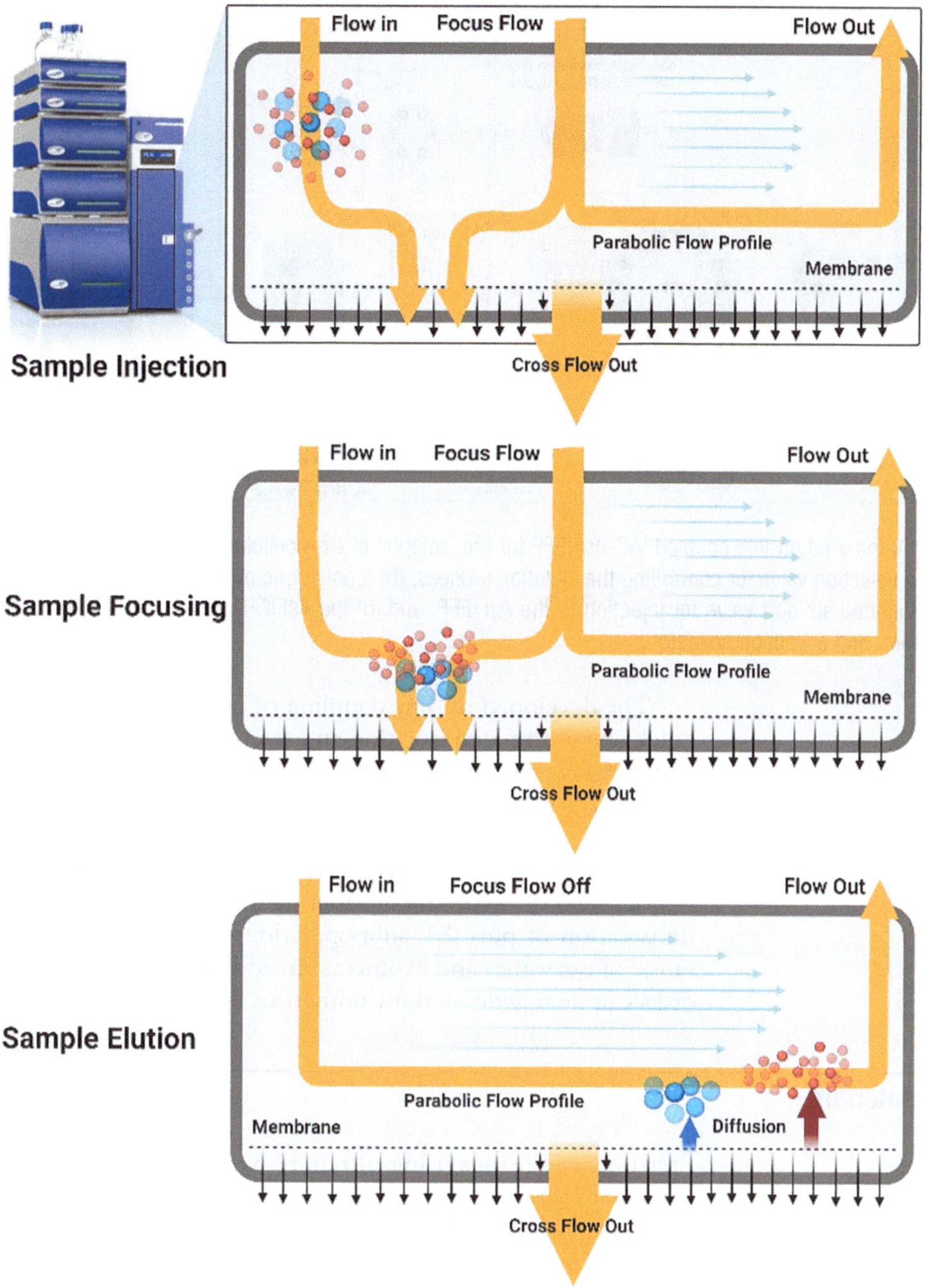

Fig. 1 Separation principle of biomacromolecules in normal mode in the AsFlFFF channel

fragile biomacromolecules. With AsFlFFF, it is also possible to exchange the buffer to a formulation buffer, which is important for fractionation of biomacromolecules for therapeutic purposes. The main disadvantages of AsFlFFF are dilution of the injected samples, overloading effects, and problems with self-association.

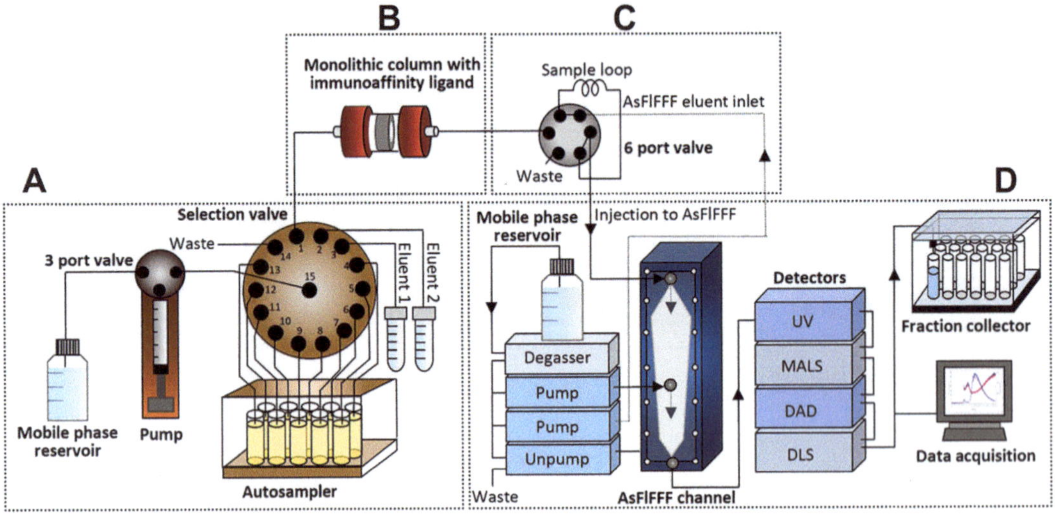

Fig. 2 Automated on-line coupled IAC-AsFlFFF for the isolation of extracellular vesicles. The system consists of (**a**) a selection valve for controlling the isolation process, (**b**) a polymeric monolithic column for the IAC, (**c**) an automated six-port valve for injection to the AsFlFFF, and (**d**) the AsFlFFF with UV, MALS, DAD, and DLS detectors, and a fraction collector

The developed on-line coupling of IAC to AsFlFFF resulted in quick and automated isolation and fractionation of nanosized subpopulations of EVs from human plasma [17]. A polymeric monolithic disk immobilized with monoclonal antibodies was utilized in IAC. The on-line IAC-AsFlFFF system is described in Fig. 2. The system is able to process reliably 18–38 samples in 24 h with only minor operator involvement, offering highly reproducible and gentle isolation of pure EV subpopulations in high yields in the size range of exosomes and exomeres. In addition, IAC polymeric CDI disks can be regenerated for numerous uses.

2 Materials

Prepare all solutions using ultrapure water (prepared by purifying deionized water) and analytical grade reagents. Prepare and store all reagents at room temperature (unless indicated otherwise). Diligently follow all waste disposal regulations when disposing waste materials.

1. Buffer: Phosphate phosphate-buffered saline (PBS, pH 7.4) is prepared by dissolving a PBS tablet in 200 mL Milli-Q (MQ) water. The solution is filtered through a 0.2 μm membrane filter (Supor-200) before use.
2. Carbonate-bicarbonate solution: 1.06 g of anhydrous Na_2CO_3 is dissolved in 100 mL Milli-Q water to prepare 0.1 M Na_2CO_3 stock solution. 0.8401 g of $NaHCO_3$ is dissolved in 100 mL Milli-Q water to prepare 0.1 M $NaHCO_3$ stock solution. The carbonate-bicarbonate solution is prepared by mixing 90 mL of

0.1 M Na_2CO_3 stock solution with 10 mL of 0.1 M $NaHCO_3$ stock solution. The pH of the solution is adjusted with NaOH to 11.3.

3. Ammonium hydroxide (NH_4OH) solution (0.15 M at pH 11.5): Dilute 1.13 mL of 25% ammonia with MQ water to a final volume of 50 mL.
4. Ethanolamine solution (2 M at pH 9): Dilute 0.603 mL of ethanolamine in a small amount of MQ water and then add 0.9 mL of fuming 37% HCl by placing the pipette tip directly inside the ethanolamine solution. Add MQ water to a final volume of 5 mL.
5. Antibodies of anti-CD9, anti-CD32, anti-CD61, anti-CD63, anti-CD81, and anti-CD82 are immobilized on 1,1′--carbonyldiimidazole (CDI) disks.

3 Methods

3.1 Immobilization of Antibodies on 1,1′-Carbonyldiimidazole Disks

Use CDI disks for immobilization of the antibodies (*see* **Note 1**). The immobilization procedure is shown in Fig. 3 [18].

Carry out all procedures at room temperature unless otherwise specified.

1. Place a CIM® CDI disk with a volume of 0.34 mL and a pore size of 1.3 μm in the disk housing.
2. Wash the disk first with 6.8 mL of Milli-Q water, then with 15 mL of PBS buffer.
3. Inject the antibody (1 mL of 0.5 mg/mL of anti-CD9, anti-CD32, anti-CD61, anti-CD63, anti-CD81, or anti-CD82) through the housing back and forth using a syringe pump and two syringes at both ends of the column with a flow rate of 0.5 mL/min for 1.5 h at room temperature (*see* **Notes 2** and **4**).
4. Repeat the cycle of injecting back and forth the solution five times daily for three consecutive days.

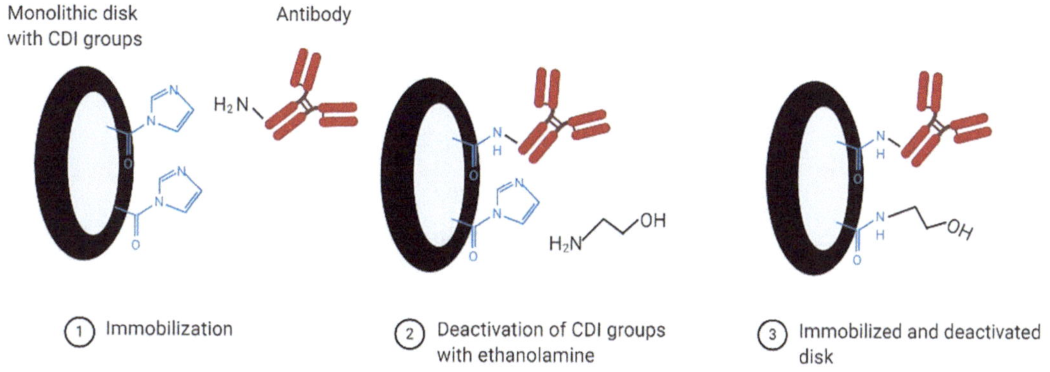

Fig. 3 Immobilization of antibodies on a polymeric monolithic CIM® CDI disk [18]

5. Leave the disk for 48 h to ensure maximum antibody immobilization.
6. Wash the disk with 3 mL of PBS, and deactivate the remaining CDI groups by passing 1 mL of 2 M ethanolamine (pH 9) through the system twice.
7. Leave the washed disk overnight with stoppers.
8. Finally wash the system with 20 mL PBS buffer to remove excess ethanolamine (*see* **Note 3**).

3.2 Isolation of Subpopulations of Extracellular Vesicles by Immunoaffinity Chromatography

The IAC process cycle for the isolation of EVs is shown in Fig. 4 and the EV isolation program in Table 1 (*see* **Note 5**) [17].

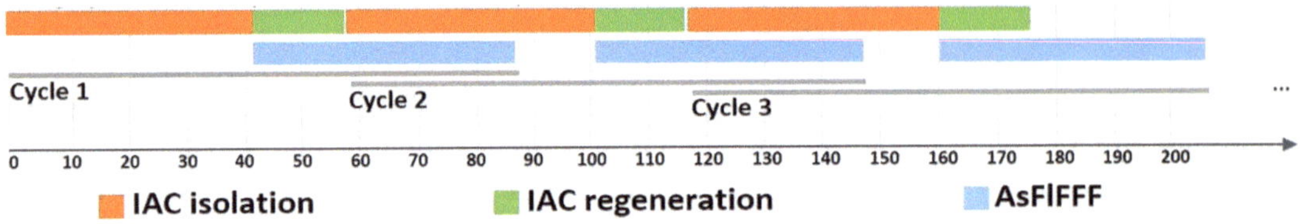

Fig. 4 Repeated successive IAC-AsFlFFF analysis cycles for the EVs. Automated injection to AsFlFFF is done when the eluate from IAC is fully transferred to the sample loop of the six-port valve [17]

Table 1
Process cycle used in IAC for the isolation of EVs [17]

EV Isolation program			
	mL	mL/min	min
Sample loading	5	6	0.833
Sample injection	5	0.25	20
PBS loading	3	6	0.5
PBS injection	3	0.25	12
Carbonate-bicarbonate loading	2	6	0.333
Carbonate-bicarbonate injection	2	0.25	8
PBS loading	3	6	0.5
PBS injection	3	1	3
Ammonium hydroxide loading	2	6	0.333
Ammonium hydroxide injection	2	1	2
PBS loading	3	6	0.5
PBS injection	3	1	3
	Total		51

3.3 Fractionation of Subpopulations of Extracellular Vesicles by Asymmetrical Flow Field-Flow Fractionation On-Line Coupled to IAC

Fractionation of the IAC isolated EVs is carried out by AsFlFFF (*see* **Note 6**). The AsFlFFF channel is connected to UV (SPD-20A Prominence, Shimadzu, Japan), multiangle light scattering (Postnova PN3070 MALS), diode array (DAD) (G1315A Agilent Technologies, Waldbronn, Germany), and dynamic light scattering (DLS) (Zetasizer Nano, Malvern Instruments, UK) detectors. Fractions are collected using a CBM-20A modular system controller (Shimadzu, Japan) and a fraction collector (FRC-10A, Shimadzu, Japan). Typical analysis profiles are seen in Figs. 5 and 6. The EV fractions do not contain contaminating lipoproteins proved by Figs. 5 and 6 where no carotenoids are detected in the DAD isoabsorbance plots of EVs.

1. Use a PBS 7.4 buffer as the mobile phase.
2. Select a Postnova AF2000 MT spacer channel with a thickness of 350 μm.
3. Install a regenerated cellulose membrane with a 10 kDa mass cutoff.
4. The AsFlFFF loop size is 500 μL.
5. Inject the sample for 5 min at a flow rate of 0.1 mL/min.

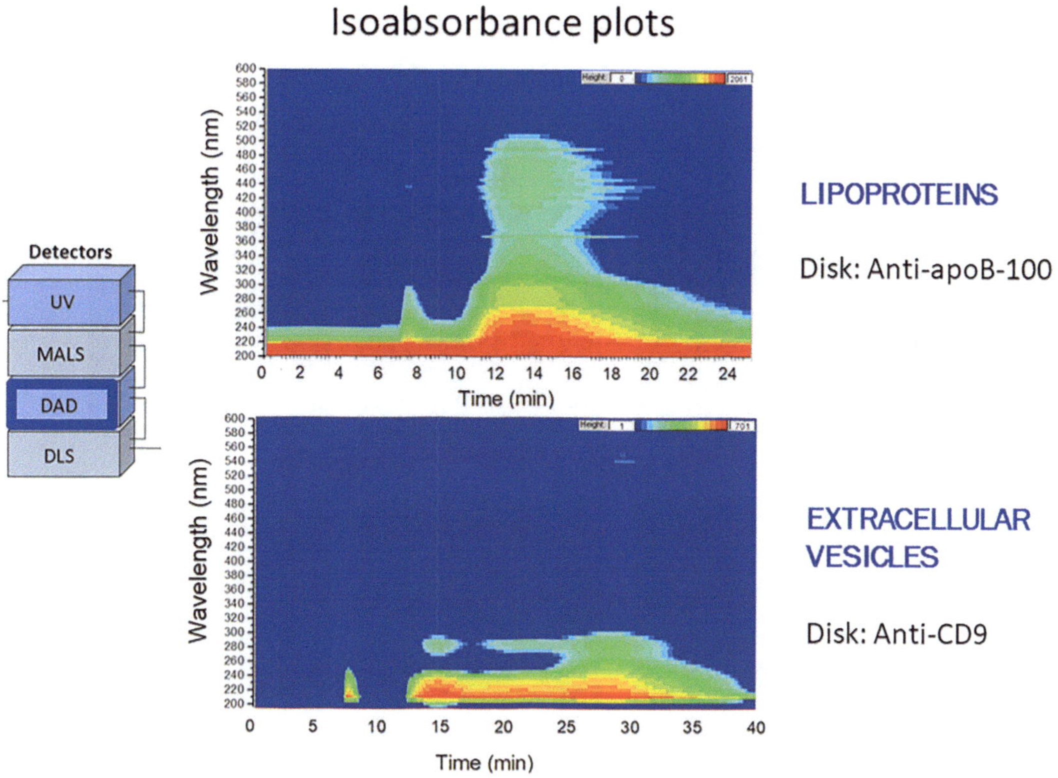

Fig. 5 The absence of carotenoids confirmed by diode array detector proves that no lipoproteins are present as contaminants in the EV fractions

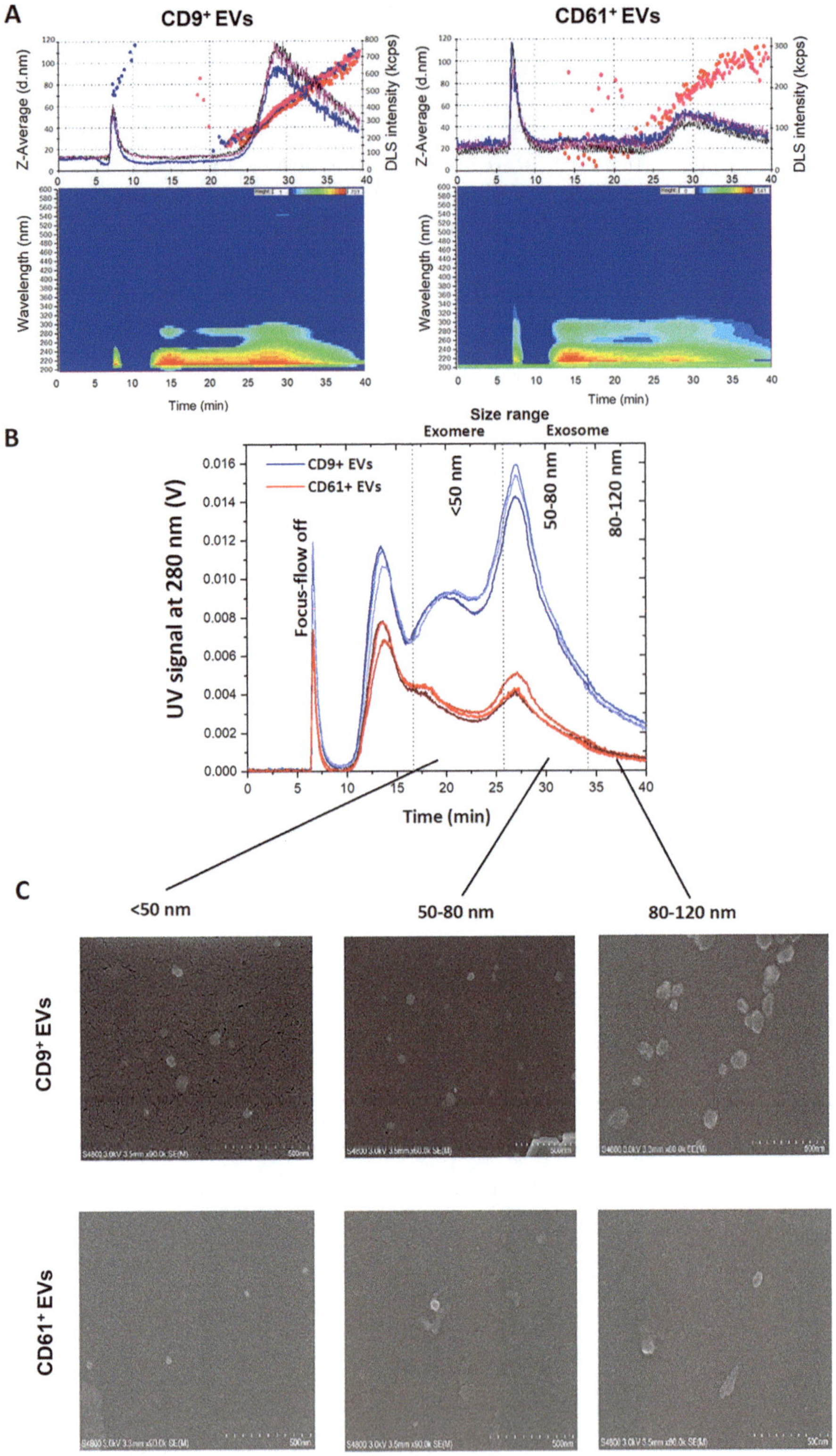

Fig. 6 IAC-AsFlFFF analysis profiles after fractionating the $CD9^+$ and $CD61^+$ EV isolates. Technical replicates ($n = 3$) of raw flow DLS data, the hydrodynamic diameter (dots as *Z*-Average) on top, and (**a**) isoabsorbance plot of the selected run at the bottom. (**b**) Overlaid UV spectra (280 nm) from the technical replicates, and (**c**) field emission scanning electron microscopy morphology of fractionated subpopulations for $CD9^+$ and $CD61^+$ EVs [17]

6. Set the detector flow at 0.5 mL/min.
7. The focusing step, including a 1 min transition time, is 6 min with a cross flow rate of 3 mL/min.
8. Separation and fractionation are done in three steps with decreasing cross flow rate (1) for 5 min with a linear decay of the cross flow from 3.0 mL/min to 0.5 mL/min, (2) for 15 min with a linear decay of the cross flow from 0.5 mL/min to 0 mL/min, and (3) for 14 min with 0 mL/min cross flow.

4 Notes

1. Selection of the appropriate antibodies is crucial.
2. During the immobilization step, the flow rate needs to be well controlled as it has a direct effect on the obtained yields.
3. The polymeric monolithic CDI disk, immobilized with antibody can be regenerated for multiple uses (even hundred), thus reducing the relative cost of the disk.
4. For higher EV yields, the polymeric monolithic disk can be loaded with a higher amount of the antibody, and for the best plasma sample volumes and flow rates, proper optimization is needed.
5. For higher yields, the isolation procedure can be repeated several times.
6. The on-line coupled IAC-AsFlFFF system can be used for the isolation of any nanosized biomacromolecules of interest and their subpopulations, as long as suitable antibodies are available for IAC.

References

1. Liangsupree T et al (2022) Raman spectroscopy combined with comprehensive gas chromatography for label-free characterization of plasma-derived extracellular vesicle subpopulations. Anal Biochem 647:114672
2. Gonda A et al (2019) Internalization of exosomes through receptor-mediated endocytosis. Mol Cancer Res 17(2):337–347
3. Termini CM, Gillette JM (2017) Tetraspanins function as regulators of cellular signaling. Front Cell Develop Biol 5(APR)
4. Meldolesi J (2018) Exosomes and Ectosomes in intercellular communication. Curr Biol 28(8):R435–R444
5. Liangsupree T et al (2022) Kinetics and interaction studies of anti-tetraspanin antibodies and ICAM-1 with extracellular vesicle subpopulations using continuous flow quartz crystal microbalance biosensor. Biosens Bioelectron 206:114151
6. Liangsupree T, Multia E, Riekkola M-L (1636) Modern isolation and separation techniques for extracellular vesicles. J Chromatogr A 2021: 461773
7. Multia E et al (2019) Fast isolation of highly specific population of platelet-derived extracellular vesicles from blood plasma by affinity monolithic column, immobilized with anti-human CD61 antibody. Anal Chim Acta 1091:160–168
8. Shao H et al (2018) New Technologies for Analysis of extracellular vesicles. Chem Rev 118(4):1917–1950

9. Koritzinsky EH et al (2017) Quantification of exosomes. J Cell Physiol 232(7):1587–1590
10. Morani M et al (2020) Electrokinetic characterization of extracellular vesicles with capillary electrophoresis: a new tool for their identification and quantification. Anal Chim Acta 1128: 42–51
11. Piotrowska M et al (1621) Capillary zone electrophoresis of bacterial extracellular vesicles: a proof of concept. J Chromatogr A 2020: 461047
12. Yohannes G et al (2011) Asymmetrical flow field-flow fractionation technique for separation and characterization of biopolymers and bioparticles. J Chromatogr A 1218(27): 4104–4116
13. Qureshi RN, Kok WT (2011) Application of flow field-flow fractionation for the characterization of macromolecules of biological interest: a review. Anal Bioanal Chem 399(4): 1401–1411
14. Zhang X et al (2018) Field-flow fractionation: a gentle separation and characterization technique in biomedicine. TrAC Trends Anal Chem 108:231–238
15. Kavurt UB et al (2015) Membranes for separation of biomacromolecules and bioparticles via flow field-flow fractionation. J Chem Technol Biotechnol 90(1):11–18
16. Lespes G, De Carsalade Du Pont V (2022) Field-flow fractionation for nanoparticle characterization. J Sep Sci 45(1):347–368
17. Multia E et al (2020) Automated on-line isolation and fractionation system for Nanosized biomacromolecules from human plasma. Anal Chem 92(19):13058–13065
18. Multia E (2021) Development of fast, reliable and automated isolation and fractionation methods for nanosized subpopulations of human biomacromolecules. Helsinki: Unigrafia, Doctoral School in Natural Sciences Dissertation Series. 2669–882X

Part II

Targeting Cell Behavior Functions of EVs

Chapter 9

A Novel Assay for Investigating the Role of Exosomes in Tumor Cell-Endothelial Cell Crosstalk

Yan Qiu, Wenli Jiang, and Ye Zeng

Abstract

Exosomes have critical role in regulating the tumor development and progression and resistance following antiangiogenesis therapies (AATs). Exosomes could be released by both tumor cells and surrounding endothelial cells (ECs). Here, we describe the methods to explore the cargo transfer between tumor cells and ECs by a novel four-compartment co-culture methods and to investigate the effect of tumor cells on angiogenic ability of ECs by Transwell co-culture methods.

Key words Tumor cells, Endothelial cells, Cell–cell interactions, Exosomes

1 Introduction

Liver cancer is one of the most common cancers, ranking sixth in incidence and fourth in mortality worldwide [1]. Hepatocellular carcinoma (HCC) is the main type of primary liver cancer, responsible for over 750,000 deaths globally each year, with China accounting for more than half of all cases [2]. Targeting tumor vasculature via VEGFR inhibitors (e.g., vandetanib), or antiangiogenic therapies (AATs), however, has not resulted in optimal outcomes for most cancer patients [3–5].

Exosomes (Exos) are lipid bilayer-enclosed vesicles with a diameter of 30–200 nm [6] and are primarily involved in cell communication. Exosomes contain a variety of functional components, including proteins, DNA, RNA, and lipids, and they serve critical roles in intercellular material and information transfer. Exosomes have been shown to play key roles in the control of the tumor microenvironment by promoting angiogenesis, signaling pathway activation, carcinogenesis, and metastasis [7].

Exosomes mediated the crosstalk between tumor cells and human umbilical vein endothelial cells (HUVECs). Hypoxia-induced exosomes generated by lung cancer, papillary thyroid

Seppo Vainio (ed.), *Cell-Secreted Vesicles: Methods and Protocols*, Methods in Molecular Biology, vol. 2668,
https://doi.org/10.1007/978-1-0716-3203-1_9,

cancer, and esophageal squamous carcinoma cells have been demonstrated to increase human umbilical vein endothelial cells (HUVECs) proliferation, migration, invasion, and angiogenic ability [8–10]. Exosomes from several human malignancies have been linked to tumor angiogenesis via changing endothelial cell properties [11, 12]. Exosomes produced from HCC cells under hypoxia have been proven in studies to trigger tube formation in HUVECs [13].

When plated on Matrigel, human umbilical vein endothelial cells (HUVECs) undergo differentiation into capillary-like tube structures in vitro. A tube-formation assay has been used to evaluate angiogenesis in vitro as previously described [14]. Here we demonstrate a series of protocols to explore the tumor-HUVECs crosstalk.

2 Materials

2.1 Differential Centrifugation

1. Refrigerated benchtop centrifuge.
2. 0.22 μm-filters.
3. Ultracentrifuge.
4. Phosphate buffer saline (PBS).

2.2 Conditional Medium Transfer Assay

1. HUVEC-NC.
2. HUVEC-miR9.
3. HepG2.
4. DMEM with high glucose.
5. FBS.
6. BSA.
7. Penicillin-streptomycin.
8. Quartile petri dish.
9. 24 mm cell culture slide.
10. Trizol.
11. mi*DETECT*™ miRNA External Control: cel-miR-39.
12. Bulge-Loop TM miRNA qRT-PCR Primer.
13. Chloroform.
14. Isopropyl.
15. Absolute ethyl alcohol.
16. DEPC H_2O.
17. qPCR RT Kit.

2.3 Co-Culture (HUVECs/HCC) Assay

1. HUVECs.
2. HUVEC culture medium.
3. HCC tumor cell lines (SMMC-7721 and Huh7).
4. DMEM with high glucose.
5. FBS.
6. Penicillin-streptomycin.
7. GW4869 (Dissolved in DMSO to 0.58 mg/mL, 1 mM; and then diluted in culture medium to 10 μM).
8. Pharmacological agents such as vandetanib (Dissolved in DMSO to 1.90 mg/mL, 4 mM; and then diluted in culture medium to 4 μM) or 3-Methyladenine (3-MA; dissolved in DMSO to 29.83 mg/mL, 200 mM; and then diluted in culture medium to 5 mM).
9. Transwell (polyester (PET) membrane, 0.4 μm pore size, 10 μm thickness, 6.5 mm diameter, 24-well plates).
10. Matrigel.
11. Paraformaldehyde or other suitable fixation solution.

3 Methods

3.1 Exosomes Isolation by Differential Centrifugation

1. Collects 50 mL of cell culture media from cells (*see* **Note 1**).
2. Centrifuge at 300 × *g* at 4 °C for 20 min to remove dead cells and transfer supernatant to fresh tube.
3. Centrifuge at 3000 × *g* at 4 °C for 20 min to remove cell debris and transfer supernatant to fresh tube.
4. Centrifuge at 10,000 × *g* at 4 °C for 30 min to pellet the microvesicles (MVs) and transfer supernatant to fresh tube. If needed, resuspend MVs in 100 μL PBS and store at 4 °C for subsequent VEGF assays.
5. Filter supernatant containing exosomes with a 0.22 μm filter.
6. Ultracentrifuge at 100,000 × *g* at 4 °C for 2 h to pellet the exosomes.
7. Resuspend the exosomes in 100 μL PBS and store at 4 °C for subsequent VEGF assays (*see* Fig. 1 and **Note 2**).

3.2 Conditional Medium Transfer During HUVECs and HCC Cells' Crosstalk

1. Sterilize the culture dish and cell crawl for 30 min before using them.
2. The cells were cultured in DMEM high glucose medium containing 10% FBS and 1% Penicillin-Streptomycin. The cells were grown in a humidified 5% CO_2 incubator at 37 °C. After confluence, the cells were trypsinized with 0.25% trypsin

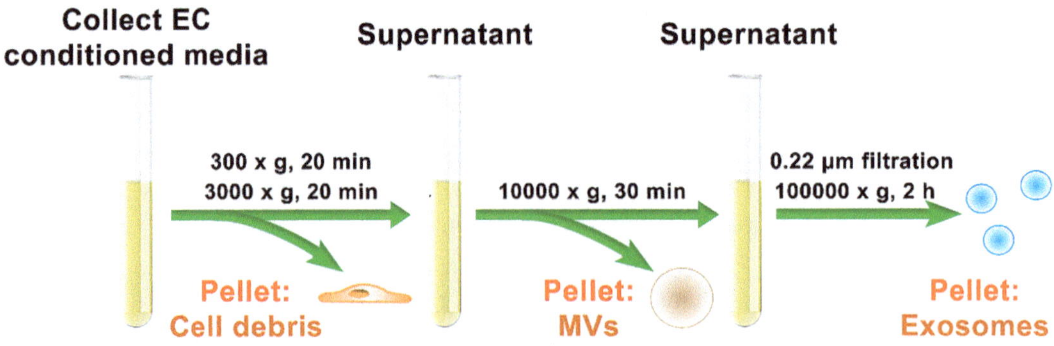

Fig. 1 Diagram of the differential centrifugation of microvesicles (MVs) and exosomes

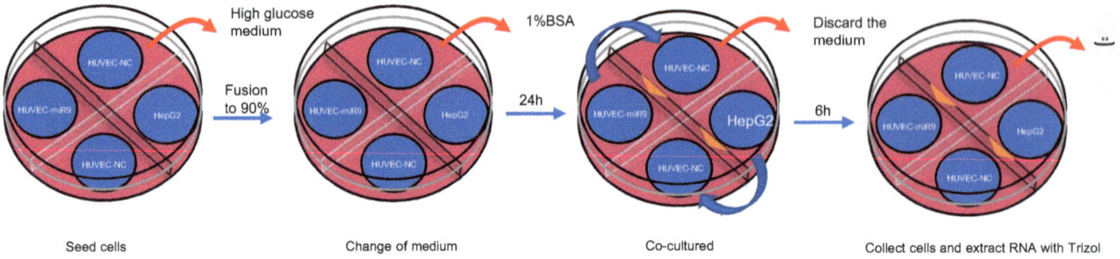

Fig. 2 Schematic diagram of the four-compartment co-culture method

and 0.05% EDTA, and then concentrated by centrifugation. To achieve a cell concentration of 5×10^5 cells/mL, the harvested cells were resuspended in fresh medium (*see* Fig. 2).

3. Inoculate 500 μL of each crawl slice, discard the nonadherent cells after 1 and a half hours, and add 2 mL complete medium to each well.
4. When the cells confluence to about 90%, replace the 1% BSA medium (BSA: DMEM = 1:30). After 24 h, the HUVEC-NC culture was discarded, and the interval between petri dishes was cut to allow the HepG2 culture to treat HUVEC-NC for 6 h (*see* **Note 3**).
5. Cells were scraped off the climbing sheet with a cell scraper and thoroughly mixed with RNA extraction lysate to allow full lysis of the sample (*see* **Note 4**).
6. Add external reference: after the sample is thoroughly lysed and mixed, add 5 μL (1 pmol) of external reference, mix thoroughly, and stand for 5–10 min (*see* **Note 5**).
7. RNA extraction: complete RNA extraction according to the process (*see* Fig. 3).
8. Add 200 μL of chloroform, stir for 15 s, and set aside for 3 min.
9. For 10 min, centrifuge at 12,000 rpm at 4 °C while transferring the upper aqueous phase to a new EP tube.

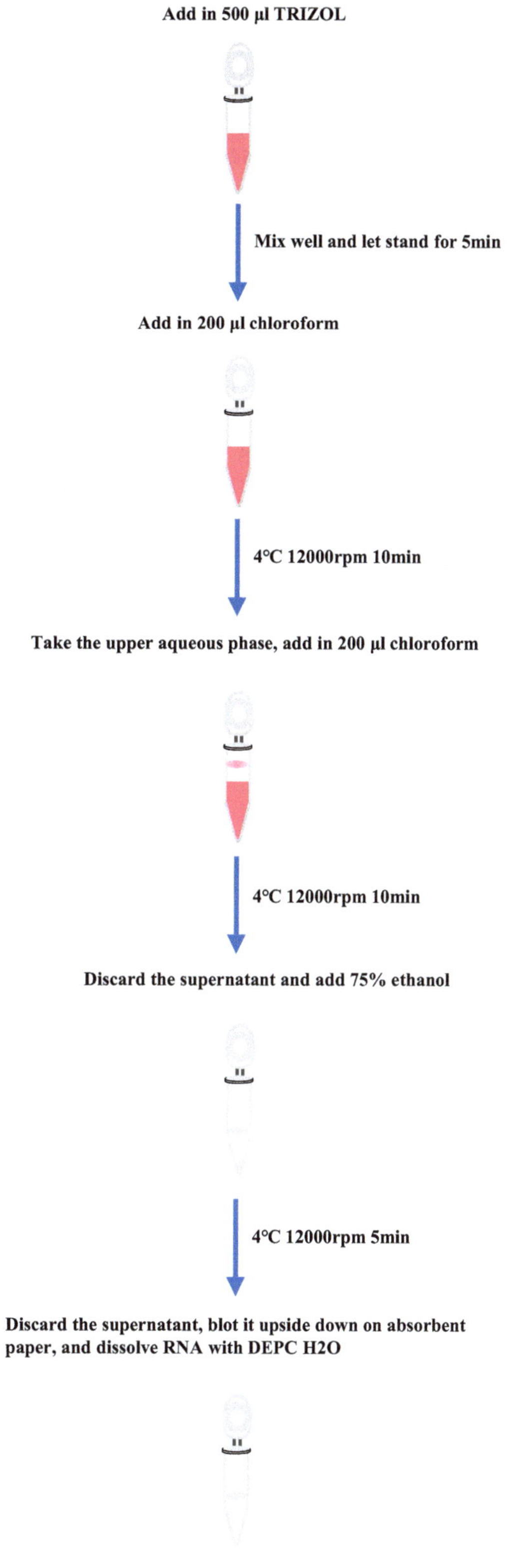

Fig. 3 TRIZOL method for RNA extraction

10. Add an equal volume of isopropyl alcohol, mix thoroughly, set aside for 20 min, centrifuge at 12,000 rpm for 10 min at 4 °C, and discard supernatant.
11. Add 75% DEPC ethanol to the sediment, gently spin and reverse wash the sediment, and centrifuge at 12,000 rpm for 10 min at 4 °C (*see* **Note 6**).
12. Discard the supernatant, place the EP tube upside down on absorbent paper, and let it dry at room temperature for 5–10 min. RNA should be dissolved in 20–50 μL of DEPC water and frozen at −80 °C (*see* **Note 7**).
13. The purity and concentration were determined by a spectrophotometer.
14. Using 1 μg total RNA as template, the reverse transcription reaction system was prepared according to the instructions of the qPCR RT Kit. The total system was 20 μL, and the first strand of cDNA was synthesized (*see* **Note 8**).
15. qRT-PCR was performed in 10 μL of quantitative reaction mixtures containing 5 μL of SYBR Premix ExTaq, 0.4 μM each of forward and reverse primers, 1 μL cDNA, and nuclease-free water.
16. The program used for qRT-PCR consisted of a denaturing cycle of 3 min at 94 °C, 40 cycles of PCR (94 °C for 10 s, 60 °C for 40 s), and a melting cycle of 60 °C for 15 s.
17. The primer sequences are listed as below:

    hsa-miR-9-5p RT: CTCAACTGGTGTCGTGGAGTCGG CAATTCAGTTGAGTCATACAGC

    hsa-miR-9-5p F: ACACTCCAGCTGGGTCTTTGGTTATC TAG

    All R:CTCAACTGGTGTCGTGGA

    U6 F:CTCGCTTCGGCAGCACA

    U6 R:AACGCTTCACGAATTTGCGT

    Cel-miR39: 5′-UCACCGGGUGUAAAUCAGCUUG-3′

3.3 Role of HUVECs-Derived Exosomes on Tumor Cells' Tubulogenesis by Co-Culture Assay

1. Seed 1×10^4 HUVECs per Transwell insert.
2. Culture the HUVECs in the Transwell insert in HUVEC culture medium containing 10% FBS and 1% penicillin-streptomycin for 48–72 h in 24-well plates.
3. Treat HUVECs with HUVEC culture medium containing 10 μM GW4869 in the presence of pharmacological agents such as 4 μM vandetanib or 5 mM 3-MA for 48 h (*see* **Note 9**).
4. Coat the bottom of a 24-well plate with Matrigel (300 μL/well) without introducing air bubbles.

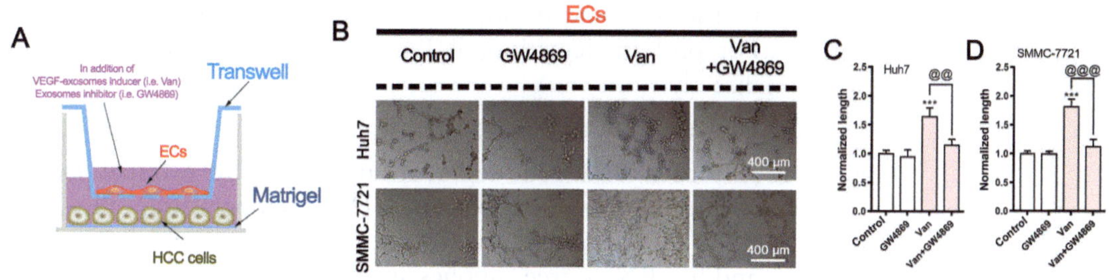

Fig. 4 EC-derived exosomes induce the tubulogenesis of HCC cells. HCC cells (5×10^4) per Matrigel-coated well in DMEM containing 10% FBS and 1% penicillin-streptomycin are seeded in 24-well plate for 24 h. ECs were treated with 10 μM GW4869 in the presence of pharmacological agents such as 4 μM vandetanib (Van) for 48 h, and then co-cultured with HCC cells for 8 h. ECs without GW4869 and Van treatments are set as control. (**a**) Diagram of the Transwell with ECs and HCC cells is represented on the left. (**b**) Tube formation. Normalized length of tubes of Huh7 cells (**c**) and SMMC-7721 (**d**) cells. Mean ± SEM, $n = 3$. One-way analysis of variance (ANOVA) with the least significant difference test. ***$P < 0.001$ versus control. @@$P < 0.01$, @@@$P < 0.001$ versus Van. (Revised from ref. [15])

5. After gelling of the Matrigel, seed 5×10^4 tumor cells per Matrigel-coated well in DMEM containing 10% FBS and 1% penicillin-streptomycin in 24-well plate for 24–48 h.
6. Change the fresh medium in 24-well plate for tumor cell culture.
7. Transfer the Transwell inserts with HUVECs and culture medium onto the 24-well plates, and co-culture with tumor cells for 8 h.
8. Remove the inserts and aspirate the media from the 24-well plates.
9. Wash in PBS three times for 10 s.
10. Fix in 4% paraformaldehyde or other suitable fixation solution. The fixation is not necessary if you can image all the tube structures within 5–10 min.
11. Observe and image the tube structures of tumor cells.
12. Acquire the images from three to four fields per well from three to four independent experiments to calculate the average length and junction of tube networks (*see* Fig. 4 and **Note 10**).

4 Notes

1. If culture medium volume is low, PBS can be added to 50 mL. The volume used for centrifugation is dependent on ultracentrifuge parameters.
2. Further detection should be performed immediately. To avoid secretory proteins contamination, differential centrifugation is more reliable than commercial isolation kits. The isolation

methods should be adjusted for blood and other body fluids. For protein detection, RIPA is mixed with exosome pellets directly.

3. Pay attention to aseptic operation.
4. In order to avoid degradation caused by external factors (including enzymes, extreme pH or temperature), all operations should strictly follow the RNA operation specifications and use RNase-free supplies and reagents.
5. The sample is generally rich in RNase, so it is necessary to thoroughly mix to inhibit RNase activity before adding external parameters. If not thoroughly mix, adding external parameters will easily lead to the degradation of external parameters.
6. 75%DEPC ethanol need to be ready to use.
7. The drying time should not exceed 10 min, otherwise it is not easy to dissolve.
8. MiR9 and miR39 genes perform RT reactions based on specific stem-loop primers.
9. GW4869 was used to inhibit the release of exosomes. GW4869 is a commonly used pharmacological agent, which inhibits the inward budding of multivesicular bodies (MVBs) and release of mature exosomes from MVBs.
10. The pseudovascular organization of cells could be analyzed by ImageJ software (version 1.51s; National Institutes of Health, Bethesda, MD, USA) using the Angiogenesis Analyzer plugin (written by Gilles Carpentier, 2012. The plugin is available at http://imagej.nih.gov/ij/macros/toolsets/Angiogenesis%20 Analyzer.txt).

References

1. Bray F, Ferlay J, Soerjomataram I, Siegel RL, Torre LA, Jemal A (2018) Global cancer statistics 2018: GLOBOCAN estimates of incidence and mortality worldwide for 36 cancers in 185 countries. CA Cancer J Clin 68(6): 394–424. https://doi.org/10.3322/caac.21492
2. Siegel RL, Miller KD, Jemal A (2018) Cancer statistics, 2018. CA Cancer J Clin 68(1):7–30. https://doi.org/10.3322/caac.21442
3. Hurwitz H, Fehrenbacher L, Novotny W, Cartwright T, Hainsworth J, Heim W et al (2004) Bevacizumab plus irinotecan, fluorouracil, and leucovorin for metastatic colorectal cancer. N Engl J Med 350(23):2335–2342. https://doi.org/10.1056/NEJMoa032691
4. Perren TJ, Swart AM, Pfisterer J, Ledermann JA, Pujade-Lauraine E, Kristensen G et al (2011) A phase 3 trial of bevacizumab in ovarian cancer. N Engl J Med 365(26):2484–2496. https://doi.org/10.1056/NEJMoa1103799
5. Hsu C, Yang TS, Huo TI, Hsieh RK, Yu CW, Hwang WS et al (2012) Vandetanib in patients with inoperable hepatocellular carcinoma: a phase II, randomized, double-blind, placebo-controlled study. J Hepatol 56(5):1097–1103. https://doi.org/10.1016/j.jhep.2011.12.013
6. Pegtel DM, Gould SJ (2019) Exosomes. Annu Rev Biochem 88:487–514. https://doi.org/10.1146/annurev-biochem-013118-111902
7. Taylor DD, Gercel-Taylor C (2011) Exosomes/microvesicles: mediators of cancer-

associated immunosuppressive microenvironments. Semin Immunopathol 33(5):441–454. https://doi.org/10.1007/s00281-010-0234-8
8. Li R, Wang Y, Zhang X, Feng M, Ma J, Li J et al (2019) Exosome-mediated secretion of LOXL4 promotes hepatocellular carcinoma cell invasion and metastasis. Mol Cancer 18(1):18. https://doi.org/10.1186/s12943-019-0948-8
9. Fang JH, Zhang ZJ, Shang LR, Luo YW, Lin YF, Yuan Y et al (2018) Hepatoma cell-secreted exosomal microRNA-103 increases vascular permeability and promotes metastasis by targeting junction proteins. Hepatology (Baltimore, MD) 68(4):1459–1475. https://doi.org/10.1002/hep.29920
10. Kumar A, Deep G (2020) Exosomes in hypoxia-induced remodeling of the tumor microenvironment. Cancer Lett 488:1–8. https://doi.org/10.1016/j.canlet.2020.05.018
11. Zeng Z, Li Y, Pan Y, Lan X, Song F, Sun J et al (2018) Cancer-derived exosomal miR-25-3p promotes pre-metastatic niche formation by inducing vascular permeability and angiogenesis. Nat Commun 9(1):5395. https://doi.org/10.1038/s41467-018-07810-w
12. Yokota Y, Noda T, Okumura Y, Kobayashi S, Iwagami Y, Yamada D et al (2021) Serum exosomal miR-638 is a prognostic marker of HCC via downregulation of VE-cadherin and ZO-1 of endothelial cells. Cancer Sci 112(3):1275–1288. https://doi.org/10.1111/cas.14807
13. Matsuura Y, Wada H, Eguchi H, Gotoh K, Kobayashi S, Kinoshita M et al (2019) Exosomal miR-155 derived from hepatocellular carcinoma cells under hypoxia promotes angiogenesis in endothelial cells. Dig Dis Sci 64(3):792–802. https://doi.org/10.1007/s10620-018-5380-1
14. Ponce ML (2009) Tube formation: an in vitro matrigel angiogenesis assay. Methods Mol Biol 467:183–188. https://doi.org/10.1007/978-1-59745-241-0_10
15. Zeng Y, Yao X, Liu X, He X, Li L, Liu X et al (2019) Anti-angiogenesis triggers exosomes release from endothelial cells to promote tumor vasculogenesis. J Extracell Vesicles 8(1):1629865. https://doi.org/10.1080/20013078.2019.1629865

Chapter 10

Nanoparticle (NP) Loading by Direct Incubation with Extracellular Vesicles-Secretor Cells: NP Encapsulation and Exosome Characterization

Ana Martín-Pardillos and Pilar Martín-Duque

Abstract

Exosomes are small vesicles released by all types of cells, and they have been postulated as a promising natural way to carry information amongst cells. Exosomes might serve as mediators for intercellular communication through the delivery of their endogenous cargo to neighbor or distant cells. Recently, this ability to transfer their cargo has open a new therapeutic approach and exosomes have been investigated as vectors for the delivery of the loaded cargo, for instance nanoparticles (NPs).

Currently, several methods to load exosomes with NPs have been described; however, the maintenance of the membrane integrity on the vesicle has to be taken into consideration, in order to choose one or another methodology. Here we describe the NP encapsulation through the incubation of the cells with the NPs and the subsequential methods to determine their cargo and to discard detrimental alterations on the loaded exosomes.

Key words Encapsulation, Nanoparticle, Integrity, Characterization, Immunoblot, Size, NTA, Protein, MP-AES

1 Introduction

Multicellular organisms have extracellular spaces, which contain a great diversity of molecules and ions, but also a large number of extracellular vesicles (EVs). EVs are constituted by several types of vesicles, including exosomes, microvesicles, and apoptotic bodies [1]. EVs differ in size and biogenesis pathway. Apoptotic bodies are the bigger vesicles (500–4000 nm) and they resulted from the membrane blebbing that occurs during apoptosis. Microvesicles are smaller than apoptotic bodies and tend to be larger than exosomes (100–2000 nm), although their size range overlaps with the exosome size (40–160 nm), therefore the mechanism of biogenesis remains the primary distinction between them, as most of the surface markers are common. Microvesicles arise through direct

Seppo Vainio (ed.), *Cell-Secreted Vesicles: Methods and Protocols*, Methods in Molecular Biology, vol. 2668, https://doi.org/10.1007/978-1-0716-3203-1_10,

outward budding and fusion with the plasma membrane, meanwhile exosomes are released by fusion of an organelle from the endocytic pathway, the multivesicular body, with the plasma membrane [2, 3].

Exosomes are able to carry various biomolecules, including nucleic acids, proteins, lipids, amino acids, and metabolites, which can reflect their cell of origin [4, 5] and have the ability to transfer their cargo to acceptor cells, thereby inducing phenotypic changes in the latter [6]. Cancer cells generally show a production of exosomes higher than in normal cells and at the same time have a strong capacity to modify both local and distant microenvironments, which is crucial for premetastatic niche formation, metastasis and tumor immune response modulation, and drug resistance induction [7, 8].

The ability to transfer their cargo to other cells has generated substantial interest in the scientific community, not only because the possibility of the message's exchanges among cells but also because it opens the possibility to use them as exosome-based therapeutics with fewer off-target effects than other bio-vehicles and with higher specificity (tropism). Therefore, exosomes have become promising tools for the treatment of cancer and other diseases, through the delivery and transfer of drugs and other compounds that remain protected from the immune system and degradation, mainly due to their encapsulation [9, 10].

Multiple techniques have been developed to load cargos into exosomes, such as incubating cargos together with exosomes or exosome-secreting cells, transfection, physical treatments (sonication, electroporation, extrusion, freeze-thaw, surfactant treatment and dialysis), and in situ synthesis [8, 9, 11]. Several of those methods allow the load of nanoparticles (NPs) however had a negative effect regarding the morphology and integrity of the charged exosomes. Nevertheless, the preincubation of the cells with the NPs as an attempt to incorporate them into the exosomal biogenesis pathway is a highly successful method, and exosomes showed a significant load, retaining the membrane integrity [11].

However, manipulation steps might change the exosomes and the tropism that they could transfer to the nanoparticles. In order to check whether the incorporation of the NPs might destroy or partially alter the exosomal membrane, we follow the minimal requirements described by the International society of extracellular vesicles (ISEV) [12]. We encourage authors to follow their protocols after the isolation of the exosomes NP loaded.

Here we describe a complete methodology for loading NPs in exosomes: from the NP incubation with the cells to the purification of exosomes and their analysis. To summarize, the first step is to determine the subcytotoxic concentration of NPs to treat cells, once it is determined we will proceed to the treatment of the cells, selecting the time for maximum internalization and later on, we will recover the supernatant to purify the exosome fraction. Finally, we will determine the concentration, morphology and integrity of the exosomes.

2 Materials

2.1 Cytotoxicity Test

1. Cell-culture treated plates.
2. Cell line to test: the selection of the cell line depends on the researcher's needs (related to the field of study).
3. Cell culture medium: chose the appropriate cell culture medium for the selected cell line and complement it with serum and antibiotics and/or glutamine if it is needed.
4. PBS-EDTA: 6.572 mM Na_2HPO4, 1.910 mM KH_2PO4, 0.139 M NaCl, 3.756 mM KCl, 1.236 mM EDTA $Na_2{\cdot}2H_2O$, 10.066 mM LiCl.
5. Trypsin: 2.5% Trypsin in PBS.
6. Centrifuge.
7. Cell counting chamber.
8. Optical microscope.
9. Multiwell plates.
10. PBS without Ca^{+2}/Mg^{+2}: 0.154 M NaCl, 5.599 mM Na_2HPO4, 1.058 mM KH_2PO4.
11. Viability test assays.

2.2 Internalization Time Evaluation

1. Cell culture medium: chose the appropriate cell culture medium for the selected cell line and complement it with serum and antibiotics and/or glutamine if it is needed.
2. PBS-EDTA: 6.572 mM Na_2HPO4, 1.910 mM KH_2PO4, 0.139 M NaCl, 3.756 mM KCl, 1.236 mM EDTA $Na_2{\cdot}2H_2O$, 10.066 mM LiCl.
3. Trypsin: 2.5% Trypsin in PBS.
4. Centrifuge.
5. Cell counting chamber.
6. Optical microscope.
7. Multiwell plates.
8. PBS without Ca^{+2}/Mg^{+2}: 0.154 M NaCl, 5.599 mM Na_2HPO4, 1.058 mM KH_2PO4.
9. Aqua regia: 75% HCl, 25% HNO_3.
10. Milli-Q water.
11. MP-AES equipment.

2.3 Recovery of Loaded Exosomes

1. Cell-culture treated plates.
2. Cell culture medium: chose the appropriate cell culture medium for the selected cell line and complement it with serum and antibiotics and/or glutamine if it is needed.
3. PBS without Ca^{+2}/Mg^{+2}: 0.154 M NaCl, 5.599 mM Na_2HPO4, 1.058 mM KH_2PO4.
4. Exosome-depleted cell culture medium complemented with exosome-depleted serum (*see* **Note 1**) and antibiotics and/or glutamine if it is needed.
5. Centrifuge: it has to reach a minimum speed up to 500*g*.
6. Centrifuge tubes.
7. Ultracentrifuge: range needed 12,000*g* to 100,000*g*.
8. Ultracentrifuge tubes (reference depends on the rotor and ultracentrifuge model).

2.4 Characterization of the Recovered Exosomes

2.4.1 Protein Quantification

1. Microtubes/microplates.
2. Plate reader.
3. Protein quantification test (preferably bicinchoninic acid assay (BCA assay)).

2.4.2 Exosomes Count by Nanoparticle Tracking Analysis (NTA)

1. PBS without Ca^{+2}/Mg^{+2}: 0.154 M NaCl, 5.599 mM Na_2HPO4, 1.058 mM KH_2PO4.
2. Nanoparticle Tracking Analysis (NTA) equipment.

2.4.3 Exosome Markers Evaluation by Immunoblot Detection

1. Cold (−20 °C) acetone.
2. Microtubes.
3. RIPA buffer: 150 mM NaCl, 1% NP-40, 0.5% sodium deoxycholate, 0.1% SDS, 50 mM Tris (pH 8).
4. Protease inhibitor solution.
5. Western-blot materials and reagents.

2.4.4 Evaluation of NP Loading

1. Aqua regia: 75% HCl, 25% HNO_3.
2. Milli-Q water.
3. MP-AES equipment.

3 Methods

3.1 Cytotoxicity Test

1. Seed the cells in a 100 mm diameter plate (*see* **Note 2**), with 10 mL cell culture medium, and keep the medium on the cell culture for 2–3 days, until to get confluency.

2. When the cells reached 90% confluency: wash the plate twice with PBS-EDTA.
3. Remove PBS-EDTA and add 2 mL trypsin to the plate.
4. Incubate the plate 5 min in the CO_2 incubator.
5. Check the morphology under the microscope, when the cells acquire round shape, add 5 mL cell culture medium and recover the medium (*see* **Note 3**).
6. Centrifuge the medium that contains the cells: 250*g* during 5 min (*see* **Note 4**).
7. Remove the supernatant and resuspend the cells in 10 mL cell culture medium.
8. Add 15 microliters of the resuspension to the cell counting chamber to count the cells (*see* **Note 5**).
9. Dilute the cells as needed to seed them in a 96-well plate (*see* **Note 6**): seed the same number of wells than concentrations you would like to test by triplicate (every concentration should be tested in three wells).
10. Allow the cells to attach to the well during 24 h.
11. After 24 h, add serial dilutions of the NPs to test: 150 μL per well (*see* **Notes** 7 and **8**).
12. Incubate the cells with the NP dilutions during different times, for example 24, 48, and 72 h.
13. At each selected time point, wash the wells twice with PBS Ca^{+2}/Mg^{+2} to eliminate the NPs (*see* **Note 9**) and determine survival using a cell viability test of your election (*see* **Note 10**).
14. For the next step, to load the cells with the NPs, a subcytotoxic concentration should be selected, i.e., a concentration that reduces less than 20% the viability of the cells (Fig. 1).

3.2 Internalization Time Evaluation

1. Trypsinize the cells as previously described.
2. Seed the cells in a 96-well plate: seed the number of cells needed to reach 80% confluency after 24-h, in 150 μL per well (*see* **Note 11**).
3. Allow the cells to attach to the well during 24 h.
4. After 24 h seed, remove the supernatant and add the subcytotoxic NP concentration previously selected, to each well (150 μL per well).
5. Evaluate the internalization of NPs every 2 h up to 72 h to determine the maximum internalization time (*see* **Note 12**):
   (a) Evaluation by optical microscopy: NP clusters might be detected by optical microscopy; as dark spots appear on the cells.

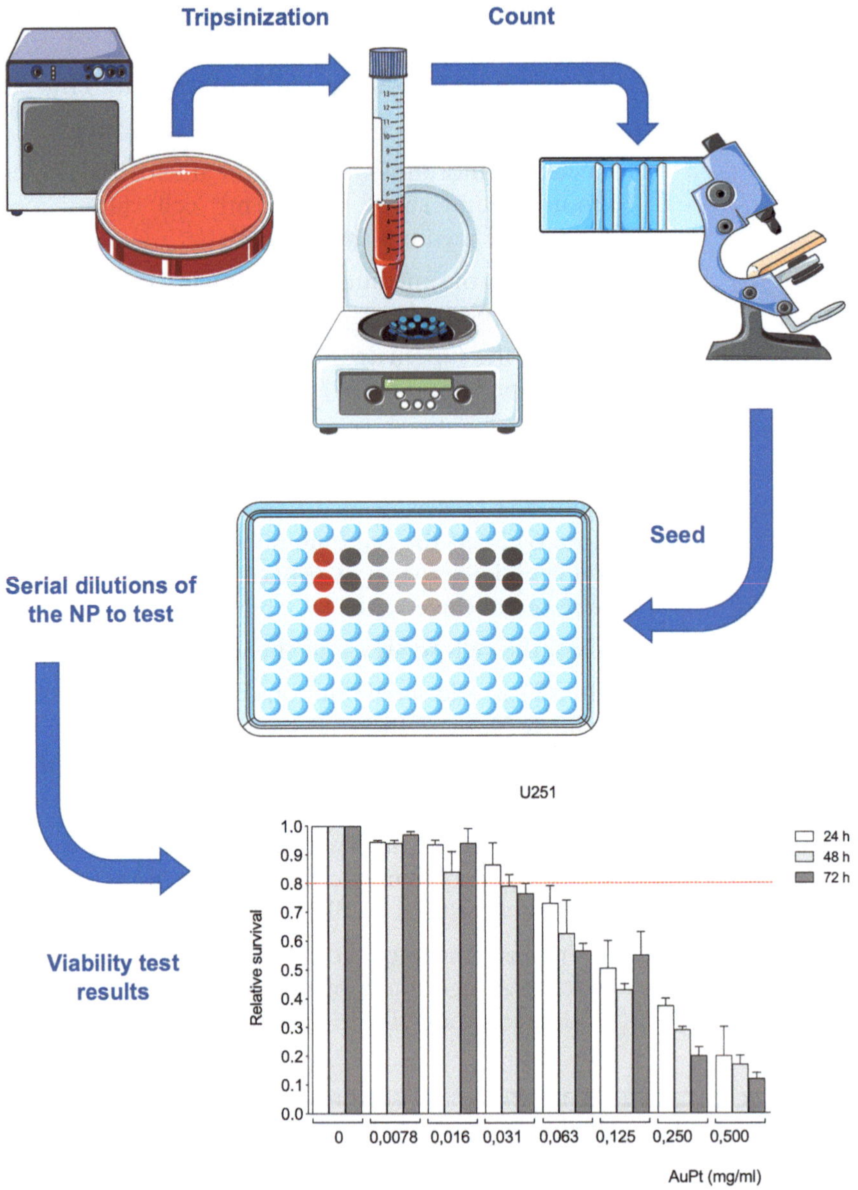

Fig. 1 Example of cytotoxicity test: procedure from trypsinization to evaluation of AuPt NP cytotoxicity in U251MG cells: relative survival of U251MG cells after 24, 48, and 72 h of incubation with 0, 0.0078, 0.016, 0.031, 0.063, 0.125, 0.250, and 0.500 mg AuPt/mL. The subcytotoxicity concentration selected to check the internalization time should be 0.031 mg AuPt/mL, due to after 72 h incubation, the relative survival reduction is less than 20%

(b) Evaluation by metal quantification: in the case of metallic NPs, additionally to the optical microscopy, metals might be detected by Microwave Plasma Atomic Emission Spectroscopy (MP-AES) (*see* **Note 13**). For that technique, cells have to be collected by trypsinization (wash the cells

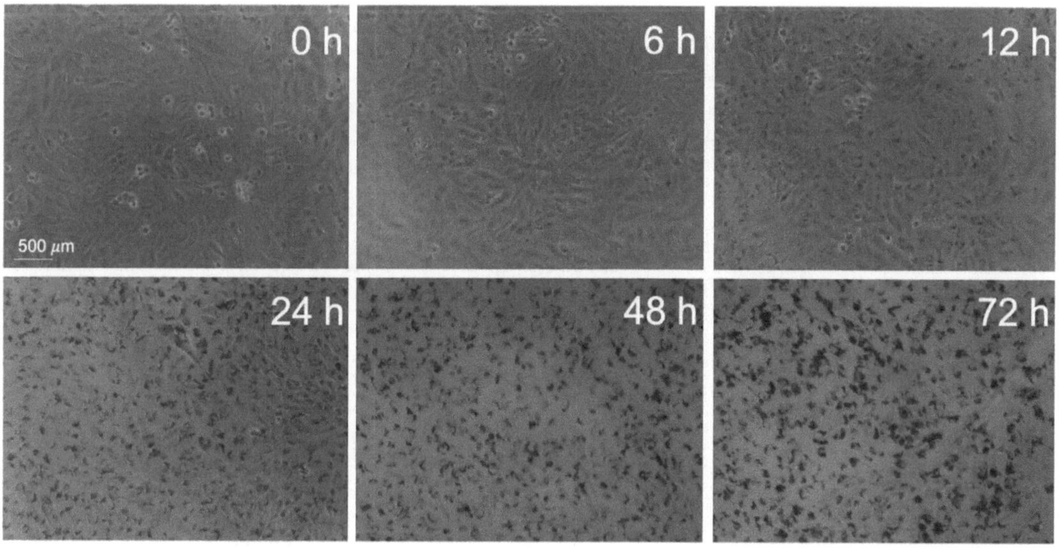

Fig. 2 Example of internalization time evaluation: incubation of U251MG cells with 0.031 mg AuPt NPs/mL during 0, 6, 12, 24, 48, and 72 h. After 12 h is possible to detect internalization of AuPt NPs in the cells; however, the internalization is low. After 24 h, the internalization is significant and it does not significantly increase with longer incubations times. Therefore, 24 h should be the selected time to incubate the cells with the NPs to get the maximum internalization. Scale bar: 500 μm

twice with PBS-EDTA, add trypsin and resuspend the pellet in cell culture medium), counted and centrifuged. Discard the supernatant and lyse the pellet in a dilution of aqua regia to proceed to the metal quantification by MP-AES (*see* **Note 14**).

6. According to the obtained results, the incubation time that allows the maximum internalization of NPs for the incubation and loaded-exosome purification should be chosen (Fig. 2).

3.3 Recovery of Loaded Exosomes

1. Seed the cells in 100 mm plates (*see* **Note 15**).
2. Incubate the cells with the selected NP concentration and during the selected time.
3. After the optimized time incubation, remove the supernatant and wash the cells twice with PBS without Ca^{+2}/Mg^{+2} to remove the NPs that are not internalized.
4. Incubate cells with exosome-depleted medium during 48 h (*see* **Note 1**): during this time, cells release exosomes, some of them loaded with NPs.
5. After 48 h, collect the supernatant.
6. Centrifuge the supernatant at 500*g* during 10 min at 4 °C, collect the supernatant and discard the pellet. Debris and death cells are the major constituents of the pellet.

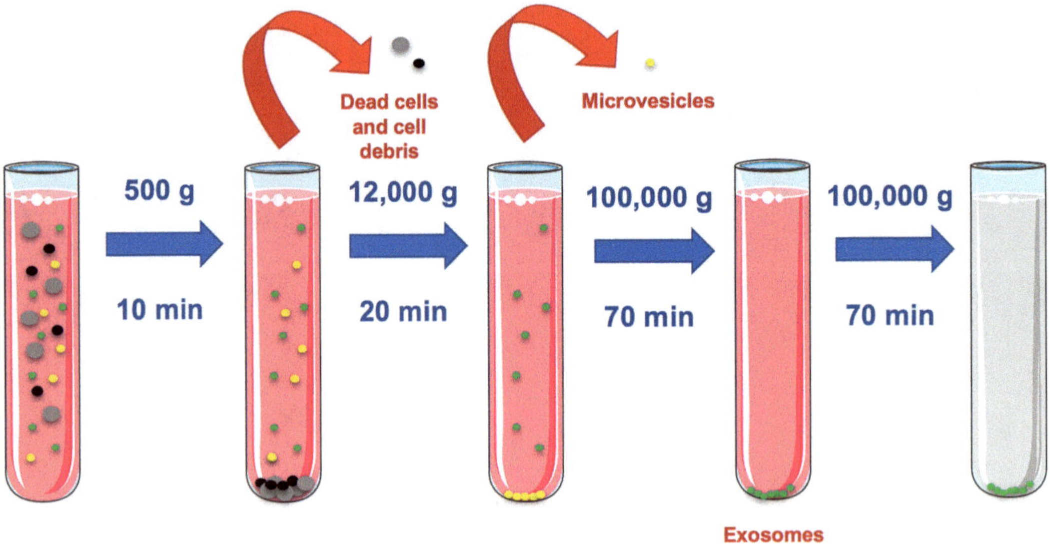

Fig. 3 Procedure summary

7. Centrifuge the supernatant at 12,000*g* during 20 min at 4 °C, collect the supernatant and discard the pellet. Microvesicles are the major population that constitute the pellet.
8. Centrifuge the supernatant at 100,000*g* during 70 min at 4 °C: keep the pellet, as the exosomes are located in the pellet. Discard the supernatant.
9. To wash the exosomes, add PBS without Ca^{+2}/Mg^{+2} in the ultracentrifuge tube and centrifuge again at 100,000*g* during 70 min at 4 °C. With this step, we will remove all the leftovers from the cell culture medium that could alter the protein quantification.
10. Discard the supernatant and resuspend the pellet in 250 μL PBS without Ca^{+2}/Mg^{+2} (*see* **Note 16**). The NP-loaded exosomes are ready for further applications and characterization (Fig. 3).

3.4 Characterization of the Recovered Exosomes

For every procedure is possible to follow you own protocol, however we include some instructions to get better results.

3.4.1 Protein Quantification

To quantify exosomal protein amounts, we recommend bicinchoninic acid assay (BCA assay).

1. In case of using the bicinchoninic acid assay (BCA assay), follow the enhanced protocol for concentrations that range from 5 to 250 μg/mL (*see* **Note 17**) and measure the absorbance at 562 nm.

3.4.2 Exosome Count by Nanoparticle Tracking Analysis (NTA)

1. Thaw the exosome sample in ice.
2. Dilute the sample in PBS without Ca^{+2}/Mg^{+2} to get 5–10 μg protein/mL. Keep the samples in ice until the measurement.
3. Follow the manufacturer instructions to use the equipment, as they are differences between different equipments.
4. Consider the size obtained by NTA to confirm that you have extracted the exosome fraction correctly.

3.4.3 Exosome Markers Evaluation by Immunoblot Detection

To evaluate alterations of the exosomal markers (*see* **Note 18**), we recommend to perform western blot, following your own protocol.

1. Load the concentration of protein previously optimized for every antibody (*see* **Notes 19** and **20**): in case that the protein would be too diluted, it is possible to concentrate the protein following this protocol:
   (a) Cool the required volume of acetone to −20 °C.
   (b) Add one volume of protein plus four volumes of cold acetone.
   (c) Vortex the tube and incubate for 1 h at −20 °C.
   (d) Centrifuge 10 min to 13,000*g*.
   (e) Discard the supernatant without disturbing the pellet.
   (f) Leave the tube open and allow the acetone remains to evaporate at room temperature during 30 min.
   (g) Add RIPA buffer with protease inhibitor and recalculate the concentration in the final volume.
2. Continue with your western-blot protocol.

3.4.4 Evaluation of NP Loading

To evaluate the internalization of NPs, metals can be detected by Microwave Plasma Atomic Emission Spectroscopy (MP-AES):

1. Lyse the total exosome extract with aqua regia at room temperature during 24 h: add 900 μL mL aqua regia to 100 μL exosome sample.
2. Dilute the sample 1/100 in Milli-Q water.
3. Proceed with your MP-AES quantification protocol following the manufacturer's instructions.
4. Calculate the ng of metal by μg exosomal protein.

4 Notes

1. Exosome-depleted cell culture medium: the serum has to be added to the cell culture medium to complement it. Before to be added to the cell culture medium, serum has to be ultracentrifugated (100,000*g* – 70 min) to remove fetal bovine serum exosomes.

2. Initial number depends on the selected cell line.
3. During trypsinization to facilitate the cellular detach, gently sharply hit the plate on a side before adding cell culture medium.
4. Centrifugation is a mechanical process that utilizes and applied centrifugal force field to separate components of a mixture by density and/or particle size. Centrifugation settings (rcf-g and time) could be adapted to the cell type, according to its sensitivity to centrifugation.
5. To calculate the number of cells in the solution, count the four squares of the cell counting chamber and average the count. Multiply the average by 10,000 to get the number of cells per milliliter.
6. Seeding concentration has to be optimized for every cell line; however, a recommendation is to seed 2500 cells per well in a final volume of 150 μL per well for tumor cells or 5000 cells per well for slower proliferating cells.
7. To test NP concentration, exclude the concentrations that generate precipitates on the cells, in this case the toxicity would be by physical interference, instead of by mechanistic toxicity.
8. Usually, NPs are dissolved in MQ water. When preparing the serial dilutions do not add more than 25% MQ water to the initial dilution, otherwise the toxicity could be consequence of the MQ water instead of the NP concentration.
9. Before to perform the viability assay test, wells must to be washed with PBS without Ca^{+2}/Mg^{+2} to remove the NPs that can interfere with the assay. Add the PBS very slowly, resting the pipette tips against the lateral wall of the well, without touching the well bottom, to avoid cell detachment.
10. The binding of proteins to NPs can lead to misinterpretation of the biological activity and false results by in vitro toxicity assays. As example, Lactate Dehydrogenase (LDH) is released as result of the loss of membrane integrity. Therefore, its determination is a classic method to quantify toxicity by cell death. However, the adsorption of LDH at the surface of several NPs has been proved [13]. The use of this kind of assays should be carefully analyzed to select the one without interactions, to avoid alteration of the results due to the binding of the NPs to the used reagents. We recommend to evaluate the interaction of the reagents with the nanoparticles co-incubating them.
11. To check NP internalization, usually around 10,000 cells per well could be a good starting point, but it will be depending on the cell line.

12. If it is possible, to check the time of maximum internalization of NPs we recommend optical microscopy, due to the easiness of the assay.
13. For the evaluation of NP internalization by metal quantification is recommended to seed the cells in 35 mm plates, the cell number is determining to get enough material to quantify the metals.
14. The quantified metal amount has to be expressed in relation to the cell count.
15. If the NP incubation time is 24 h, seed the number of cells needed to reach 75% of confluence after that time (usually 5–7, five million of cells per plate). For longer incubation times, seed the number of cells per plate needed to reach 50% confluence after the 24-h seed.
16. Resuspend in smaller volumes if you want a more concentrated sample.
17. Preferably, use the bicinchoninic acid assay (BCA assay) to quantify exosomal protein. Incubation of the samples with the working reagent at 60 °C during 30 min (enhanced protocol), before measuring the absorbance, is preferable for samples with low protein concentration, as usually occurs with exosomal extracts.
18. In case of performing western-blot to evaluate the alteration of the exosomal markers, we recommend to detect proteins from tetraspanin family (CD9, CD81, CD63).
19. We recommend do not load less than 10 μg protein.
20. The exosome markers could also be quantified according to the total number of cells that produce exosomes, for example load the amount of exosome protein produced by 1,000,000 cells, instead of a fixed protein amount. In that case, after recovering the supernatant, it is necessary to count the cells as previously described.

Acknowledgments

We gratefully acknowledge financial support from the ERC Advance Grant CADENCE (grant no- ERC-2016-ADG-762684). P.M-D also thanks Instituto de Salud Carlos III (PI19/01007 and DTS21/00130). Authors thank Prof Jesús Santamaría for the support on this work. The authors declare no conflict of interest.

References

1. György B, Szabó TG, Pásztói M, Pál Z, Misják P, Aradi B, László V, Pállinger E, Pap E, Kittel A, Nagy G, Falus A, Buzás EI (2011) Membrane vesicles, current state-of-the-art: emerging role of extracellular vesicles. Cell Mol Life Sci 68(16):2667–2688. https://doi.org/10.1007/s00018-011-0689-3
2. Akers JC, Gonda D, Kim R, Carter BS, Chen CC (2013) Biogenesis of extracellular vesicles (EV): exosomes, microvesicles, retrovirus-like vesicles, and apoptotic bodies. J Neuro-Oncol 113(1):1–11. https://doi.org/10.1007/s11060-013-1084-8
3. Hessvik NP, Llorente A (2018) Current knowledge on exosome biogenesis and release. Cell Mol Life Sci 75(2):193–208. https://doi.org/10.1007/s00018-017-2595-9
4. Wortzel I, Dror S, Kenific CM, Lyden D (2019) Exosome-mediated metastasis: communication from a distance. Dev Cell 49(3): 347–360. https://doi.org/10.1016/j.devcel.2019.04.011
5. Kalluri R, LeBleu VS (2020) The biology, function, and biomedical applications of exosomes. Science 367(6478):eaau6977. https://doi.org/10.1126/science.aau6977
6. Mathieu M, Martin-Jaular L, Lavieu G, Théry C (2019) Specificities of secretion and uptake of exosomes and other extracellular vesicles for cell-to-cell communication. Nat Cell Biol 21(1):9–17. https://doi.org/10.1038/s41556-018-0250-9
7. Zhang L, Yu D (2019) Exosomes in cancer development, metastasis, and immunity. Biochim Biophys Acta Rev Cancer 1871(2): 455–468. https://doi.org/10.1016/j.bbcan.2019.04.004
8. Li SP, Lin ZX, Jiang XY, Yu XY (2018) Exosomal cargo-loading and synthetic exosome-mimics as potential therapeutic tools. Acta Pharmacol Sin 39(4):542–551. https://doi.org/10.1038/aps.2017.178
9. Fu S, Wang Y, Xia X, Zheng JC (2020) Exosome engineering: current progress in cargo loading and targeted delivery. NanoImpact 20:100261. https://doi.org/10.1016/j.impact.2020.100261
10. Sancho-Albero M, Encabo-Berzosa MDM, Beltrán-Visiedo M, Fernández-Messina L, Sebastián V, Sánchez-Madrid F, Arruebo M, Santamaría J, Martín-Duque P (2019) Efficient encapsulation of theranostic nanoparticles in cell-derived exosomes: leveraging the exosomal biogenesis pathway to obtain hollow gold nanoparticle-hybrids. Nanoscale 11(40): 18825–18836. https://doi.org/10.1039/c9nr06183e
11. Sancho-Albero M, Encinas-Giménez M, Sebastián V, Pérez E, Luján L, Santamaría J, Martin-Duque P (2022) Transfer of photothermal nanoparticles using stem cell derived small extracellular vesicles for in vivo treatment of primary and multinodular tumours. J Extracell Vesicles 11(3):e12193. https://doi.org/10.1002/jev2.12193
12. Witwer KW, Goberdhan DC, O'Driscoll L, Théry C, Welsh JA, Blenkiron C, Buzás EI, Di Vizio D, Erdbrügger U, Falcón-Pérez JM, Fu QL, Hill AF, Lenassi M, Lötvall J, Nieuwland R, Ochiya T, Rome S, Sahoo S, Zheng L (2021) Updating MISEV: evolving the minimal requirements for studies of extracellular vesicles. J Extracell Vesicles. 10(14): e12182. https://doi.org/10.1002/jev2.12182
13. Forest V, Pourchez J (2017) Preferential binding of positive nanoparticles on cell membranes is due to electrostatic interactions: a too simplistic explanation that does not take into account the nanoparticle protein corona. Mater Sci Eng C Mater Biol Appl 70(Pt 1): 889–896. https://doi.org/10.1016/j.msec.2016.09.016

Chapter 11

Two Complementary Strategies to Quantitate Extracellular Vesicle Uptake Using Bioluminescence and Non-Lipidic Dyes

Víctor Toribio and María Yáñez-Mó

Abstract

The study of the molecular mechanisms controlling extracellular vesicle uptake by a target cell is an aspect of great interest within the EV community due to EV relevance in intercellular communication for tissue homeostasis or different disease progressions such as cancer or Alzheimer's. Since the EV field is relatively young, standardization of techniques for even basic aspects such as their isolation and characterization is still under development and debate. So it is for the study of EV uptake, where the currently most used strategies have critical limitations. Newly designed techniques should try to discern the uptake events from the surface EV binding or to improve the sensitivity and fidelity of the assays. Here, we describe two different complementary methods to measure and quantify EV uptake that we believe, help to overcome certain limitations of the currently used techniques. One is based on a mEGFP-Tspn-Rluc construct, to sort these two reporters into EVs. The use of bioluminescence signal to measure EV uptake allows for a better sensitivity, discerns EV binding from uptake, and allows kinetics measurement in alive cells, being compatible with a high-throughput screen format. The second one is a flow cytometry assay based in EV staining with a maleimide conjugated with a fluorophore, a chemical compound that covalently binds to proteins within sulfhydryl residues, being a good alternative to lipidic dyes and compatible with flow cytometry sorting of cell populations that have captured the labeled EVs.

Key words Extracellular vesicles, Uptake, Luciferase, Quantitative, Non-lipidic staining

1 Introduction

Extracellular vesicles have arisen as an important pillar in intercellular communication both at paracrine and at endocrine levels. They are relevant under physiological and pathological conditions, since they can induce major phenotypic, metabolic, or epigenetic changes in the receptor cell, leading to cell differentiation/dedifferentiation, tissue remodeling, or cellular chemotaxis [1, 2]. Thus, in recent decades the biomedical scientific community has been highly interested in how EVs are taken up by receptor cells. The majority of the reports indicate that EV uptake is an active process,

Seppo Vainio (ed.), *Cell-Secreted Vesicles: Methods and Protocols*, Methods in Molecular Biology, vol. 2668,
https://doi.org/10.1007/978-1-0716-3203-1_11,

being the main described pathways clathrin-dependent endocytosis, caveolin-dependent endocytosis, endocytosis mediated by lectins, phagocytosis, or macropinocytosis [3]. Depending on the receptor cell and the EV origin, the pathway may differ, and different pathways may also cooperate in this process.

The current strategies to study EV uptake are either based on protein fusion reporters built from EV-enriched proteins (mostly from the tetraspanin superfamily), or based on EV staining with lipophilic dyes, such as PKH. The most usually used reporters are fluorescent proteins such as EGFP [4] or phluorin, a variant of EGFP that is brightly fluorescent only around physiological pH, so that EVs will only display green fluorescence upon secretion, being of great help to study EV secretion or distribution in vivo [5]. These strategies have the advantage of being traceable by fluorescence microscopy and can be used in animal models, but they pose several limitations for quantification, which is difficult and time-consuming. In addition, tagged molecules will present poor sensitivity and subpopulation selectivity due to the fact that each EV carries very few reporter molecules, even under overexpression conditions of an EV-enriched protein as the carrier. On the contrary, EV staining can yield a stronger signal, and every EV would be stained independently of its marker composition but a higher background signal arises from non-specific labeling events. Lipophilic fluorescent dyes can aggregate producing micellar structures that can be confused with bona-fide EVs, and they can diffuse between hydrophobic membranes, leading to staining of the target cells that would be counted as false uptake events [6].

Here, we describe two different methods optimized to quantitate EV uptake, one using a Renilla luciferase (Rluc) fusion–protein reporter [7] and another using a non-lipophilic EV staining strategy. The first method uses a double-tagged mEGFP-Tetraspanin-Renilla luciferase construct. Both tags (mEGFP and Rluc) are, respectively, fused to the tetraspanin intracellular N and C terminal ends, to be sorted into EVs lumen (Fig. 1a). The dual tag allows to perform classical analysis of EV uptake and subcellular localization by confocal microscopy (Fig. 1b) by using the fluorescent reporter. However, for accurate and sensitive quantification, luciferase bioluminescence signal has several advantages over fluorescence: it can be easily quantified and displays higher signal than a fluorescent reporter with the same number of molecules with negligible background noise, thus improving signal-to-noise ratio. In addition, in our assay bioluminescence signal will come only from EVs that have been already effectively taken up by target cells, distinguishing uptake from binding events, something that is not easily feasible with fluorescent reporters. This effect is achieved by the use of the chemically protected Rluc substrate EnduRen™ (Promega). EnduRen™ is cytopermeable and can be preloaded into alive target cells.

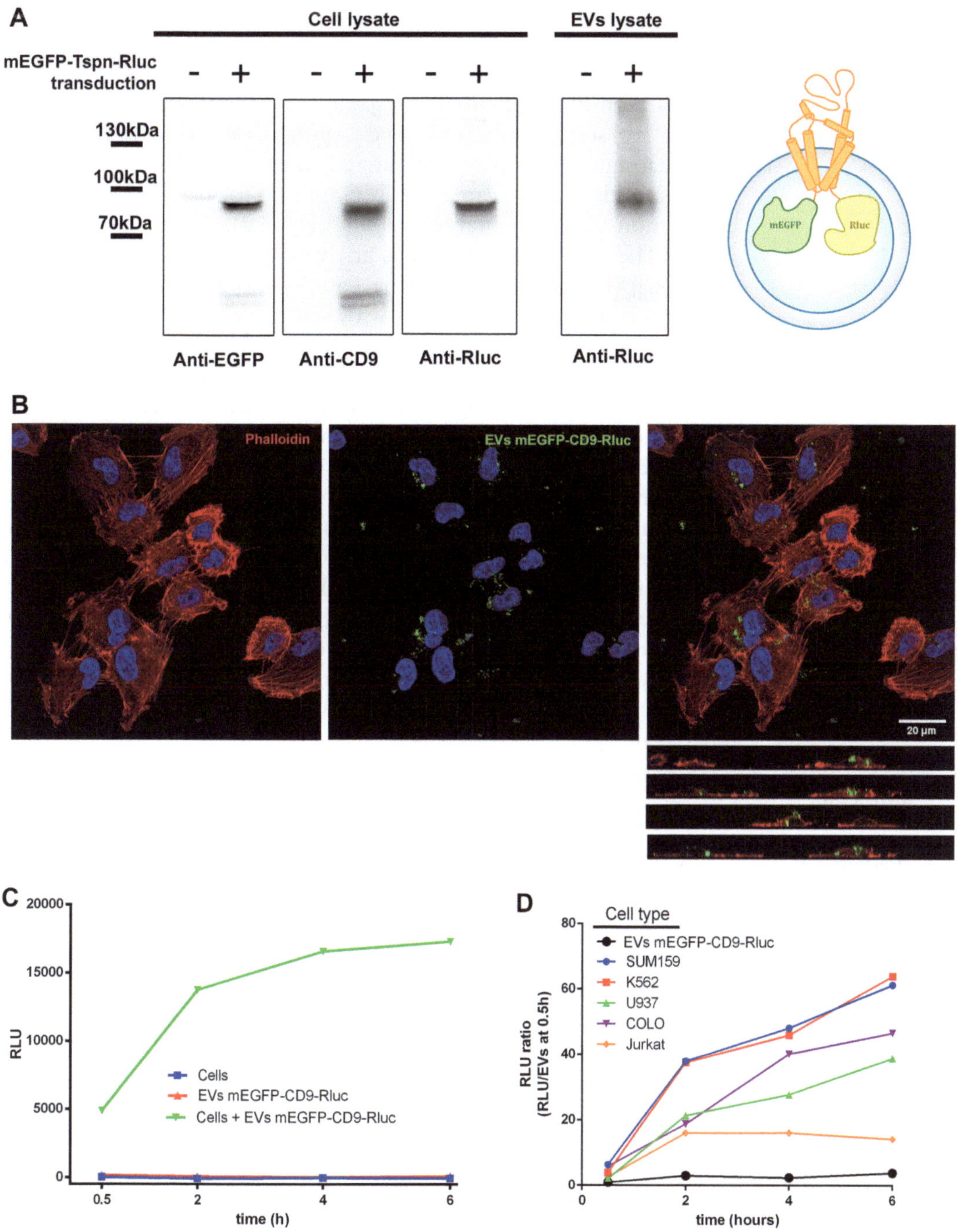

Fig. 1 Renilla luciferase-based EV uptake assay. (**a**) mEGFP-CD9-Rluc is sorted into EVs. Western blot characterization of the mEGFP-CD9-Rluc construct in cell lysates and in EV lysates isolated by SEC with anti-EGFP, anti-CD9, or anti-Rluc antibodies. mEGFP-CD9-Rluc appears as a single band around 90 kDa. (**b**) mEGFP-CD9-Rluc EVs uptake by confocal microscopy. Confocal microscopy analysis of SUM159 cells after 2 h of mEGFP-CD9-Rluc EV uptake. Target cells were stained with Phalloidin 647 in order to detect cortical actin (Red) and EVs were detected by their green EGFP fluorescence. Orthogonal views are shown in bottom right panel to illustrate that EVs are inside the cells. (**c**) Standard EV uptake luciferase assay. Target SUM159 cells

Intracellular esterases would deprotect EnduRen™ converting it into coelenterazine H, the canonical Rluc substrate. No luminescence signal is detected when incubating either cells alone or isolated mEGFP-Tspn-Rluc carrying EVs alone with EnduRen™, (Fig. 1c) so that both preloaded target cells and EVs are needed for actual signal to display. The tagged construct is only carried in the EVs, so this system can be easily employed to analyze the uptake by any non-transfected target cell line (Fig. 1d).

The second technique was firstly described by Roberts-Dalton, et al. [8] and is based on the compound maleimide, which can stablish covalent interactions with sulfhydryl/thiol residues (-SH) only present in free cysteines within proteins. Sulfhydryl residues appear not to be present in the glycocalyx, so binding to glycans is not expected. Although maleimide is chemo-selective for thiols from pH 6.5–7.5, above pH 7.5, thiol chemoselectivity is lost and the reagent starts reacting with free amines [9]. Thus, pH range and avoiding amine or sulphate-based buffers are important considerations of the technique. Maleimide is commercially sold conjugated to different fluorophores such as Alexa 633 or 488, so it is a powerful tool to stain pre-isolated EVs (Fig. 2a) and a great alternative to fluorescent lipophilic dyes, since it does not aggregate and does not diffuse between lipid membranes, still allowing for an easy quantification by flow cytometry (Fig. 2c, d) and traceability by fluorescence microscopy techniques (Fig. 2e).

2 Materials

2.1 Buffers and Solutions

1. Phosphate-buffered saline, pH 7.4 (PBS): 10 mM Na_2HPO_4, 1.8 mM KH_2PO_4, 137 mM NaCl, 2.7 mM KCl.

2.2 Cell Line Culture and Media

1. SUM159 breast cancer cell line is cultured in Dulbecco's modified Eagle's medium (DMEM) supplemented with 5% FBS, 1% penicillin/streptomycin, non-essential amino acids (80 mg/mL), insulin (5 μg/mL), and hydrocortisone (1 μg/mL).

   SKMEL147 melanoma cell line is cultured in DMEM supplemented with 10% FBS, 1% penicillin/streptomycin.

   K562 and U937 cell lines were cultured in RPMI supplemented with 10% FBS, 1% penicillin/streptomycin, and 2 mM L-glutamine.

Fig. 1 (continued) or mEGFP-CD9-Rluc EVs alone were pre-incubated with EnduRen™ for 2 h. Then, EnduRen™ preloaded cells were incubated or not with mEGFP-CD9-Rluc EVs and bioluminescence signal was measured at different time points. (**d**) EV uptake luciferase assay using different receptor cell lines. The EV uptake kinetics of SUM159, K562, U937, COLO, and Jurkat cell lines are shown. Data were normalized to signal of EVs alone at 0.5 h time point

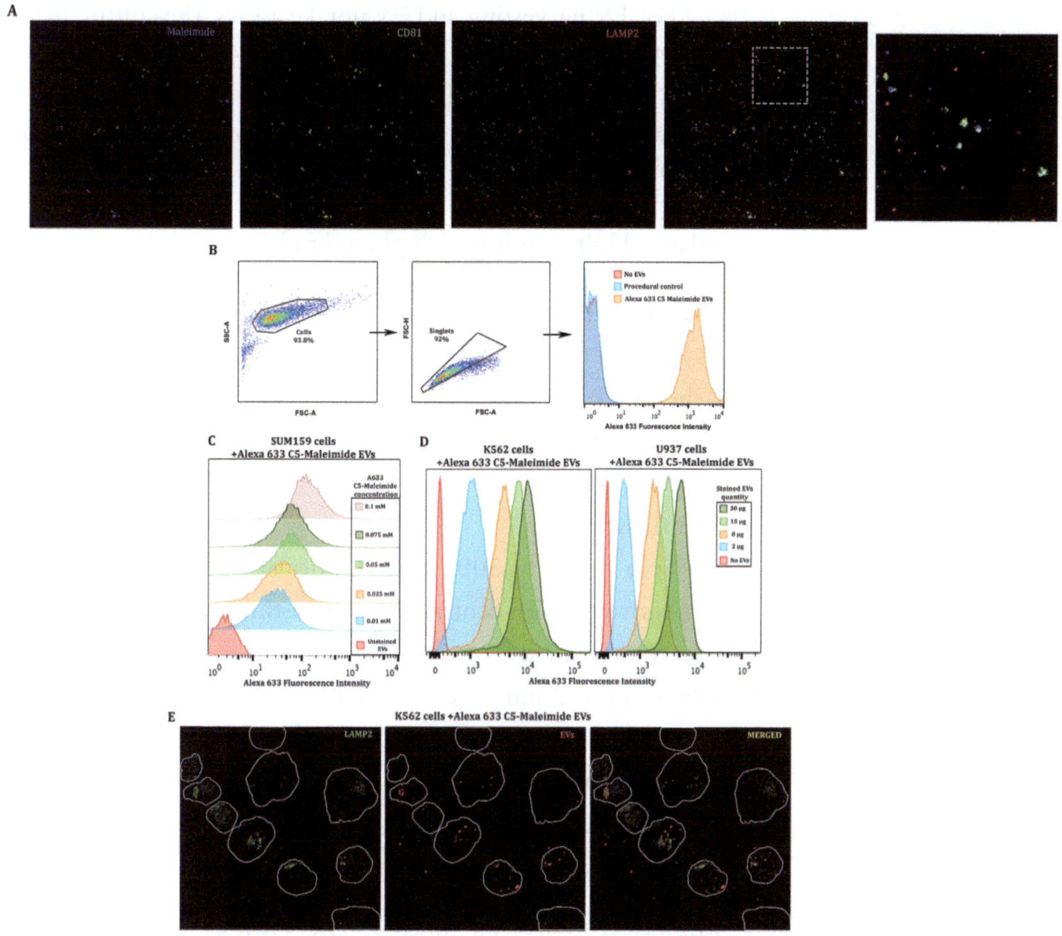

Fig. 2 Flow cytometry EV uptake assay using Alexa Fluor™ C5-Maleimide-stained EVs. (**a**) Alexa Fluor™ C5-Maleimide EV staining characterization by confocal microscopy. 2.5 μg of SKMEL147 EVs, isolated by differential ultracentrifugation and stained with Alexa 633 C5-Maleimide, was immobilized on coverslips using a CytoSpin™. Samples were fixed with 4% PFA, blocked with PBS + 1% BSA, and stained with primary antibodies anti-LAMP2 (Abcam, rabbit) and anti-CD81 (5A6, mouse) followed by secondary antibodies GαM Alexa 488 and DαR Alexa 555 diluted. Dotted white line indicates the area of the close-up image shown in the right, in which the colocalization of the maleimide staining with the two different EV markers is observed. (**b**) Flow cytometry EV uptake assay gating strategy and controls. Target cells were incubated with maleimide-stained EVs for 2 h. The live-cell population is identified by their SSC-A vs FSC-A profile in a dotplot (left panel). Within that population, single-cell events are selected using the FSC-H vs FSC-A profile (center panel). The overlay of the histogram plots of the negative control (cells with no EVs), the procedural control (PBS + Alexa633 C5-Maleimide treated as EVs), and a positive control are shown in the right panel. (**c**) Alexa 633 C5-Maleimide EV staining optimization. The same amount of EVs were stained with different concentrations of Alexa 633 C5-Maleimide. Target SUM157 cells were incubated with stained EVs for 2 h and the far-red signal was measured by flow cytometry. (**d**) Dose–response EV uptake assay. Two different cell lines K562 or U937 were incubated with different μg of EVs stained with 25 μM of Alexa 633 C5-Maleimide. After 2 h of EV uptake, cells were analyzed by flow cytometry. (**e**) EV uptake detection by confocal microscopy. K562 were incubated with Alexa 633 C5-Maleimide EVs (red) for 2 h and after fixation and permeabilization, stained for the late endosomal marker LAMP2 (green). Cell contours were delimited using the corresponding wide field image

For EV harvesting and EV uptake assays, the culture media is supplemented with EV-depleted FBS (*see* **Note 1**).

2. Cell culture plates, p100 and p150.
3. p96-well assay white plate with clear bottom with lid and tissue culture treated.
4. Cell lifting solution: 0.25% trypsin, 0.02% EDTA in Hanks' balanced salt solution with phenol red.

2.3 Consumable Materials and Instruments

1. Micropipettes and micropipettes tips.
2. Eppendorf tubes and 15 mL tubes.
3. Benchtop centrifuge.
4. Flow cytometry tubes (5 mL, 12 × 75 mm).
5. 0.22 μm filters.
6. 50 mL syringe.
7. ExoSpin Columns 3000 MW (Optional) (*see* **Note 2**).
8. 2 mL syringe with central Luer and a three-way valve.
9. CL-2B Sepharose.
10. Column frit, 2 μm pore (optional).
11. Nylon 20 denier brown (natural) pantihose (optional).
12. Column support.
13. Pierce BCA protein assay kit.
14. NTA analysis device (e.g., Nanosight).

2.4 EV Uptake Reporters and Staining Reagents

1. EnduRen™ Live Cell Substrate (*see* **Note 3**).
2. Alexa Fluor™ C5-Maleimide (*see* **Note 4**).

3 Methods

3.1 EV Uptake Luciferase Assay

3.1.1 Cell Culture and Previous Considerations

- This protocol was optimized with EVs produced by the breast cancer cell line SUM159 (*see* **Note 5**). For EV production, SUM159 cells stably transfected with the reporter construct were cultured in p150 culture plates with 20 mL of culture media supplemented with 5% EV-depleted FBS for 5–7 days. Cells were seeded at a density to reach confluence around the day of conditioned media collection.
- The protocol was optimized with EVs isolated by differential ultracentrifugation. Other EV isolation protocols such as size exclusion chromatography should work but a reconcentration of the EVs may be required.

- After isolation, characterize EVs following MISEV current recommendations [10]. Ensure your EVs present EV markers and check for the lack of non-EV markers. Determine EV number by NTA and protein quantity by BCA.

3.1.2 Performance of the Assay

- The day before the assay count and seed around 20,000–30,000 of receptor cells per well in a white-bottom p96 well plate, in 100 μL of complete culture medium supplemented with EV-depleted FBS and leave ON at 37 °C (*see* **Note 6**).
- All conditions should be performed at least in duplicates (two wells per condition), in order to reduce variability or to discard artifacts (*see* **Note 7**).
- Remove the culture media and replace it with 100 μL of a 60 μM EnduRen™ solution diluted in culture media supplemented with EV-depleted FBS.
- Negative controls: Seed two wells for the control of cells loaded with EnduRen™ but do not add GFP-CD9-Rluc EVs, and leave two other empty wells for the control of EGFP-Tspn-Rluc EVs with EnduRen™ but without target cells (*see* **Note 8**). For the EVs alone negative control, no removal of culture media is performed at this point.
- Incubate cells for 2 h at 37ªC for EnduRen™ loading and activation by target cells.
- Remove the supernatant. Wash once with culture media supplemented with EV-depleted FBS to get rid of the EnduRen™ remnants.
- Add 100 μL of a solution of purified EVs diluted in culture media supplemented with EV-depleted FBS per well (*see* **Note 9**).
- Leave the plate in the incubator at 37 °C and in presence of CO_2 until your first time point measurement (*see* **Note 10**).

3.1.3 Quantification

- Measurements are done with a luminescence microplate reader (*see* **Note 11**).
- The mean is calculated from each technical replicate. The background for each time point measurement is obtained from the EVs alone control or from the cells alone control if the EV background signal is shown. Usually, background level is subtracted from all the measurements. In EV uptake kinetics experiments, the data can be normalized, if needed, to the first time point of the appropriated control (EV background, untreated cells, and non-transduced cells) (*see* **Note 12**).

3.2 EV Uptake Flow Cytometry Assay

3.2.1 Cell Culture and Previous Considerations

- This protocol was optimized with EVs produced by the melanoma cell line SKMEL147 or the breast cancer cell line SUM159. For EV production, cells were cultured in p150 culture plates with 20 mL of culture media supplemented with 10% or 5% EV-depleted FBS, respectively, for 5–7 days. Cells were seeded at density that reached confluence around the day of conditioned media collection.
- The protocol has been optimized with EVs isolated by differential ultracentrifugation or by size exclusion chromatography (SEC).
- After isolation, characterize EVs following MISEV current recommendations [10]. Ensure your EV preparations present of EV markers and check also for the lack of some non-EV markers. Determine EV number by NTA and protein quantity by BCA.

3.2.2 EV Staining

- Add Alexa Fluor™ C5-Maleimide at a final concentration of 25 μM (*see* **Note 13**) to a solution with 200 μg of isolated EVs in 200 μL of filtered PBS (*see* **Note 14**). As a procedural control, incubate filtered PBS with Alexa Fluor™ C5-Maleimide and treat it as an EV sample.
- EVs can be stained ON at 4 °C or 2 h at RT in Eppendorf tubes without agitation.
- After incubation, the unbound remnants of Alexa Fluor™ C5-Maleimide have to be removed by either ExoSpin or SEC:

  Exo-Spin Columns (3000 MW) (Invitrogen)

  - One possibility is to use commercial Exo-Spin columns with 3000 Da of molecular weight exclusion. The resin is pre-hydrated in 650 μL filtered PBS for 15 min at RT (*see* **Note 15**).
  - Remove the inferior cap of the spin column and put it into a 2 mL collector tube and centrifuge the column at 750*g* for 2 min at RT to remove excess of interstitial fluid (*see* **Note 16**).
  - Discard the collector tube and immediately add the stained exosome sample (20–100 μL) directly to the center of the resin bed at the top of the column (*see* **Note 17**).
  - Place the spin column in a 1.5 mL elution tube (*see* **Note 18**).
  - Centrifuge the spin column at 750*g* for 2 min at RT. The exosome sample will be eluted into the elution tube, and the Alexa Fluor™ C5-Maleimide remnants will be retained in the column.

Size Exclusion Chromatography (SEC) [11]

- Remove the plunger from a 2 mL syringe, which will be our column. Place a column frit as a filter inside the syringe, near the Luer side. An alternative is to use a pantihose inserted inside the Luer (*see* **Note 19**). Once placed one of the two types of filters into the column, fill the column with filtered PBS to hydrate the filters and to remove the bubbles inside them.
- Put the column in the column support vertically with the Luer facing downwards. Add some filtered PBS and put the three-way valve in the Luer.
- Add CL-2B Sepharose until it reaches 1 mL of settled resin. Open the valve to let the Sepharose settle (*see* **Note 20**).
- Equilibrate the column with various volumes of filtered PBS (*see* **Note 21**).
- Before adding the stained EV sample, let the PBS volume elute. Then, add the whole sample (around 200–400 μL) taking care of not disturbing the front of the resin.
- Start collecting fractions in Eppendorfs of around 100 μL (*see* **Note 22**). When the sample has entirely entered, the resin adds more filtered PBS over the column and continues collecting fractions (*see* **Note 23**).
- Pool together the EV-enriched fractions for the assay (*see* **Note 24**).

3.2.3 Performance of the Assay

- For adherent cells, between 100,000–250,000 cells are seeded in p24 wells the day before the assay in 300–500 μL of culture media supplemented with EV-depleted FBS (*see* **Note 25**).
- Alexa Fluor™ C5-Maleimide-stained EVs are added to cells in the culture plate and left for 2 h in the incubator at 37 °C in presence of CO_2 and in the darkness (*see* **Note 26**).
- Control conditions: cells non-incubated with EVs (negative control) or incubated with a PBS + Alexa Fluor™ C5-Maleimide sample treated equally as the EV samples (procedural control) (Fig. 2b).
- After the incubation time, cells are washed once with PBS and lifted up with 100 μL of trypsin and neutralized with 100 μL of culture media (*see* **Note 27**).

3.2.4 Quantification

- To analyze EV uptake, we measure the fluorescence on the channel corresponding to the used fluorophore by flow cytometry. Use the negative control to stablish the threshold for positive signal and the procedural control to measure the background of the free dye.

- Cells are gated by their FSH and SSC profile and subsequently by their FSC-A/FSC-H profile to gate on cellular singlets (Fig. 2b). A negative control sample (cells with no EVs, cells with unstained EVs or cells with free dye) is used to stablish the threshold to determine a positive fluorescence signal. The geometric mean of the fluorescence of the population of each condition is obtained. Data can be normalized, if needed, to a stablished control (*see* **Note 28**).

4 Notes

1. Fetal bovine serum (FBS) (Sigma) is EV-depleted by 100,000*g* ultracentrifugation ON at 4 °C. After the centrifugation, the supernatant is filtered with a 0.22 μm filter.
2. (Invitrogen, 15382476).
3. (Promega, E6482).
4. The assay was optimized with the Alexa 633-conjugated maleimide (Thermo Fisher, A20342) since far-red fluorophores yield better sensitivity at a wavelength that presents less cellular autofluorescence. However, the assay can be performed with any fluorophore of preference.
5. A stable cell line of your choice expressing the EGFP-Tspn-Rluc construct is needed as EV producer.
6. For non-adherent cells, this step can be done the same day the assay is performed. In this case, cells are centrifuged and resuspended in culture media supplemented with EV-depleted FBS (*see* **Note 1**).
7. Leave at least one empty well of separation between different conditions to avoid bioluminescence signal leaking from the most intense samples to the less intense samples.
8. EVs for this control can be added to the plate the day before, to have a control of bacterial contamination. Although the EV isolation method is done in sterile conditions, some contaminations may appear, bacteria or fungi in our EV sample will give us luminescence signal when EnduRen™ is added, so plating this EV control the day before the assay will allow us to identify if we have contamination thus avoiding consequent artifacts.
9. Add around 4–8 μg or 5×10^9 EVs per well, to have a good signal, although this has to be stablished for each EV producer cell line, Tspn construct, and isolation method used. The limit of detection in our conditions was around $9.35 \pm 2.46 \times 10^8$ EVs/well [7].
10. A kinetic analysis can be performed by repeatedly measuring the same plate at different time point. Between different time point measurements, place the plate back into the incubator.

11. In our case, we used a Tecan GENios Microplate reader, 3 reads per well, 5000 ms of exposure, and 150 of gain. These settings are subjected to changes depending on the plate reader and the quantity of EVs used.
12. The luminescence assay allows us to analyze and measure EV uptake kinetics in alive cells (Fig. 1c, d). A small number of target cells are required per condition since it is performed in a p96-well plate. The reporter used is EV specific and the method cleanly distinguishes uptake from binding, since EnduRen™-incubated EVs display no signal (Fig. 1c, d). The main handicap of this procedure is that EVs have to be isolated from producing cells that express the construct, limiting the EV origin availability.
13. Alexa Fluor™ stock is resuspended DMSO at 5 mM and maintained at −20 °C.
14. For lower concentrations of protein, the concentration of maleimide can be reduced proportionally.
15. It is important to turn upside down the tube in order to get the whole resin hydrated. It is also important that there are no bubbles in the resin bed. Hydrated columns can be left at 4 °C for up to 3 days.
16. It is important to check the orientation of the column in the rotor if it is an angular one.
17. Take care to not disturb the resin surface.
18. Place the tube in the rotor maintaining orientation established during the first centrifugation.
19. In case of using a pantihose as filter (a small rectangular piece of around 0.5 × 1.5 cm), once hydrated in filtered PBS and free of interstitial air bubbles, roll the piece up around the long side using the help of a pair of tweezers and insert the humidified pantihose roll into the Luer from the outside.
20. Avoid the formation of bubbles inside the resin.
21. Precast columns can be left for various days with PBS at 4 °C. For longer periods of time, it is advised to store them in distilled H_2O + 0.05% azide.
22. Two drops each, each drop being around 50 μL.
23. Around 10 fractions are collected. EVs usually elute around fractions 5, 6, and 7 although this has to be checked for each cell line or sample type. Unbound maleimide will elute later in the soluble fractions.
24. To identify the EV-positive fractions, it is recommended to use a Dotblot assay using 1 μL of each SEC fraction into a nitrocellulose membrane and revealing the membrane with an EV marker antibody.

25. For cells that grow in suspension, the seeding can be done the same day of the assay.
26. Use around 1.75 10^9 or 2 μg of stained EVs per 250,000 cells. These parameters have to be optimized for each producer and receptor pair of cells.
27. For non-adherent cells, a 1–2 min of trypsin incubation step can be done to get rid of surface-bound EVs.
28. The assay based on C5-Maleimide EV staining is an end-point assay, a bigger number of receptor cells is required, and the EV staining is less specific. Although trypsin treatment will reduce the percentage of cells displaying fluorescence because of EV binding without active uptake, we cannot rule out the possibility that some binding events are included in our quantitation. Otherwise, EVs can be produced by any kind of cell and flow cytometry allows to a single-cell EV uptake measurement or sorting, having the possibility to identify or analyze different cellular populations. To sum up, both assays are complementary and can give different information when characterizing EV uptake, being very interesting alternatives to the EV uptake assays that are currently being performed.

References

1. Yanez-Mo M, Siljander PR, Andreu Z, Zavec AB, Borras FE, Buzas EI et al (2015) Biological properties of extracellular vesicles and their physiological functions. J Extracell Vesicles 4: 27066
2. Alberro A, Iparraguirre L, Fernandes A, Otaegui D (2021) Extracellular vesicles in blood: sources, effects, and applications. Int J Mol Sci 22(15):8163
3. Mulcahy LA, Pink RC, Carter DRF (2014) Routes and mechanisms of extracellular vesicle uptake. J Extracell Vesicles 3(1):24641
4. Carpintero-Fernández P, Fafián-Labora J, O'Loghlen A (2017) Technical advances to study extracellular vesicles. Front Mol Biosci 4:79
5. Verweij FJ, Revenu C, Arras G, Dingli F, Loew D, Pegtel DM et al (2019) Live tracking of inter-organ communication by endogenous exosomes in vivo. Dev Cell 48(4):573–589.e4
6. Simonsen JB (2019) Pitfalls associated with lipophilic fluorophore staining of extracellular vesicles for uptake studies. J Extracell Vesicles 8(1):1582237
7. Toribio V, Morales S, López-Martín S, Cardeñes B, Cabañas C, Yáñez-Mó M (2019) Development of a quantitative method to measure EV uptake. Sci Rep 9(1):10522
8. Roberts-Dalton HD, Cocks A, Falcon-Perez JM, Sayers EJ, Webber JP, Watson P et al (2017) Fluorescence labelling of extracellular vesicles using a novel thiol-based strategy for quantitative analysis of cellular delivery and intracellular traffic. Nanoscale 9(36): 13693–13706
9. Hermanson GT (2013) Chapter 3 – The reactions of bioconjugation. In: Hermanson GT (ed) Bioconjugate techniques, 3rd edn. Academic Press, Boston, pp 229–258
10. Théry C, Witwer KW, Aikawa E, Alcaraz MJ, Anderson JD, Andriantsitohaina R et al (2018) Minimal information for studies of extracellular vesicles 2018 (MISEV2018): a position statement of the International Society for Extracellular Vesicles and update of the MISEV2014 guidelines. J Extracell Vesicles 7(1):1535750
11. Monguió-Tortajada M, Morón-Font M, Gámez-Valero A, Carreras-Planella L, Borràs FE, Franquesa M (2019) Extracellular-vesicle isolation from different biological fluids by size-exclusion chromatography. Curr Protoc Stem Cell Biol 49(1):e82

Chapter 12

Integrin-Mediated Exosomal Homing to Organs

Eun Jeong Park and Motomu Shimaoka

Abstract

Exosomes, the small extracellular vesicles of 40–150 nm in size, are secreted by nearly all types of cells and play a dynamic role in intercellular and interorgan communications. These vesicles secreted by source cells contain a variety of biologically active materials such as microRNAs (miRNAs) or proteins, thereby utilizing these cargoes in modifying molecular functionalities of the target cells in the remote tissues. Consequently, several key functions of microenvironmental niches in the tissues are regulated in an exosome-dependent manner. The precise mechanisms by which the exosomes bind and home to different organs remained largely unknown. In recent years, integrins, a large family of cell adhesion molecules, have been revealed to play a crucial role in guiding homing of exosomes to target tissues, as integrins regulate tissue-specific homing of cells. In this regard, it is imperative to experimentally determine the roles played by integrins on the exosomes in their tissue-specific homing. This chapter presents a protocol to investigate exosomal homing regulated by integrins in in vitro and in vivo settings. We focus on β7 integrin, as its role in mediating the gut-specific homing of lymphocytes has been well established.

Key words Exosome, Integrin, Ligand, Cell, Adhesion, Homing

1 Introduction

Exosomes are the nanosized extracellular vesicles secreted from nearly all cell types through the pathways of endocytosis and exocytosis [1–3]. In the endosomal pathway, intraluminal vesicles (ILVs; exosomes) are generated by inbound sprouting of membranes of late endosomal organelles termed multivesicular bodies (MVBs) [4, 5]. At this stage, cytoplasmic and MVB-membraneous molecules are incorporated into the exosomes. Further, some MVBs are fused with the plasma membrane via an exocytic pathway to release the exosomes into the extracellular space, while the other MVBs can be degraded after their fusion with lysosomes [5, 6]. The processes of exosomal biogenesis are distinct from those of other extracellular vesicles, including microvesicles or apoptotic bodies [7, 8]. The mechanism by which biological substances including proteins, RNAs, or lipids are sorted into the exosomes remains

Seppo Vainio (ed.), *Cell-Secreted Vesicles: Methods and Protocols*, Methods in Molecular Biology, vol. 2668,
https://doi.org/10.1007/978-1-0716-3203-1_12,

largely unknown. The source cells exposed to pathological conditions such as inflammation or hypoxia may secrete the exosomes that contain significantly altered cargoes. The exosomal cargo contents do not always resemble those of their source cells. In considerable cases, the exosomes produced from stressed cells may play an important role in the cytoprotective mechanisms, by possessing the RNAs or proteins effective to dampen stresses [9]. Thus, the exosomes exert variable effects relevant to the pathologic microenvironment, thereby modifying the distant or neighboring tissues via transferring the bioactive cargoes [10–12].

Of note, the exosomes possess an ability to undergo transfer of biologically active substances to recipient cells to reshape microenvironmental niches in tissues [13, 14]. To modulate the functionalities of recipient cells of the microenvironmental niches at remote organs, exosomes themselves must travel there via systemic circuits, such as blood vessels. Integrins on the exosomes have been shown to regulate tissue-specific homing of exosomes [12, 15, 16]. Integrins represent a large family of transmembrane cell adhesion molecules [17]. Cells use integrins to mediate the dynamic performance for directional migration and homing. Thus, attention has been paid to determine whether and how exosomes depend on integrins to reach target cells in remote organs. Accumulating evidence demonstrates that the exosomes exploit the integrins to mediate homing to the target tissues as do cells [15, 18, 19]. Therefore, it is of importance to better understand the mechanisms by which exosomes migrate and home to specific tissues in vivo. This chapter describes the protocols to study integrin-mediated exosomal homing via in vitro and in vivo assessments.

2 Materials

2.1 Cell Culture and Exosome Isolation

1. TK-1 ATCC ® CRL-2396™, Mus musculus T-cell lymphoma (ATCC, Manassas, VA, USA).
2. Cell culture incubator maintained at 37 °C and 5% CO_2.
3. 100 mm × 20 mm cell culture dishes.
4. RPMI-1640 medium supplemented with 10% heat-inactivated fetal bovine serum (FBS) without bovine exosomes (exosome-free 10% FBS) (*see* **Note 1**), 10 mM 4-(2-hydroxyethyl)-1-piperazineethanesulfonic acid (HEPES), 2 mM L-glutamine, and 1% penicillin–streptomycin (*see* **Note 2**).
5. Phorbol-12-myristate-13-acetate (PMA).
6. L-70 ultracentrifuge (Beckman Coulter, Brea, CA, USA) or similar device.
7. SW32Ti rotor (Beckman Coulter).
8. SW41Ti rotor (Beckman Coulter).

9. Ultracentrifuge tube (38.5 mL) (Beckman Coulter).
10. Ultracentrifuge tube (13.2 mL) (Beckman Coulter).
11. Syringe filter unit (0.22 μm) (Merck, Darmstadt, Germany).
12. Pierce bicinchoninic acid (BCA) protein assay kit (Thermo Fisher Scientific, Waltham, MA, USA) or a similar kit to determine protein concentration.

2.2 Nanoparticle Tracking Analysis (NTA)

1. NanoSight LM10-HS equipped with a 405-nm laser (Malvern Panalytical Ltd, Malvern, UK).
2. NTA software (version 2.3) (Malvern Panalytical Ltd).

2.3 Flow Cytometry

1. Exosomes (*see* **Notes 3** and **4**):
   (a) TK1 β7-scramble (scr) exosomes.
   (b) TK1 β7-knockdown (KD) exosomes.
2. 1× phosphate-buffered saline (PBS) (without calcium and magnesium).
3. Bovine serum albumin (BSA) (Sigma, St. Louis, MO, USA).
4. Latex beads (4 μm), aldehyde sulfate, 4% w/v (Thermo Fisher Scientific).
5. 1 M glycine.
6. Poly-L-lysine (0.01% solution, upon being needed) (*see* **Note 5**).
7. Test tube rotator wheel.
8. Antibodies:
   (a) Unconjugated antibody to CD9 (KMC8) (BD Biosciences, San Jose, CA, USA).
   (b) Unconjugated antibody to CD63 (NVG-2) (Biolegend, San Diego, CA, USA).
   (c) Unconjugated antibody to CD81 (Eat-2) (Biolegend).
   (d) Phycoerythrin (PE)-conjugated antibody to β7 integrin (M293) (BD Biosciences).
   (e) Allophycocyanin (APC)-conjugated antibody to rat IgG2a (Biolegend).
   (f) PE-conjugated antibody to α4 integrin (PS/2) (Southern Biotech, Birmingham, AL, USA).
   (g) PE-conjugated antibody to Armenian hamster IgG (Biolegend).
   (h) PE-conjugated antibody to rat IgG2a (Biolegend).
   (i) PE-conjugated antibody to rat IgG2b (Biolegend).

9. 0.5% BSA in PBS buffer.
10. Falcon polystyrene 5 mL tubes for flow cytometry.
11. Flow cytometer (for example, BD Accuri™ C6, BD Biosciences).

2.4 Analysis of Integrin-Mediated Exosomal Binding to Ligand

1. Recombinant mouse mucosal addressin cell adhesion molecule-1 Fc (MAdCAM-1 Fc) (R&D Systems, Minneapolis, MN, USA).
2. Latex beads (4 μm), aldehyde sulfate, 4% w/v (Thermo Fisher Scientific).
3. TK1 β7-scramble (scr) exosomes and TK1 β7-scramble (scr) exosomes.
4. 2′,7′-Bis-(2-carboxyethyl)-5-(and-6)-carboxyfluorescein acetoxymethyl ester (BCECF-AM) (Thermo Fisher Scientific).
5. HEPES-buffered saline (HBS).
6. 0.5% BSA in HBS.
7. 1 M glycine.
8. Divalent cation solutions: 1 M $MgCl_2$, 1 M $CaCl_2$, and 1 M $MnCl_2$.
9. 0.5 M ethylenediaminetetraacetic acid (EDTA) solution.
10. PBS containing 2% FBS (exosome-free) (*see* **Note 1**).
11. Falcon polystyrene 5 mL tubes for flow cytometry.
12. Flow cytometer (for example, BD Accuri™ C6).

2.5 In Vivo Exosome Homing Assay in a Competitive Context

1. PKH26GL and PKH67GL kits (Sigma).
2. L-70 ultracentrifuge or similar device.
3. Amicon ultracentrifugal filter column (UFC9010) (Sigma).
4. C57BL/6J mice (CLEA Japan, Tokyo, Japan).
5. 30-G needles.
6. 1 mL syringes.
7. Basemold (Falma, Tokyo, Japan).
8. Tissue-Tek O.C.T. Compound (Sakura-Finetek, Torrance, CA, USA).
9. Cryostat microtome (for example, CM1850, Leica Biosystems, Wetzlar, Germany).
10. 4′,6-Diamidino-2-phenylindole (DAPI) with NucBlue (Thermo Fisher Scientific).
11. BZ-X700 fluorescent microscope or similar device (Keyence, Itasca, IL, USA).

3 Methods

3.1 Exosome Isolation

In this section, we show the methods of isolating exosomes using ultracentrifugations, from culture media of mouse TK1 cells (*see* **Note 3**). TK1 cells highly expressing integrins α4 and β7 were used to generate the β7-scr and β7-KD exosomes [15] (*see* **Note 4**).

1. Culture TK1 cells in RPMI-1640 medium containing 10% fetal bovine serum (FBS) and penicillin (100 U/mL)/streptomycin (100 μg/mL) at 60–70% confluency.
2. Wash the cells with PBS twice and count cells using a hemocytometer.
3. Incubate for 72 h at 37 °C with 5% CO_2 in RPMI-1640 medium supplemented with 10% heat-inactivated fetal bovine serum (FBS) without bovine exosomes (exosome-free 10% FBS) (*see* **Note 1**), 10 mM 4-(2-hydroxyethyl)-1-piperazineethanesulfonic acid (HEPES), 2 mM L-glutamine, 1% penicillin-streptomycin, and 100 nM PMA. The addition of retinoic acid to the culture media at a concentration of 200 nM can be considered when the exosomal increase in β7 integrin is required (*see* **Note 6**).
4. Centrifuge culture medium at 1000 × *g* for 10 min to remove cells and then at 2000 × *g* for 20 min to remove apoptotic bodies.
5. Centrifuge supernatant at 16,500 × *g* for 20 min to remove microvesicles.
6. Filtrate supernatant by using 0.22 μm filter unit.
7. Centrifuge at 100,000 × *g* for 70 min.
8. Resuspend pellet in 12 mL of PBS.
9. Centrifuge at 100,000 × *g* for 70 min.
10. Resuspend pellet in 0.2 mL of PBS.
11. Measure concentration of proteins by using a Pierce bicinchoninic acid (BCA) protein assay kit.

3.2 Characterization of Exosome Size by Nanoparticle Tracking Analysis (NTA)

1. Measure particle size and number of exosomes via NTA using the NanoSight LM10-HS equipped with a 405-nm laser [20, 21] (*see* **Note 7**).
2. Prepare 1 mL of exosome samples at a concentration of 0.2 ~ 1.0 × 10^9 particles/mL using PBS and infuse approximately 0.5 mL into the chamber [22] (*see* **Note 8**).
3. Find the microscopic optimum areas in the NanoSight LM10-HS.

4. Record a video for 60 s in length using the NTA software (version 2.3) with the following parameters: 10 for camera level, 4 for threshold, and at ambient room temperature.
5. Repeat recording videos twice more at different regions to acquire comprehensive data.
6. Analyze the particle sizes and numbers using captured report files.

3.3 Analysis of Exosomal Expressions of Integrins and Tetraspanins Using Flow Cytometry

3.3.1 Immobilization of Exosomes to Beads

1. Mix 5 μg TKl exosomes (as measured by a BCA protein assay) with 5-μL latex beads in a 1.5 mL microcentrifuge tube and incubate the mixture for 15 min at room temperature [23]. The beads are 4 μm in diameter at a concentration of 1.2×10^6/μL.
2. Add PBS to obtain a final volume of 1 mL and incubate on a test tube rotator wheel overnight at 4 °C.
3. Add 110 μL of 1 M glycine (100 mM final) to block any non-specific binding, mix gently, and incubate at room temperature for 30 min.
4. Microcentrifuge for 5 min at 7900 × *g* at room temperature. Remove and discard supernatant.
5. Resuspend the bead pellet in 1 mL PBS/0.5% BSA and centrifuge for 5 min at 7900 × *g* at room temperature. Remove and discard supernatant.
6. Resuspend the bead pellet with 0.5 mL 0.5% BSA in PBS.
7. Wash the beads twice.
8. Resuspend the beads in 1 mL 0.5% BSA in PBS.
9. A final concentration of 4–5 × 10^7 beads per mL can be adequate to analyze expression of exosomal tetraspanins.

3.3.2 Determination of Exosomal Marker Tetraspanin (CD9, CD63, and CD81) Expressions

1. To detect tetraspanin proteins on the surface of exosomes using flow cytometry, incubate 10 μL exosome-conjugated beads with 10 μg/mL anti-CD9 antibody, 10 μg/mL anti-CD63 antibody, or 10 μg/mL anti-CD81 antibody diluted in 0.5% BSA in PBS for 1 h at 4 °C.
2. Wash twice with 0.5 mL 0.5% BSA in PBS.
3. Incubate the bead pellets with 10 μg/mL APC-conjugated secondary antibody in 0.5% BSA in PBS for 1 h at 4 °C.
4. Wash twice in 0.5% BSA in PBS.
5. Analyze antibody-stained exosome conjugated with beads using a flow cytometer (BD Accuri™ C6). Adjust forward scatter (FSC) and side scatter (SSC) to observe both single and doublet beads. Gate on single beads to analyze fluorescence (Fig. 1a). Compare the fluorescence levels obtained with a specific antibody to that of the isotype control (Fig. 1b).

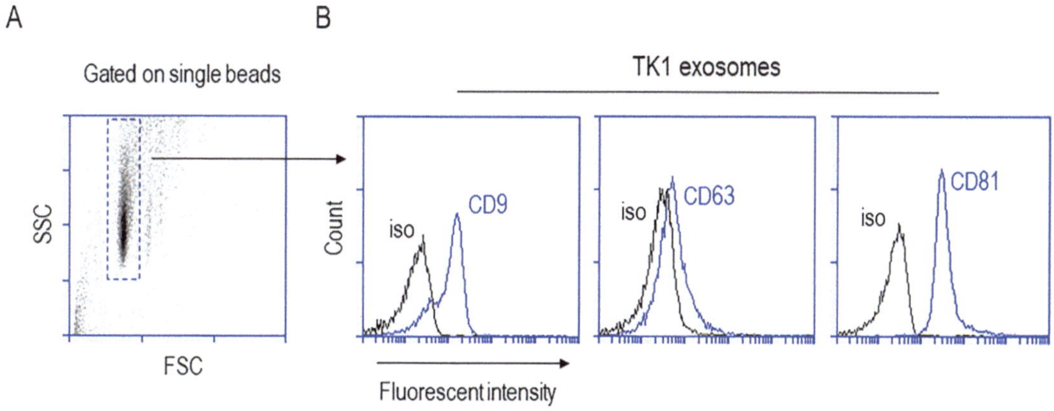

Fig. 1 Expression of tetraspanins on exosomes isolated from culture media of TK1 cells. The exosomes were immobilized on beads and stained with the indicated antibody to tetraspanins followed by the secondary antibody fluorescently labeled. The single beads were gated (**a**) and analyzed for the tetraspanin expressions using flow cytometry (**b**). FSC forward scatter, SSC side scatter, and iso isotype control

3.3.3 Examination of Integrin Expression on Exosomes

1. The TK1 β7- knockdown (KD) exosomes and TK1 β7-scramble (scr) exosomes were first generated using culture supernatant of TK1 β7-KD cells and TK1 β7-scr cells, respectively [15] (also *see* **Note 4**). Both cells were tested for their altered expressions of β7 and α4 antibody using flow cytometry and the reduction in β7- and α4-expression was found in β7-KD cells (Fig. 2a).
2. After immobilization of exosomes to beads as stated above in Subheading 3.3.1, incubate 10 μL exosome-conjugated beads with 10 μg/mL phycoerythrin (PE)-labeled anti-β7 integrin antibody or 10 μg/mL PE-labeled anti-α4 integrin antibody diluted in 0.5% BSA in PBS for 1 h at 4 °C.
3. Wash twice in 0.5% BSA in PBS and resuspend in 0.2 mL 0.5% BSA in PBS (if sample number is 4).
4. Analyze antibody-stained exosomes coated on beads using a flow cytometer (BD Accuri™ C6). Adjust the FSC and SSC to observe both singlet and doublet beads. Gate on singlet beads to analyze fluorescence. Compare the fluorescence levels obtained with a specific antibody to that of the isotype control (Fig. 2b).

3.4 Analysis of Integrin-Mediated Exosomal Binding to Ligand

3.4.1 Immobilization of MAdCAM-1 to Beads

1. Conjugate MAdCAM-1 protein to beads by mixing 5 μg MAdCAM-1 with 10 μL 4 μm latex beads in a microfuge microtube [15].
2. Add 1 mL of HBS buffer and incubate in a tube rotator at 4–8 °C for 16 h.
3. Add 1 M glycine solution at 100 mM, mix well, and incubate at room temperature for 30 min (*see* **Note 9**).

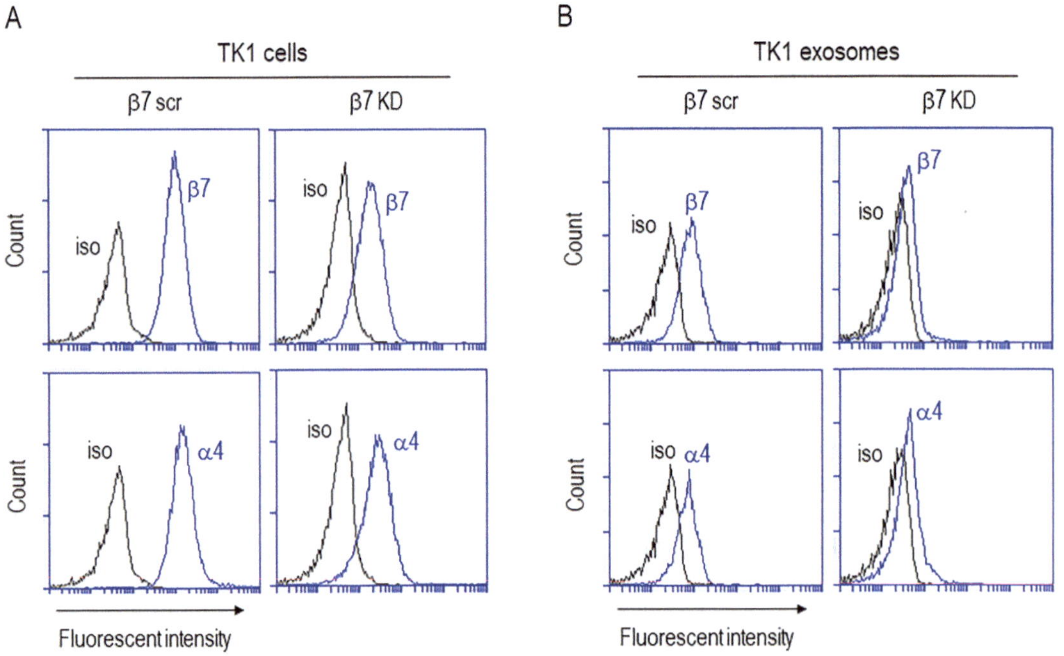

Fig. 2 Expression of integrins on cells and exosomes. (**a**) Cellular expression levels of indicated integrins are shown. (**b**) The exosomes were isolated from culture media of β7-scr and β7-KD of TK1 cells and tested for their expression of β7 and α4 integrin as shown in Subheading 3.3.3. Iso isotype control, scr scramble, and KD knockdown

4. Centrifuge at 2000 × *g* in a microfuge for 3 min and discard supernatant.
5. Add 1 mL of 0.5% BSA in HBS buffer (0.5% BSA/HBS) to the pellet (bead-conjugated MAdCAM-1), mix well, and centrifuge at 2000 × *g* for 3 min.
6. Repeat this wash step twice more.
7. Resuspend pellet with 0.25 mL 0.5% BSA/HBS (if sample number is 5).
8. Keep samples on ice until use.

3.4.2 Fluorescent Labeling of Exosome

1. To fluorescently label exosomes, mix 25 μg of the exosomes in 0.2 mL HBS buffer with BCECF-AM at a concentration of 5 μM in a microfuge tube.
2. Incubate at 37 °C for 30 min and tap tubes every 10 min.
3. Add 1 mL of HBS buffer, mix well, and centrifuge at 10,000 × *g* for 70 min.
4. Resuspend the pellet with 10 μL of HBS buffer (2.5 μg/μL) and keep on ice until use.

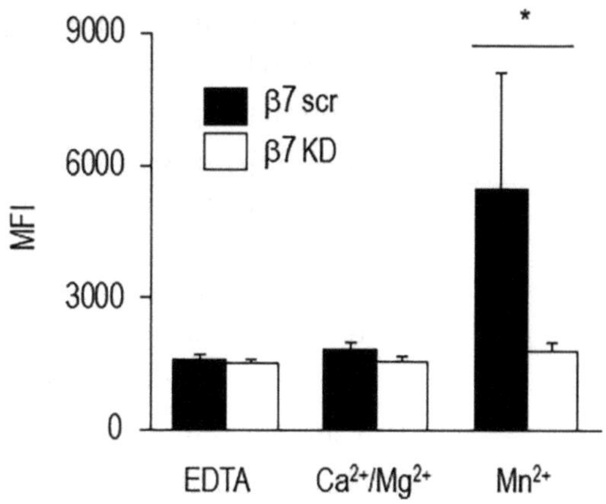

Fig. 3 Binding of TK1 β7-scr and β7-KD exosomes to MAdCAM-1. Differential levels of exosomal binding were tested under the conditions of 1 mM EDTA, 1 mM Ca^{2+}/Mg^{2+}, or 1 mM Mn^{2+}. Data represent the mean ± standard error of the mean (SEM) for MFI values obtained from three to five independent experiments. MFI mean fluorescent intensity, scr scramble, and KD knockdown

3.4.3 Analysis of Exosomal Binding to Integrin Ligand

1. Mix the BCECF-AM-labeled exosomes (5 μg in 2 μL) with bead-conjugated MAdCAM-1 (1 μg in 50 μL) in the presence of either 1 mM EDTA or divalent cation (1 mM $MgCl_2$/1 mM $CaCl_2$ or 1 mM $MnCl_2$) (*see* **Note 10**).
2. Incubate at 37 °C for 30 min.
3. Add 0.2 mL PBS containing 2% FBS per sample and mix well.
4. Determine the binding levels using a flow cytometer (BD Accuri™ C6).
5. Adjust the FSC and SSC to observe both single and doublet beads. Gate on single beads to analyze fluorescence.
6. Compare the fluorescence intensity obtained with β7-scr and β7-KD TK1 exosomes under different incubation conditions. The representative results are shown in Fig. 3.

3.5 Competitive In Vivo Exosomal Homing Assay

In this section, we show the methods of labeling TK1 exosomes using PKH dyes (*see* **Note 11**).

3.5.1 Fluorescent Labeling of Exosomes

1. Prepare equal amount (100 μg per each) of two exosome samples (TK1 β7-scr exosomes and TK1 β7-KD exosomes).
2. Resuspend 50 μg of each sample with 1 mL of Diluent C in an ultracentrifuge tube (13.2 mL).
3. A total of four tubes (two β7-scr exosomes and two β7-KD exosomes) will be prepared.
4. Add 6 μL PKH67 or PKH26 dye into each tube containing 1 mL exosomes in Diluent C.

5. Mix thoroughly with gentle pipetting and incubate at room temperature for 5 min.
6. Stop staining by mixing with 2 mL 10% BSA/PBS.
7. Add 5.5 mL serum-free RPMI-1640 media to make a total volume of 8.5 mL per tube.
8. Place carefully and slowly 1.5 mL 971 mM sucrose solution onto the bottom of each tube, making a sucrose cushion. No turbulence is required.
9. Perform centrifugation at 100,000 × *g* at 4 °C for 3 h.
10. Carefully remove the supernatant solution.
11. Resuspend the pellet of exosomes in 1 mL PBS by gently pipetting.
12. Transfer the sample to an Amicon ultracentrifugal filter column (UFC9010) and add 9 mL PBS to the column.
13. Centrifuge the tubes at 3000 × *g* at 4 °C for 40 min to concentrate the sample volume up to 0.5 mL.
14. Take the concentrate, place into a 1.5 mL microfuge tube, and put on ice until use.

3.5.2 Competitive In Vivo Exosomal Homing Assay (See Fig. 4)

1. Mix equal amounts (0.1 mL each) of the two samples of fluorescently labeled exosomes (PKH67-labeled TKI β7-scr exosomes and PKH26-labeled TKI β7-KD exosomes or PKH26-labeled TKI β7-scr exosomes and PKH67-labeled TKI β7-KD exosomes).
2. Intravenously inject 0.2 mL of the mixed exosomes into mice using 30-G needles and 1 mL syringes (*see* **Note 12**).
3. Euthanize the recipient mice at 16 h after injection and prepare frozen sections of tissues using an O.C.T. compound and a Cryostat microtome.
4. Locate cells by staining the nuclei using DAPI (NucBlue).
5. Quantify two different fluorescent spots (green and red) by using a BZ-X700 fluorescent microscope to quantify the exosomes competitively homed to different tissues.

4 Notes

1. Exosome-free FBS was used. This exosome-free FBS was generated by eliminating extracellular vesicles via ultracentrifugation (110,000 × *g* for 18 h) followed by filtration through a 0.22 μm filter unit [24].
2. Store at 4 °C and use for no more than 3 months.

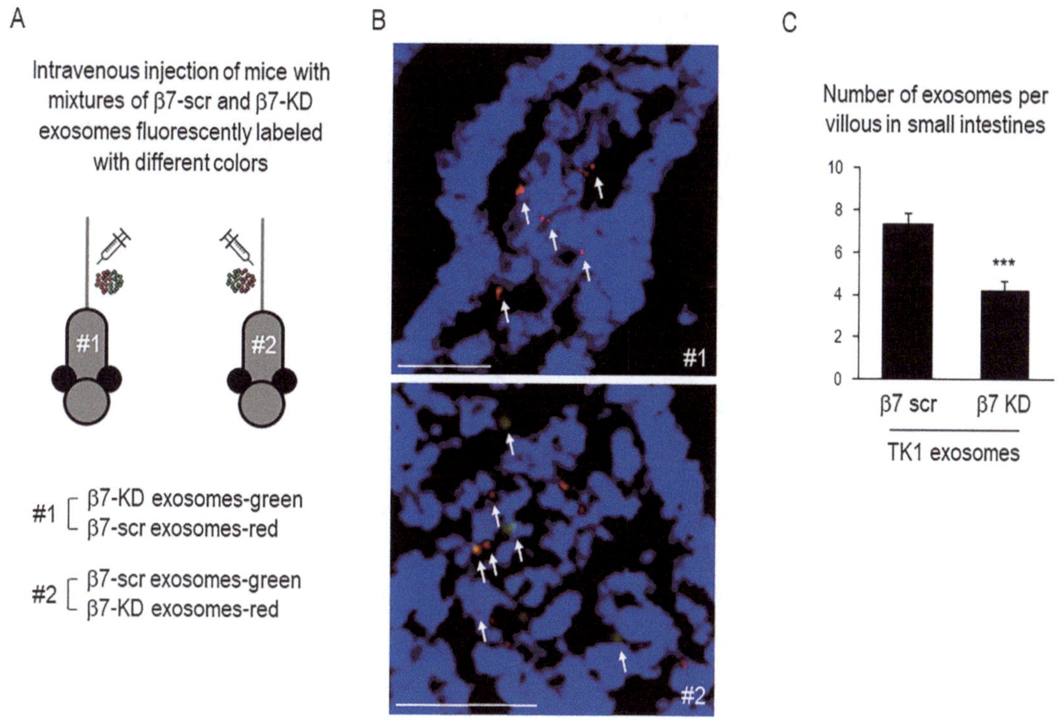

Fig. 4 Competitive in vivo exosomal homing assay and representative results. (**a**) The intravenous injection of two mice (#1 and #2) with mixtures of same numbers of β7-scr and β7-KD exosomes fluorescently labeled with different dyes (green and red) is shown. (**b**) Two representative images of villi in small intestinal tissues are shown. Sixteen hours after injections, frozen sections of intestine tissues separated from the recipient mice were subjected to nuclear staining with DAPI and fluorescent microscopic analysis. Arrows indicate some of the detected exosomes fluorescently labeled. Scale bars, 20 μm. (**c**) Significant reduction in the number of TK1 β7-KD exosomes homed into villi of small intestines when compared with that of TK1 β7-scr exosomes (*see* **Note 13**). Data represent the mean ± SEM obtained from 43 villi of three to five mice. ***, $P < 0.001$; scr scramble, and KD knockdown

3. The methods to isolate exosomes using ultracentrifugation were described in previous reports [23, 25–27].
4. TK-1 cells were transfected with β7 scramble (scr; wildtype) or β7 short hairpin RNAs (shRNAs; knockdown) (Santa Cruz, Dallas, TX) using Amaxa Nucleofector II (Lonza, Basel, Switzerland) according to the manufacturer's instructions. The stably transfected cells were analyzed for their expressions of β7 integrins (Fig. 2).
5. Should one be unable to attach sufficient amounts of exosomes to the aldehyde/sulfate latex beads, consider coating beads with poly-L-lysine (0.01% solution). Poly-L-lysine is widely used in coating solutions to enhance cell attachment and adhesion to both plastic and glass surfaces [28]. Incubate aldehyde/sulfate latex beads with poly-L-lysine (0.01% solution)

overnight at 4 °C. Wash twice with sterile PBS before use. Exosomes may strongly attach to aldehyde/sulfate latex beads treated with poly-L-lysine (0.01% solution) [27].

6. Retinoic acid has been known to enhance the expression of β7 integrins on surface of the cells and consequently their gut-homing capacity [29]. Although current protocol does not include the addition of retinoic acid to culture media, use of retinoic acid can be considered for further investigation.
7. Dynamic light scattering (Horiba, Kyoto, Japan) is an alternative to consider for measuring particle size and concentration [30, 31].
8. Do not make air bubbles on sample infusion to the device because they can lower the quality of video live capture.
9. The glycine is used to block the unbound surface on the beads from the unnecessary antibody binding [32].
10. $MnCl_2$ is known to directly activate integrin proteins, whereas $CaCl_2/MgCl_2$ stabilize integrins in the inactive state [33].
11. Alternatively, the staining of exosomes can be also performed using Exo-Glow kit (System Biosciences, Palo Alto, CA, USA) according to the manufacturer's instructions as described previously [15].
12. By analyzing the ratio of two alternately labeled exosomes homed to specific tissue, the competitive homing assay has an advantage to compare the homing capacity of two cohorts of exosomes distributed to tissues. Before intravenous injection to mice, the two exosome samples must be adjusted to identical particle numbers and protein concentrations.
13. The relatively high expression of the major ligand for β7 integrins, MAdCAM-1, in the small intestines compared to other organs is responsible for differential homing capacities of β7-scr and β7-KD exosomes [34].

References

1. Pan BT, Johnstone RM (1983) Fate of the transferrin receptor during maturation of sheep reticulocytes in vitro: selective externalization of the receptor. Cell 33(3):967–978. https://doi.org/10.1016/0092-8674(83)90040-5
2. Raposo G, Stoorvogel W (2013) Extracellular vesicles: exosomes, microvesicles, and friends. J Cell Biol 200(4):373–383. https://doi.org/10.1083/jcb.201211138
3. Gonda A, Kabagwira J, Senthil GN, Wall NR (2019) Internalization of exosomes through receptor-mediated endocytosis. Mol Cancer Res 17(2):337–347. https://doi.org/10.1158/1541-7786.MCR-18-0891
4. Thery C, Zitvogel L, Amigorena S (2002) Exosomes: composition, biogenesis and function. Nat Rev Immunol 2(8):569–579. https://doi.org/10.1038/nri855
5. Hessvik NP, Llorente A (2018) Current knowledge on exosome biogenesis and release. Cell Mol Life Sci 75(2):193–208. https://doi.org/10.1007/s00018-017-2595-9
6. Eitan E, Suire C, Zhang S, Mattson MP (2016) Impact of lysosome status on extracellular vesicle content and release. Ageing Res Rev 32:65–

74. https://doi.org/10.1016/j.arr.2016.05.001

7. Akers JC, Gonda D, Kim R, Carter BS, Chen CC (2013) Biogenesis of extracellular vesicles (EV): exosomes, microvesicles, retrovirus-like vesicles, and apoptotic bodies. J Neuro-Oncol 113(1):1–11. https://doi.org/10.1007/s11060-013-1084-8
8. Pegtel DM, Gould SJ (2019) Exosomes. Annu Rev Biochem 88:487–514. https://doi.org/10.1146/annurev-biochem-013118-111902
9. Villarroya-Beltri C, Baixauli F, Gutierrez-Vazquez C, Sanchez-Madrid F, Mittelbrunn M (2014) Sorting it out: regulation of exosome loading. Semin Cancer Biol 28:3–13. https://doi.org/10.1016/j.semcancer.2014.04.009
10. Valadi H, Ekstrom K, Bossios A, Sjostrand M, Lee JJ, Lotvall JO (2007) Exosome-mediated transfer of mRNAs and microRNAs is a novel mechanism of genetic exchange between cells. Nat Cell Biol 9(6):654–659. https://doi.org/10.1038/ncb1596
11. Park EJ, Appiah MG, Myint PK, Gaowa A, Kawamoto E, Shimaoka M (2019) Exosomes in sepsis and inflammatory tissue injury. Curr Pharm Des 25(42):4486–4495. https://doi.org/10.2174/1381612825666191116125525
12. Myint PK, Park EJ, Gaowa A, Kawamoto E, Shimaoka M (2020) Targeted remodeling of breast cancer and immune cell homing niches by exosomal integrins. Diagn Pathol 15(1):38. https://doi.org/10.1186/s13000-020-00959-3
13. Ludwig AK, Giebel B (2012) Exosomes: small vesicles participating in intercellular communication. Int J Biochem Cell Biol 44(1):11–15. https://doi.org/10.1016/j.biocel.2011.10.005
14. Gupta R, Radicioni G, Abdelwahab S, Dang H, Carpenter J, Chua M, Mieczkowski PA, Sheridan JT, Randell SH, Kesimer M (2019) Intercellular communication between airway epithelial cells is mediated by exosome-like vesicles. Am J Respir Cell Mol Biol 60(2):209–220. https://doi.org/10.1165/rcmb.2018-0156OC
15. Park EJ, Prajuabjinda O, Soe ZY, Darkwah S, Appiah MG, Kawamoto E, Momose F, Shiku H, Shimaoka M (2019) Exosomal regulation of lymphocyte homing to the gut. Blood Adv 3(1):1–11. https://doi.org/10.1182/bloodadvances.2018024877
16. Ji P, Yang Z, Li H, Wei M, Yang G, Xing H, Li Q (2021) Smart exosomes with lymph node homing and immune-amplifying capacities for enhanced immunotherapy of metastatic breast cancer. Mol Ther Nucleic Acids 26:987–996. https://doi.org/10.1016/j.omtn.2021.10.009
17. Denucci CC, Mitchell JS, Shimizu Y (2009) Integrin function in T-cell homing to lymphoid and nonlymphoid sites: getting there and staying there. Crit Rev Immunol 29(2):87–109. https://doi.org/10.1615/critrevimmunol.v29.i2.10
18. Hoshino A, Costa-Silva B, Shen TL, Rodrigues G, Hashimoto A, Tesic Mark M, Molina H, Kohsaka S, Di Giannatale A, Ceder S, Singh S, Williams C, Soplop N, Uryu K, Pharmer L, King T, Bojmar L, Davies AE, Ararso Y, Zhang T, Zhang H, Hernandez J, Weiss JM, Dumont-Cole VD, Kramer K, Wexler LH, Narendran A, Schwartz GK, Healey JH, Sandstrom P, Labori KJ, Kure EH, Grandgenett PM, Hollingsworth MA, De Sousa M, Kaur S, Jain M, Mallya K, Batra SK, Jarnagin WR, Brady MS, Fodstad O, Muller V, Pantel K, Minn AJ, Bissell MJ, Garcia BA, Kang Y, Rajasekhar VK, Ghajar CM, Matei I, Peinado H, Bromberg J, Lyden D (2015) Tumour exosome integrins determine organotropic metastasis. Nature 527(7578):329–335. https://doi.org/10.1038/nature15756
19. Shimaoka M, Kawamoto E, Gaowa A, Okamoto T, Park EJ (2019) Connexins and Integrins in exosomes. Cancers (Basel) 11(1). https://doi.org/10.3390/cancers11010106
20. Jang SC, Kim OY, Yoon CM, Choi DS, Roh TY, Park J, Nilsson J, Lotvall J, Kim YK, Gho YS (2013) Bioinspired exosome-mimetic nanovesicles for targeted delivery of chemotherapeutics to malignant tumors. ACS Nano 7(9):7698–7710. https://doi.org/10.1021/nn402232g
21. Van Deun J, Mestdagh P, Sormunen R, Cocquyt V, Vermaelen K, Vandesompele J, Bracke M, De Wever O, Hendrix A (2014) The impact of disparate isolation methods for extracellular vesicles on downstream RNA profiling. J Extracell Vesicles 3. https://doi.org/10.3402/jev.v3.24858
22. Iwai K, Minamisawa T, Suga K, Yajima Y, Shiba K (2016) Isolation of human salivary extracellular vesicles by iodixanol density gradient ultracentrifugation and their characterizations. J Extracell Vesicles 5:30829. https://doi.org/10.3402/jev.v5.30829
23. Thery C, Amigorena S, Raposo G, Clayton A (2006) Isolation and characterization of exosomes from cell culture supernatants and biological fluids. Curr Protoc Cell Biol Chapter 3(Unit 3):22. https://doi.org/10.1002/0471143030.cb0322s30

24. Shelke GV, Lasser C, Gho YS, Lotvall J (2014) Importance of exosome depletion protocols to eliminate functional and RNA-containing extracellular vesicles from fetal bovine serum. J Extracell Vesicles 3(1):24783. https://doi.org/10.3402/jev.v3.24783
25. Crescitelli R, Lasser C, Szabo TG, Kittel A, Eldh M, Dianzani I, Buzas EI, Lotvall J (2013) Distinct RNA profiles in subpopulations of extracellular vesicles: apoptotic bodies, microvesicles and exosomes. J Extracell Vesicles 2. https://doi.org/10.3402/jev.v2i0.20677
26. Holder B, Jones T, Sancho Shimizu V, Rice TF, Donaldson B, Bouqueau M, Forbes K, Kampmann B (2016) Macrophage exosomes induce placental inflammatory cytokines: a novel mode of maternal-placental messaging. Traffic 17(2):168–178. https://doi.org/10.1111/tra.12352
27. Kawamoto E, Park EJ, Shimaoka M (2021) Methods to study integrin functions on exosomes. Methods Mol Biol 2217:265–281. https://doi.org/10.1007/978-1-0716-0962-0_15
28. Mazia D, Schatten G, Sale W (1975) Adhesion of cells to surfaces coated with polylysine. Applications to electron microscopy. J Cell Biol 66(1):198–200. https://doi.org/10.1083/jcb.66.1.198
29. Iwata M, Hirakiyama A, Eshima Y, Kagechika H, Kato C, Song SY (2004) Retinoic acid imprints gut-homing specificity on T cells. Immunity 21(4):527–538
30. Sanz-Rubio D, Martin-Burriel I, Gil A, Cubero P, Forner M, Khalyfa A, Marin JM (2018) Stability of circulating exosomal miRNAs in healthy subjects. Sci Rep 8(1):10306. https://doi.org/10.1038/s41598-018-28748-5
31. Park EJ, Myint PK, Appiah MG, Worawattananutai P, Inprasit J, Prajuabjinda O, Soe ZY, Gaowa A, Kawamoto E, Shimaoka M (2021) Ligand-competent fractalkine receptor is expressed on exosomes. Biochem Biophys Rep 26:100932. https://doi.org/10.1016/j.bbrep.2021.100932
32. Dismuke WM, Liu Y (2015) Current methods to purify and characterize exosomes. In: Mesenchymal stem cell derived exosomes. Elsevier, Amsterdam, pp 63–92
33. Dransfield I, Cabañas C, Craig A, Hogg N (1992) Divalent cation regulation of the function of the leukocyte integrin LFA-1. J Cell Biol 116:219–226
34. Schippers A, Leuker C, Pabst O, Kochut A, Prochnow B, Gruber AD, Leung E, Krissansen GW, Wagner N, Muller W (2009) Mucosal addressin cell-adhesion molecule-1 controls plasma-cell migration and function in the small intestine of mice. Gastroenterology 137(3):924–933. https://doi.org/10.1053/j.gastro.2009.05.039

Chapter 13

Methods and Protocols for Using Extracellular Vesicles as Delivery Vehicles in Neuronal Research

Manjusha Vaidya, Nasif Sayeed, Caroline Hobson, Sandeep Sreerama, Jonhoi Smith, Riya Shah, and Kiminobu Sugaya

Abstract

Extracellular vesicles (EVs) transport nucleic acids, proteins, and lipid molecules for intercellular communication. The biomolecular cargo from EVs can modify the recipient cell genetically, physiologically, and pathologically. This innate ability of EVs can be harnessed to deliver the cargo of interest to a specific organ or a cell type. Importantly, due to their ability to cross the blood–brain barrier (BBB), the EVs can be used as delivery vehicles to transport therapeutic drugs and other macromolecules to inaccessible organs such as the brain. Therefore, the current chapter includes laboratory techniques and protocols focusing on the customization of EVs for neuronal research.

Key words Extracellular vesicles, Exosomes, Cargo delivery, Localization

1 Introduction

Extracellular vesicles (EVs) are lipid membrane-bound micro- and nano-vesicles carrying nucleic acids, proteins, and other macromolecules for intercellular communication. EVs are secreted by normal as well as diseased cells having endomembrane system [1]. The EVs transport single- and double-stranded DNA, microRNA, long noncoding RNA, mRNA, signaling proteins (cytokines, chemokines, growth factors, enzymes, ligands, receptors), and lipid rafts [2]. They are a heterogenous population comprising exosomes (30–150 nm diameter), micro-vesicles (100–1000 nm diameter), and apoptotic bodies (800–5000 nm diameter). Each subtype is formed by a unique mechanism thereby conferring distinct cargo, size, and a secretion route for the EVs. Once secreted by the parent cell, the EVs travel through anatomical compartments via transcytosis [3]. The recipient cell uptakes the EVs and the vesicular cargo is released into the cell. A cell exposed to pathological conditions secretes EVs packing disease-specific alterations of the cargo

Seppo Vainio (ed.), *Cell-Secreted Vesicles: Methods and Protocols*, Methods in Molecular Biology, vol. 2668,
https://doi.org/10.1007/978-1-0716-3203-1_13,

[4]. The alterations in DNA and RNA sequences or mutated forms of proteins potentially become biomarkers for the disease. The functionally active cargo of the EVs has a pleiotropic effect on the recipient cell. For example, EVs from a cancer cell carry tumor-specific macromolecules that modify the tumor microenvironment and participate in angiogenesis, metastasis, and immunity evasion [5]. EVs are the inherent carriers of a molecular signature of their parent cells and are able to cross the blood–brain barrier (BBB) to exert a paracrine effect on distant cells [6]. Drawing on these abilities, EVs can be utilized as delivery vehicles in fundamental and translational research. They can be modified to load specific biomolecules and be engineered to target the organs or the cell types of interest. The chapter includes the techniques to alter the EVs to make them more efficient in carrying out the transport function to the brain and neurons. It also includes the methods to target the macromolecules to exosomes. While electroporation and exosome–liposome hybrid can be used for delivering nucleic acids and proteins, this chapter focuses on the simple techniques and successful strategies followed in Sugaya-lab [7–9]. The engineered exosomes are useful for cargo delivery to the brain and neurons under in vivo conditions as well.

We have described the cell culture, EVs/exosome isolation, immunoaffinity purification, and characterization methods in a previously published book chapter [9]. The chapter also elaborates on quantitative and qualitative analyses of the EVs/exosomes. Therefore, in this chapter, we have directly described the protocols for adding macromolecules of interest to the vesicles, targeting the EVs to the brain and neurons, followed by the techniques to confirm the cargo delivery. Figure 1 outlines the topics covered in the current chapter. Because some techniques are specific to exosomes and others to EVs in general, we have used the terms as applicable. Additional information and useful suggestions are included in the notes at the bottom of the chapter.

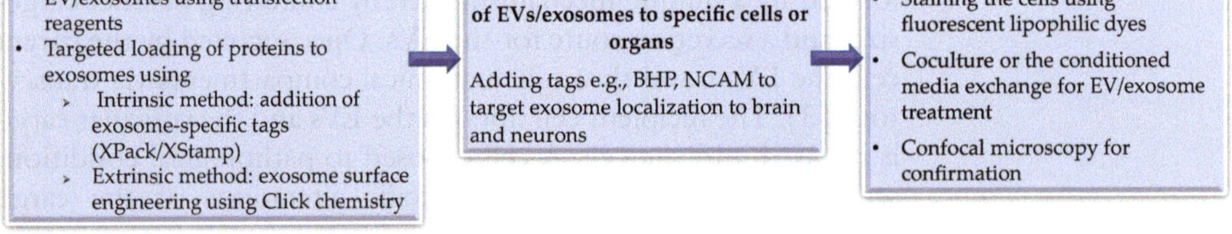

Fig. 1 Steps in extracellular vesicles (EV)/exosome-mediated delivery of biomolecular cargo

2 Materials

It is crucial to prepare all the solutions in the protocol using ultrapure, deionized water. While working with the nucleic acids, nuclease-free water must be used in every step. All the consumables used during experiments must be sterile and free of nucleases.

2.1 Loading Nucleic Acids and Proteins to Exosomes

2.1.1 Loading Nucleic Acids onto Exosomes Using Transfection Reagents

1. HEK293 cells cultured in a tissue culture-treated 6-,12-, or 24-well plate (*see* **Note 1**).
2. 10 μL Exo-Fect solution from Exo-Fect™ Exosome Transfection Kit (SBI catalog #EXFT10A-1).
3. 20 μL suspension of GFP plasmid DNA (final concentration 5 μg).
4. 70 μL sterile 1× PBS (*see* **Note 2**).
5. 50 μL HEK293 exosome suspension in 1× PBS.
6. Total transfection reaction should be 150 μL.

2.1.2 Targeted Loading of Proteins to Exosomes Using the Intrinsic Method: Addition of Exosome-Specific Tags (XPack/XStamp)

SBI-developed N terminal peptide sequence tags XPack and XStamp that target the peptide, or any protein sequence fused with them, to the inner and the outer surface of the exosomal membrane, respectively.

Commercially Available Plasmids and Cell Line Cultures

1. RFP- or GFP-packed exosomes precipitated and purified from HEK293 cells transfected with special lentivector (XPack CMV-XP-RFP-EF1α-Puro Expression Lentivector-SBI catalog # XPAK531PA-1 or XPack MSCV-XP-GFP-EF1α-Puro Expression Lentivector- SBI catalog # XPAK730PA-1) (*see* **Note 3**).
2. Neural stem cell suspension culture (*see* **Note 1**).
3. HEK293 cells in a 6-well, 12-well, or 24-well, tissue culture-treated plate.
4. HEK cell culture media (*see* **Note 3**).
5. Neural stem cell media (*see* **Note 3**).

Cloning the Gene of Interest (e.g., NGB) into a Vector to Target Its Protein Product to Exosome

NGB (human NGB neuroglobin, Gene ID: 58157, NCBI Reference Sequence: NC_000014.9) codes for an oxygen-binding heme protein of the globin family and has a neuroprotective function under ischemic or hypoxic conditions [10]. Overexpressed protein prevents neuronal apoptosis and decreases free radicles thereby improving mitochondrial function essential in neurovegetative diseases like Alzheimer's, Parkinson's, and Huntington's [11]. Therefore, exosome-mediated delivery of neuroglobin can be an effective treatment method. Here, we elaborate on the cloning of the gene in XPack Lentivector. The gene can be cloned into XStamp™ Cloning and Expression Lentivector (SBI catalog # XSTP710PA-1), using an exactly similar method, with specifically designed cloning primers.

1. Neuroglobin (NGB) (NM_021257) human-tagged ORF clone, NGB (Myc-DDK-tagged)-human neuroglobin (NGB) as template DNA (OriGene catalog # RC207482). ORF length: 456 base pairs.
2. Cloning primers: XPack-NGB-XhoI-F: 5′-ATCCTCGAGTATGGAGCGCCCGGAG-3′ and XPack-NGB-EcoRI-R: 5′-GCGGAATTCTTACTCGCCATCCCA-3′.
3. XPack MSCV-XP-MCS-EF1α-puro cloning lentivector (SBI catalog # XPAK710PA-1).
4. Cloning oligoprimers: XPack-NGB-F (with restriction enzyme XhoI cutting sequence) and XPack-NGB-R (with restriction enzyme EcoRI cutting sequence). Each primer, forward and reverse, diluted to 10 μM from 100 mM stock.
5. Two repeats of PCR amplification reaction mix (50 μL each) containing.
   (a) 0.25 μL DNA polymerase.
   (b) 10 μL 5× polymerase buffer (colorless or green) (*see* **Note 4**).
   (c) 4 μL magnesium chloride ($MgCl_2$) solution, 25 mM.
   (d) 1 μL dNTP mix (10 mM each of dATP, dTTP, dCTP, and dGTP).
   (e) 2 μL of template DNA (DNA concentration 500 ng/μL).
   (f) 1 μL of forward primer.
   (g) 1 μL of reverse primer.
   (h) 30.75 μL of molecular grade water.
6. PCR product purification after thermocycling (*see* **Note 5**).
   (a) 500 μL DNA binding buffer PB (Qiagen catalog # 19066) for 100 μL PCRs combined post-thermocycling.
   (b) QIAquick PCR purification and gel extraction spin column (Qiagen catalog #28115).
   (c) 750 μL column wash buffer PE (Qiagen catalog # 19065).
   (d) 30–40 μL DNA elution buffer EB (Qiagen catalog # 19086) or molecular grade water.
7. Agarose gel electrophoresis.
   (a) 750 mg molecular biology grade agarose, for 1.5% gel.
   (b) 50 mL 1× Tris–acetate–EDTA (TAE) buffer.

   (c) 5 μL Invitrogen™ SYBR™ Safe DNA Gel Stain (Fisher Scientific catalog # S33102).
   (d) 5–8 μL 1 kb Plus DNA Ladder.
8. PCR product purification after agarose gel electrophoresis.
   (a) Gel solubilization and DNA binding buffer QG (Qiagen catalog # 19063) The volume added should be three times the weight of the gel, e.g., for 100 mg of a gel-piece, 300 μL of QG buffer to be added.
   (b) QIAquick PCR Purification- and Gel Extraction spin column (Qiagen catalog #28115).
   (c) 750 μL column wash buffer PE (Qiagen catalog # 19065).
   (d) 30–40 μL DNA elution buffer EB (Qiagen catalog # 19086) or molecular grade water.
9. Restriction enzyme (XhoI/EcoRI) digestion of NGB-PCR product.
   (a) 30 μL of purified PCR product.
   (b) 3 μL FastDigest enzyme XhoI (Thermo Scientific™ catalog # FD0694).
   (c) 3 μL FastDigest enzyme EcoRI (Thermo Scientific™ catalog # FD0274).
   (d) 6 μL FastDigest buffer (10×) (Thermo Scientific™ FastDigest Colorless Buffer catalog # B64 <u>or</u> FastDigest Green Buffer (10×) catalog # B72) (*see* **Note 4**).
   (e) 18 μL molecular grade water.
   (f) Total digestion reaction volume 60 μL.
10. Restriction enzyme (XhoI/EcoRI) digestion of XPack MSCV-XP-MCS-EF1α-puro cloning lentivector.
   (a) 6 μL of XPack MSCV-XP-MCS-EF1α-puro cloning lentivector.
   (b) 3 μL FastDigest enzyme XhoI (Thermo Scientific™ catalog # FD0694).
   (c) 3 μL FastDigest enzyme EcoRI (Thermo Scientific™ catalog # FD0274).
   (d) 6 μL FastDigest buffer (10×) (Thermo Scientific™ FastDigest Colorless Buffer catalog # B64 <u>or</u> FastDigest Green Buffer (10×) catalog # B72) (*see* **Note 4**).
   (e) 42 μL molecular grade water.
   (f) Total digestion reaction volume 60 μL.
11. PCR product and vector DNA purification after enzymatic digestion.

(a) 60 μL digestion reaction.

(b) 500 μL DNA binding buffer PB (if the reaction is cleaned right after digestion) or buffer QG (if the reaction is subjected to agarose gel electrophoresis after digestion). The volume added should be three times the weight of the gel, e.g., for 100 mg of gel piece, 300 μL of QG buffer to be added.

(c) QIAquick PCR Purification and Gel Extraction spin column (Qiagen catalog #28115).

(d) 750 μL column wash buffer PE (Qiagen catalog # 19065).

(e) 30–40 μL DNA elution buffer EB (Qiagen catalog # 19086) or molecular grade water.

12. Ligation of NGB-PCR product with XPack MSCV-XP-MCS-EF1α-puro cloning lentivector.

(a) 12 μL PCR product, digested and cleaned (DNA concentration 400 ng/μL).

(b) 4 μL vector, digested and cleaned (DNA concentration 25 ng/μL).

(c) 2 μL T4 DNA ligase.

(d) 2 μL T4 DNA ligase buffer.

(e) Total ligation reaction mix should be 20 μL.

13. Chemically competent E. coli transformation with ligated product and subsequent culture of bacterial colonies.

(a) 20 μL (entire) ligation reaction.

(b) E. coli (strain Stbl3) chemically competent cells (*see* **Note 6**).

(c) 500 μL of SOC or LB media without antibiotics.

(d) LB agar with ampicillin (100 μg/mL), 1 plate per transformation.

(e) LB broth with ampicillin (100 μg/mL), 11 mL per 50 mL tube. One clone to be cultured per tube.

14. Colony PCR

(a) 1 mL overnight culture of each clone. Minimum 3 clones should be tested. One colony equals one clone. The remaining 10 mL culture to be kept on ice for plasmid DNA extraction.

(b) 0.25 μL DNA polymerase.

(c) 10 μL 5× polymerase buffer (colorless or green) (*see* **Note 4**).

(d) 4 μL magnesium chloride solution, 25 mM.

(e) 1 μL dNTP Mix (10 mM each of dATP, dTTP, dCTP, and dGTP).

(f) 2 μL of template DNA (bacterial suspension).

(g) 1 μL of forward primer.

(h) 1 μL of reverse primer.

(i) 30.75 μL of molecular grade water.

(j) Total reaction volume 25 μL.

15. Extraction of plasmid DNA from overnight bacterial culture of NGB clone.

(a) 10 mL overnight culture of each clone.

(b) 250 μL resuspension buffer P1 with RNase A (Qiagen catalog # 19051).

(c) 250 μL lysis buffer P2 (Qiagen catalog # 19052).

(d) 250 μL neutralization buffer N3 (Qiagen catalog #19064).

(e) QIAprep 2.0 spin columns (Qiagen catalog # 27115).

(f) 750 μL wash buffer PE with ethanol (Qiagen catalog # 19065).

(g) 50–100 μL elution buffer EB (Qiagen catalog # 19086).

(h) Verify the clone by Sanger sequencing.

Transfection of HEK293 cell line with NGB-XPack clone.

Refer to our previously published materials list for Lipofectamine® 2000-mediated transfection of HEK293 cells and precipitation as well as CD63 immunocapture of exosomes ($CD63^+$ exosomes) [9].

Immunocytochemistry (ICC) of Transfected HEK293 to Detect NGB Protein

1. HEK293 cells grown in 24-well plate at ~85–90% confluency [9].

2. Fixing cells.

(a) Ice-cold, 100% methanol.

(b) 1× PBS.

3. Immunostaining.

(a) NGB antibody (anti-neuroglobin mouse monoclonal antibody).

(b) Fluorescein (FITC)-conjugated secondary antibody; fluorescein (FITC)-AffiniPure donkey anti-mouse IgG (H + L) (λEx/λEm: 495 nm/519 nm).

(c) Blocking buffer containing 1× PBS, 5% normal serum (from the animal in which the secondary antibodies were raised), 0.3% Triton™ X-10 [e.g., for 20 mL blocking buffer: 1 mL serum, 10 mL PBS (vortexed together), and 60 μL Triton)].

(d) 1× PBS.

(e) Antibody dilution buffer containing 1× PBS, 1% bovine albumen serum (BSA), 0.3% Triton™ X-10. [e.g., for 20 mL blocking buffer: 60 μL Triton added to 20 mL PBS along with 0.2 gm bovine serum, albumen)].

(f) DAPI staining solution (λEx/λEm: 360 nm/460 nm).

Dot Blot of Transfected HEK293 Exosomes to Detect NGB Protein

1. Protein sample preparation from exosomes.
   (a) CD63+ exosomes from NGB-XPack transfected HEK293 cells (approximately all the exosomes collected from 10 mL conditioned media of 90% confluent cells in a T75 flask).
   (b) 1× Sample buffer (M-PER + protease/phosphatase inhibitors or RIPA) (Thermo Scientific™ M-PER™ mammalian protein extraction reagent—Catalog # PI78501, Thermo Scientific™ RIPA lysis, and extraction buffer—Catalog # PI89901).
   (c) 5% sodium dodecyl sulfate (SDS) (instead of M-PER).
2. Dot blot.
   (a) NGB primary antibody (anti-neuroglobin mouse monoclonal antibody).
   (b) TBS containing 20 mM Tris–HCl, 150 mM NaCl pH 7.5.
   (c) 5% BSA in TBST.
   (d) Polyvinylidene difluoride (PVDF) or nitrocellulose membrane.
   (e) Exosome extracted protein.
   (f) 100% methanol.
   (g) 5% skim milk powder dissolved in TBST.
   (h) Double-distilled water.
   (i) Primary antibody.
   (j) Goat anti-mouse IgG secondary antibody HRP conjugated.
   (k) Enhanced chemiluminescence (ECL) reagent (Thermo Scientific™ Pierce™ ECL Western Blotting Substrate (Catalog # PI32109)).

2.1.3 Targeted Loading of Proteins to Exosomes Using Extrinsic Method: Exosome Surface Engineering via Click Chemistry

Copper-catalyzed azide alkyne cycloaddition is a method to conjugate a ligand to an exosomal surface [12]. The goal of surface engineering HEK293 exosomes using Click chemistry is to attach anti-Il1RAP antibody to the surface of the GFP-packing exosomes to be delivered ultimately to neuronal cells. We have used HEK293 cells as recipients for confirmation of the delivery process. Interleukin 1 receptor accessory protein variations lead to microglial dysfunction in Alzheimer's disease that in turn affects Aβ clearance [13]. Antibody-based blocking of IL1RAP activity can be achieved by delivering the antibody using exosomes as vehicles.

1. Cells and exosomes.
   (a) HEK293 cells.
   (b) GFP-packed HEK293 exosomes.
   (c) HEK293 exosomes (control).
2. Immunofluorescence assay of exosomal samples.
   (a) 1× PBS.
   (b) Poly-L-ornithine solution diluted 1:10 in 1× PBS.
   (c) 4% paraformaldehyde (PFA).
   (d) Blocking/permeabilizing solution (0.1% Triton, 5% BSA in 1× PBS).
   (e) Primary antibody against CD9, at dilution 1:500.
   (f) Secondary antibodies (rhodamine TRITC AffiniPure goat anti-mouse IgG H + L, at dilution 1:200).
3. Quantification of exosomal proteins.
   (a) Radioimmunoprecipitation assay (RIPA) buffer containing protease and phosphatase inhibitors (RIPA Abcam catalog # ab156034, Thermo Scientific™ Halt™ Protease and Phosphatase Inhibitor Cocktail (100×) catalog # PI78440).
   (b) BSA stock.
   (c) Thermo Scientific™ Pierce™ BCA Protein Assay Kit containing BCA Reagents A and B (catalog # PI23227).
   (d) Molecular grade water.
4. Modification of Exosomal Surface.
   (a) 2 mL of 1× PBS (pH 7.4 without calcium chloride and magnesium chloride).
   (b) 70 mg, 0.6 mmol, 1 equivalent of Tetrazine n-hydroxysuccinimide (NHS) ester.
   (c) 58 mg, 0.6 mmol, 1 equivalent of 4-pentynoic acid.
   (d) Sodium bicarbonate to adjust pH.
   (e) 92 mg, 0.6 mmol, 1 equivalent of 1-ethyl-3-(3-carbodiimide dimethylaminopropyl).
   (f) Sephadex G-25 column conditioned with 1× PBS.
5. Modification of anti-Il1RAP.
   (a) 2 mL of 1× PBS (pH 7.4 without calcium chloride and magnesium chloride).

(b) 70 mg, 0.6 mmol, 1 equivalent of TCO-PEG4 n-hydroxysuccinimide (NHS) ester.

(c) 58 mg, 0.6 mmol, 1 equivalent of 4-pentynoic acid.

(d) Sodium bicarbonate to adjust pH.

(e) 92 mg, 0.6 mmol, 1 equivalent of 1-ethyl-3-(3-carbodiimide dimethylaminopropyl).

(f) 20 μL anti-IL1RAP, Rabbit polyclonal antibody to IL1RAP, isotype IgG, unconjugated.

(g) Sephadex G-25 column conditioned with 1× PBS.

6. Click chemistry.

(a) Control exosomes and GFP-packing HEK293 exosomes, each in 150 μL of 1× PBS.

(b) 7 μL of 0.32 M copper (II) sulfate pentahydrate.

(c) 35.5 μL of 1.44 M L-ascorbic acid (pH buffered to 7.4 with sodium bicarbonate).

(d) 16.4 μL of 0.27 M bathophenanthrolinedisulfonic acid disodium salt trihydrate.

(e) Sephadex G-25 column conditioned with 1× PBS.

7. Dot blot to confirm presence of antibody on exosomal surface.

(a) PVDF membrane.

(b) 20 μL of the click chemistry final reaction sample of control exosomes.

(c) 20 μL of the click chemistry final reaction sample of GFP-packed exosomes.

(d) 20 μL HEK 293 exosomes in RIPA buffer.

(e) 10 μL anti-IL1RAP (nonmodified) antibody.

(f) 1× TBST.

(g) Blocking buffer composed of 5% bovine serum albumin (BSA) in Tris-buffered saline with 0.1% Tween-20 (TBST).

(h) Thermo Scientific™ SuperSignal™ West Femto maximum sensitivity substrate enhanced chemiluminescence (ECL) solution (2 mL SuperSignal West Femto Stable Peroxide Buffer and 2 mL SuperSignal West Femto Luminol/Enhancer Solution) (Thermo Scientific™, Catalog # PI34096).

(i) ChemiDocTM MP imaging system chemiluminescence application with rapid auto-exposure, no illumination, and 647SP filter.

2.2 Enhanced and Targeted Localization of EVs/Exosomes to Specific Cells or Organs

2.2.1 Adding Peptide Tags BHP1 (Brain Homing Peptide 1) and NCAM (Neural Adhesion Molecule) to Target EV/Exosome Localization

1. Cell culture.
   (a) HEK293 cells.
   (b) Neural stem cells (NSC)—differentiated.
   (c) Adipose-derived mesenchymal stem cells (ADSC).
   (d) Corning® Matrigel® Matrix (Corning Life Sciences catalog # 354234).
   (e) Media comprising DMEM-F12, 10% exosome-depleted FBS, and 1% (v/v) of penicillin/streptomycin for NSC.
   (f) Media comprising DMEM-F12, 10% exosome-depleted mesenchymal culture-specific FBS and 1% (v/v) of penicillin/streptomycin for ADSC.
2. Co-transfection.
   (a) Plasmid DNA of XPack CMV-XP-GFP-EF1α-Puro Expression Lentivector (SBI catalog # XPAK530PA-1).
   (b) Plasmid DNA of XStamp™-NCAM Lentivector for Targeting Exosomes to neurons (SBI catalog # XSTP721PA-1).
   (c) Plasmid DNA of XStamp™-BHP1 Lentivector for Targeting Exosomes to Brain Cells (SBI catalog # XSTP722PA-1).
   (d) Invitrogen™ Lipofectamine™ 3000 Transfection Reagent (Fisher Scientific catalog # L3000075).
   (e) Gibco™ Opti-MEM™ I Reduced Serum Medium (Fisher Scientific catalog # 31-985-070).
3. Exosome precipitation from nontransfected and XPack CMV-XP-GFP-EF1α-Puro Expression Lentivector-transfected HEK293 cells [9].
4. Exosome treatment of NSC and ADSC.
   (a) GFP-packed HEK293 exosomes with NCAM tag.
   (b) GFP-packed HEK293 exosomes with BHP tag.
   (c) NSC media.
   (d) ADSC media.
5. Cell fixation.
   (a) 4% paraformaldehyde.
6. Microscopy.
   (a) DAPI staining solution (Miltenyi catalog #130-111-570).

2.3 Visualization of EV/Exosome Uptake and Internalization [14]

2.3.1 Staining the Cells Using Fluorescent Lipophilic Dyes

1. Cell culture.
   (a) GBM and NSC in suspension cultures [9].
   (b) NSC media comprising DMEM-F12, 0.5 U/mL heparin, 20 ng/mL EGF, 20 ng/mL bFGF, 2% B-27 supplements, penicillin, and streptomycin (100 U/mL each).
2. Lipophilic staining.

(a) Lipophilic dye DiO.
(b) Lipophilic dye DiI.

2.3.2 Coculture, Conditioned Media Exchange, or a Direct EV/Exosome Treatment

1. Coculture.
   (a) DiO-stained NSC.
   (b) DiI-stained GBM.
2. Conditioned media exchange.
   (a) Conditioned media from DiO-stained NSC.
   (b) Conditioned media from DiI-stained GBM.
3. EV precipitation [9].
   (a) PEG/NaCl.

2.3.3 Confocal Microscopy for Confirmation of EV/Exosome Uptake and Internalization

1. Fixing cells.
   (a) 4% PFA.
   (b) DAPI staining solution.
2. Slide preparation.
   (a) Transparent slide sealant.

3 Methods

3.1 Loading of Nucleic Acids and Proteins to Exosomes

3.1.1 Loading of Nucleic Acids to Exosomes (Fig. 2)

1. To a 1.5 mL microcentrifuge tube, add the Exo-Fect solution, the plasmid DNA and the exosomes, and the PBS. Invert the tube 4–6 times to mix the components. Do not vortex.
2. Subject the tube to a quick and low centrifugation to collect the mixture at the bottom.
3. Incubate the mixture at 37 °C in a horizontal or a rotisserie shaker for 10 min and immediately place the tube on ice.
4. After 2 min incubation on ice, add 30 μL of the ExoQuick-TC reagent (provided in the kit) to stop the reaction and mix by inverting the tube 6–8 times. Do not vortex.
5. Incubate the transfected exosome sample on ice for 30 min.
6. Centrifuge the sample at 10,000 × *g* for 3 min, preferably at 4 °C.
7. Remove the supernatant carefully without disturbing the exosomal pellet and resuspend the transfected exosomes in 300 μL ice-cold 1 × PBS.
8. Remove media from the well of ~80% confluent HEK293 cell culture.
9. Add 1 mL of fresh HEK media along with 150–300 μL of transfected exosomal suspension to each well of a 12-well plate and incubate at 73 °C with 5% CO_2.

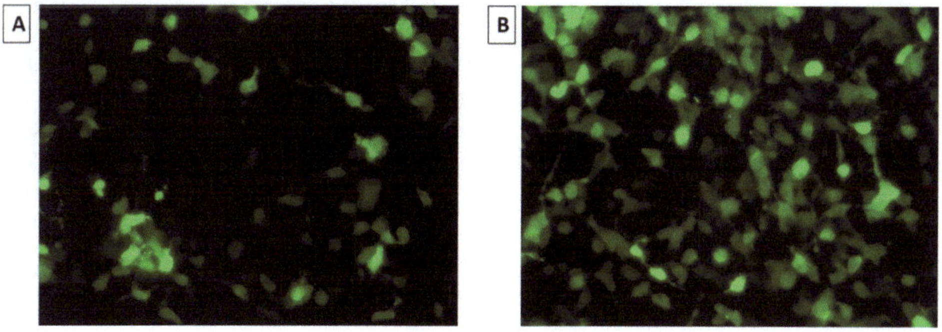

Fig. 2 HEK293 cells expressing GFP. Imaged after 48 h of incubation with GFP plasmid-transfected exosomes. (**a**) 150 μL of exosomal suspension added to the well. (**b**) 300 μL of exosomal suspension added to the well

10. After 24–48 h, visualize the cells with fluorescent microscopy at excitation wavelength of 488 nm and optimal detection at 510 nm. Figure 2 shows the HEK293 cells expressing GFP upon receiving.

3.1.2 Intrinsic Loading Proteins of Interest, e.g., RFP, NGB, to Exosomes

Cloning NGB into XPack MSCV-XP-MCS-EF1α-Puro Cloning Lentivector (Fig. 3)

1. PCR amplification of the template.
   (a) Completely thaw all the components of PCR on ice.
   (b) Except polymerase enzyme, vortex all the components and spin down briefly to collect them at the bottom. Polymerase buffer and MgCl2 should be vortexed thoroughly.
   (c) Flick the tube of polymerase 3–4 times and spin very briefly to collect the enzyme.
   (d) Make a master mix for two reactions (total volume 100 μL) and dispense 50 μL each, in two PCR tubes.
   (e) Setting up the thermocycling program as follows, complete the PCR, and proceed to either a direct purification of the PCR product or gel electrophoresis of the product followed by elution. Thermocycling program: 95 °C: 5 min, [95 °C: 30 s, 55 °C: 30 s, 72 °C: 2 min] × 35 cycles, 72 °C: 10 min.
2. PCR product purification after thermocycling.

   (Use a conventional tabletop centrifuge and perform all the centrifugation steps at room temperature at 17,900 × *g*).

   (a) Combine both the reactions in a single 1.5 μL microtube and add 500 μL buffer PB. Pipette gently but mix thoroughly.
   (b) Load the mix onto QIAquick PCR Purification spin column fitted into a collection tube.
   (c) Centrifuge the column for 1 min and discard the flow-through.

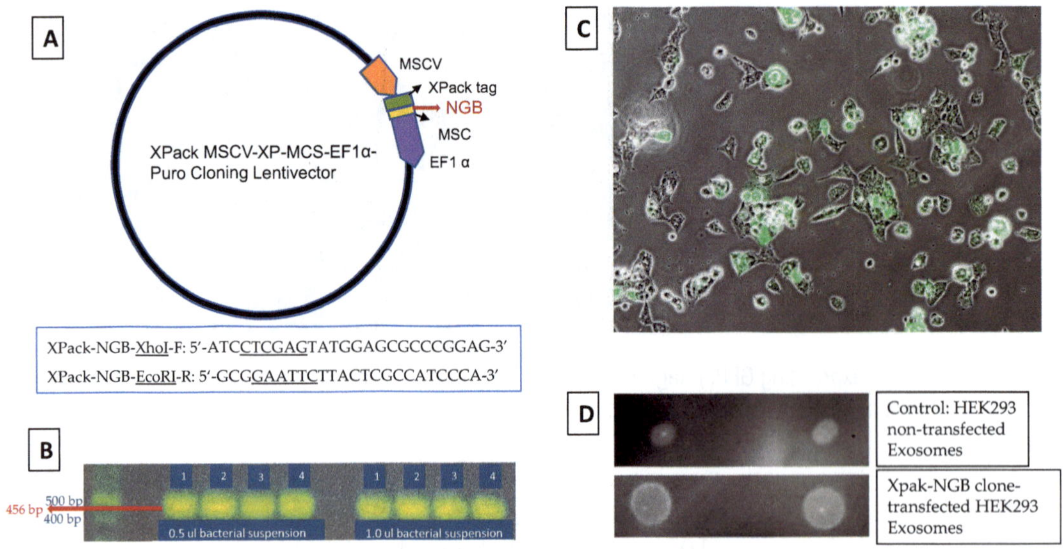

Fig. 3 Targeted delivery of NGB protein using exosome-packing tag (**a**) NGB ORF cloned into XPack MSCV-XP-MCS-EF1α-Puro Cloning Lentivector. (**b**) Colony PCR of the NGB clones—gel electrophoresis of the PCR products of 4 clones. First four clones show the PCR product with a template volume of 0.5 μL, and the last four with a template volume of 1.0 μL of bacterial suspension from overnight culture. (**c**) ICC of HEK293 cells: transfected with a clone of NGB-XPAK510PA-1. (**d**) Dot blot of CD63-purified exosomes to detect NGB protein

(d) Add 750 μL of buffer PE and repeat centrifugation as before.

(e) Discard the flow-through of the wash buffer and centrifuge the empty column again to remove traces of the wash buffer.

(f) Place the column on a new nuclease-free microcentrifuge tube and add 30–40 μL of buffer EB or molecular grade water.

(g) Incubate the column at room temperature for 2–3 min and centrifuge at highest speed for 2 min to elute the DNA from the column. The PCR product is ready for digestion with restriction endonucleases XhoI and EcoRI.

3. Agarose gel electrophoresis.

(a) To cast a 1.5% agarose gel, mix 50 mL 1× TAE buffer, 5 μL SYBR Safe DNA gel stain, and 750 mg of agarose in a 250 mL conical flask.

(b) Microwave the mixture until agarose is completely dissolved.

(c) Add the agarose mix to a gel casting tray with a comb that accommodates 25–50 μL of PCR sample.

(d) Cool the agarose mix slightly by swirling the conical flask a few times. Without letting the mix thicken, add it to the gel casting tray and wait until the gel is set.

(e) If colorless buffer is used, add ~8 μL of DNA loading dye to the PCR sample and mix by pipetting. If green buffer is used for PCR, do not add DNA loading dye.

(f) Load the samples into the wells of the gel. Also add 5–8 μL of 1 kb plus DNA ladder in the first and the last well and electrophorese at 80–90 V. After a visual confirmation of the PCR product of the expected size (456 base pairs) under UV illuminator, the DNA is ready to be eluted from the gel for digestion with restriction endonucleases XhoI and EcoRI.

4. PCR product purification after gel electrophoresis.

(a) Keep the gel on UV illuminator. Avoiding long exposure of the DNA sample to UV, cut out the part of the gel containing PCR product.

(b) Weigh the piece of the gel keeping it in a microtube.

(c) Add buffer QG and incubate at 37 °C until the gel is completely dissolved. Invert the tube intermittently to mix the buffer and the gel.

(d) Apply 750 μL of gel-buffer mix to the QIAquick PCR Purification spin column fitted into a collection tube.

(e) Process the sample further using exact steps described in subheading "Cloning NGB into XPack MSCV-XP-MCS-EF1α-Puro Cloning Lentivector (Fig. 3)" **step 2**-b to g.

5. Restriction enzyme (XhoI/EcoRI) digestion of NGB-PCR product and of XPack MSCV-XP-MCS-EF1α-Puro Cloning Lentivector.

(a) Use all of the clean PCR product (30 μL or more if additional buffer EB is added to the elution column) for digestion.

(b) In a separate microcentrifuge, use 6 μL of cloning vector for digestion.

(c) To each tube, add Fast Digest restriction endonucleases, and the buffer.

(d) Bring the volume to 60 μL by adding an appropriate amount of molecular grade water.

(e) Mix the reaction sample by gentle pipetting and short spin the tubes to collect the sample.

(f) Incubate the samples at 37 °C for 15 min and immediately proceed to gel electrophoresis. If the agarose gel is not ready, stop the digestion reaction using heat inactivation at 65 °C for 15 min. Instead of gel electrophoresis, the digested samples can be purified directly from the reaction mix.

6. PCR product and vector DNA purification after enzymatic digestion.
   (a) Add 500 μL buffer PB to each digestion reaction and follow the purification steps described in subheading "Cloning NGB into XPack MSCV-XP-MCS-EF1α-Puro Cloning Lentivector (Fig. 3)" **step 2**-b to g.
7. PCR product and vector DNA purification after gel electrophoresis.
   (a) If the digestion reactions are electrophorized, cut out the gel pieces containing DNA and add buffer QG. Then follow the protocol described in subheading "Cloning NGB into XPack MSCV-XP-MCS-EF1α-Puro Cloning Lentivector (Fig. 3)" **step 2**-b to g.
8. Ligation of NGB-PCR product and the XPack vector.
   (a) Combine PCR product, vector DNA, ligase, and the buffer in a 1.5 mL microtube.
   (b) Mix by pipetting, short spin to collect the reaction mix, and incubate overnight at 4 °C. The ligation reaction is ready for the bacterial transformation.
9. Chemically competent E. coli transformation with ligated product and subsequent culture of bacterial colonies.
   (a) Thaw the chemically competent Stbl3 E. coli cells (stored at −80 °C) on ice for 10 min.
   (b) Add entire ligation mix on the top of the cells. Without mixing or pipetting, incubate the reaction on ice for 30 min.
   (c) Heat-shock the bacteria at 42 °C for 1 min, in a water bath (*see* **Note 7**).
   (d) Immediately, keep the cells on ice for 2 min.
   (e) Add 500 μL of SOC or LB.
   (f) Incubate the cells at 37 °C for 30 min on a rotating shaker with a speed of 180 rpm.
   (g) Spin down the cells at 15,000 × *g* very briefly and remove the supernatant.
   (h) With a gentle pipetting, resuspend the cell pellet in fresh 50–100 μL of the same media.
   (i) Spread the cell suspension on an LB agar-ampicillin plate using a cell spreader.
   (j) Incubate the plate at 37 °C overnight.

(k) The next day, pick up the colonies with a sterile pipette tip and start the culture of each colony in a separate 50 mL tube containing 11 mL LB-ampicillin.

(l) Incubate the culture tubes at 37 °C overnight.

(m) The next day, reserve 1 mL culture for colony PCR and the rest of the culture for plasmid DNA extraction.

10. Colony PCR.

(a) In a 1.5 mL microtube, spin down the bacterial culture at 17,900 × *g* for 1 min and discard the supernatant.

(b) Resuspend the cell pellet in 1 mL molecular grade water and repeat centrifugation. Discard the supernatant again.

(c) Resuspend the pellet in 500 μL of molecular grade water.

(d) For each clone, set up two PCR, one with 0.5 μL bacterial suspension in place of template, and the other with 1.0 μL. Use the same primers and components described earlier. Instead of 50 μL final volume, half the reaction volume using half the amount of each component.

(e) Amplify the plasmid DNA from the bacterial clones using same thermocycling conditions as before, 95 °C: 5 min, [95 °C: 30 s, 55 °C: 30 s, 72 °C: 2 min] × 35 cycles, 72 °C: 10 min.

(f) Perform gel electrophoresis to visualize the PCR product. The product of the correct size confirms the clone to be positive for NGB. The clone can be used for cell transfections to express NGB protein that will be packed in exosomes.

11. Extraction of plasmid DNA from an overnight bacterial culture of NGB clone.

(a) Centrifuge the remaining 10 mL overnight bacterial culture at 5400 × *g* for 10 min, at 4 °C.

(b) Resuspend the cell pellet in 1 mL of molecular grade water and divide equally in separate microtubes. All the tubes are processed simultaneously, in the same way.

(c) Centrifuge the tubes at 6800 × *g* for 1 min at room temperature.

(d) Resuspend the bacterial pellet in 250 μL of buffer P1.

(e) Add 250 μL of buffer P2 and mix by inverting the tube 4–6 times. Do not use micropipette to mix. Do not exceed the time of lysis over 5 min.

(f) Add 350 μL of buffer N3 and mix again by inverting the tube 6 times.

(g) Centrifuge the tubes at 17,900 × *g* for 10 min, at 4 °C.

(h) Apply 750–800 μL of supernatant to QIAprep 2.0 spin columns and centrifuge the columns at 17,900 × *g* for 1 min.

(i) Discard the flow-through and apply 750 μL of buffer PE followed by centrifugation at 17,900 × *g* for 1 min.

(j) Discard the flow-through and repeat centrifugation to remove traces of buffer PE and ethanol.

(k) Transfer the column to a new microtube and add 50 μL buffer EB carefully at the center of the column. Avoid touching the microtip to the silica membrane in the column.

(l) Incubate the column at room temperature for 5 min.

(m) Centrifuge at 17,900 × *g* for 2 min to elute DNA from the column.

(n) The NGB clone is ready to be used for transfection.

Transfection of HEK293 Cell Line with NGB-XPack Clone

1. Refer to our previously published protocol for Lipofectamine® 2000-mediated transfection of HEK293 cells and precipitation as well as CD63 immunocapture of exosomes [9].

Immunocytochemistry (ICC) of Transfected HEK293 to Detect NGB Protein Expressed in Cytoplasm

1. Fixing cells.
   (a) Aspirate and discard the culture media from each well.
   (b) Add ice-cold 100% methanol to cover the cells completely.
   (c) Allow cells to fix for 15 min at −20 °C.
   (d) Without touching the micropipette tip to the bottom of the well, aspirate methanol (*see* **Note 8**).
   (e) To remove alcohol and rehydrate the cells, add 1 mL of 1× PBS to each well.
   (f) Keep the plate on shaker for 5 min at room temperature, with gentle rotation.
   (g) Without disturbing the cells, remove PBS and repeat the step two more times.
   (h) After the last wash, add 1 mL of 1× PBS until the cells are ready for immunostaining.
2. Immunostaining.
   (a) Vortex the antibodies to mix and collect with a short spin.
   (b) Dilute the primary antibody in dilution buffer according to the manufacturer's suggestion, or experimental requirements.
   (c) Carefully remove the PBS from the well.
   (d) Add enough volume of blocking buffer to cover the cells and incubate at room temperature for 1 h.

(e) Tilting the plate slightly, collect the buffer to the edge of the well and aspirate without touching the cells at the bottom.

(f) Apply diluted primary antibody to the cells and incubate overnight at 4 °C. Keep a few wells with no primary antibody treatment, as control samples.

(g) The next day, rinse the cells three times with 1× PBS with 5-min incubation in PBS each time.

(h) Dilute the secondary antibody in antibody dilution buffer using manufacturer's suggestion.

(i) Overlay the secondary antibody on the cells, including the controls, and incubate for 1–2 h at room temperature. Keep the samples in dark to avoid photobleaching of the fluorophore.

(j) Rinse off the antibody with PBS as step "g."

(k) Add a drop of DAPI solution and incubate in dark for 1–2 min.

(l) Wash the cells twice with PBS and finally add enough volume to keep the cells immersed for longer hours during fluorescent microscopy.

(m) The fixed and ICC-stained can be stored at 4 °C for up to 4 weeks.

Dot Blot for the Detection of NGB Protein in HEK293 Exosomes

1. Extraction of total exosomal proteins.
   (a) To 100 μL exosomal suspension of control, and the NGB-packing exosomes, add 67 μL of 20% PEG.
   (b) Centrifuge at 14,000*g* for 5 min, at 4 °C to pellet exosomes.
   (c) Resuspend the exosomes in 100 μL RIPA buffer containing protease and phosphatase inhibitors.
   (d) Vortex the samples at 4 °C for 40 min.
   (e) Centrifuge the samples at 4 °C at 14,000*g* for 20 min.
   (f) Collect 100 μL supernatant containing protein fraction for quantification.
2. Membrane preparation. All the PVDF or nitrocellulose membrane treatments are conducted in a box slightly bigger than the membrane size.
   (a) On the membrane, draw grid by pencil to indicate the region you are going to blot.
   (b) Activate the PVDF membrane by incubating in 100% methanol and nitrocellulose membrane in TBS for 5 min at room temperature for optimal protein binding.

(c) Wash the membrane in double-distilled water twice, for 5 min each time, to make sure traces of methanol are gone, in case of PVDF membrane.

(d) Using narrow-mouth pipette tip, spot 2 μL of samples onto the PVDF membrane at the center of the grid. To minimize the area of sample diffusion beyond 3–4 mm diameter, apply it slowly and carefully.

(e) Let the membrane dry at room temperature. Takes usually 20–30 min.

3. Blocking and washing.

   (a) Block nonspecific sites by soaking the membrane in 5% skim milk powder in TBST solution for 1 h, at room temperature with gentle shaking.

   (b) Wash the membrane 3 times in TBST, each time for 4 min.

   (c) Repeat the wash with TBS for the same duration.

4. Primary antibody incubation and wash.

   (a) Incubate the membrane with NGB primary antibody for 1 h, at room temperature. The antibody is diluted 1:500 with solution of 5% BSA in TBST. (The dilution factor may change depending upon the antibody or requirement of the experiment.) Intermittently, check the blot to make sure that the membrane is covered in the antibody solution.

   (b) Wash the membrane three times with TBST, each time keep in TBST for 5 min.

5. Secondary antibody incubation and wash.

   (a) Incubate the membrane with secondary antibody conjugated with HRP for 30 min at room temperature. The antibody is diluted 1:10,000 in 5% BSA and TBST solution.

   (b) Wash the membrane three times with TBST (for 15 min the first time, and twice for 5 min).

   (c) Wash the membrane once with TBS for 5 min.

6. Detection and imaging.

   (a) Mix solutions A and B of the ECL detection reagent in 1:1 ratio.

   (b) Using tweezers, tap the edge of the membrane on a Kimwipe to remove drippings of TBS and transfer the membrane to a clean box.

   (c) Overlay the detection reagent to cover the membrane (blot) evenly.

(d) Incubate the membrane in dark, at room temperature, for 5 min.

(e) Once again, lift the membrane with tweezers, drain off the dripping reagent on a Kimwipe, and image the blot.

(f) Image the blot with an imager that detects chemiluminescence. We have used ChemiDoc imaging system from BIO-RAD (Catalog # 17001401) in Fig. 3d.

Transfection of HEK293 Cells with XPack CMV-XP-RFP-EF1-α-Puro Expression Lentivector, Collection of RFP-Filled Exosomes, Treatment of Neural Stem Cells with RFP Exosomes, and Subsequent Confocal Microscopy [9] (Fig. 4).

3.1.3 Extrinsic Loading of Proteins to Exosomes via Exosome Surface Engineering (Click Chemistry) (Fig. 5)

Click Chemistry

1. HEK293 cells and exosomes. [The protocols for HEK293 cell culture, Lipofectamine-mediated transfection with XPack-GFP plasmid (System Biosciences SBI catalog # XPak-530-PA-1), exosome precipitation, and purification are described in previously published chapter [9]].
2. Immunofluorescence assay on EV samples to confirm the presence of exosomes.

(a) Coverslips were pretreated with poly-L-ornithine by keeping them submerged in the solution at 37 °C for 24 h.

(b) The next day, the coverslips were washed 3 times with 1× PBS. Adding the PBS on the top of the coverslips, gently shaking it a couple of times and aspirating it should suffice.

(c) Add 10 μL of exosomal samples to a coverslip and incubate for 24 h at 4 °C to allow them to bind to the coated coverslips (*see* **Note 9**).

(d) By gently adding PBS to the container of coverslips and submerging them with PBS, wash the coverslips 3 times.

(e) Fix the exosomes by submerging the coverslips in 4% PFA.

(f) Incubate at room temperature for 20 min.

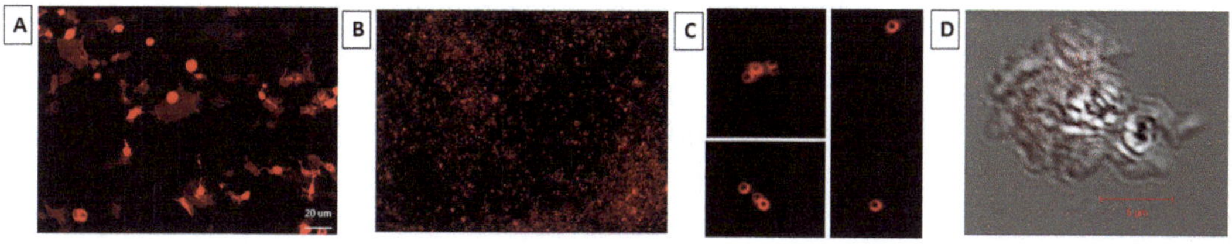

Fig. 4 Targeted delivery of RFP using exosome-packing tag (**a**, **b**) HEK293 transfected with XPAK-RFP, (**b**) CD63^{+} exosomes packing RFP, and (**c**) neural stem cell neurospheres treated with RFP (confocal micrograph)

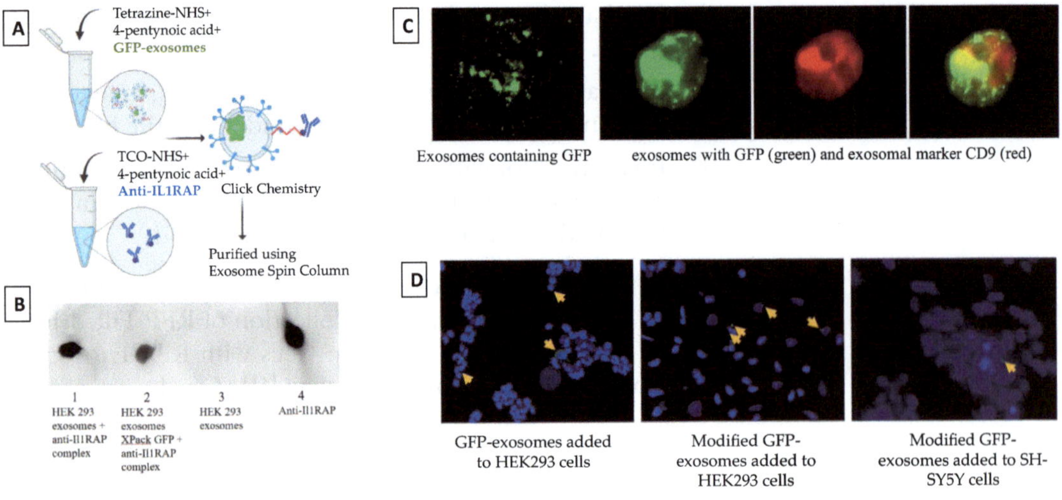

Fig. 5 Click chemistry-mediated surface-engineered exosomes deliver anti-Il1RAP antibody. (**a**) Diagrammatic representation for surface modification of exosomes. (**b**) Exosomal dot blot confirming the presence of anti-Il1RAP antibody associated with engineered exosomes. (**c**) Fluorescent microscopic image of GFP-packed exosomes immuno-stained for exosomal marker protein CD9. (**d**) Exosome uptake analysis and functional confirmation of exosomal cargo delivery

(g) Treat the exosomal samples with blocking/permeabilizing solution (0.1% Triton, 5% BSA in PBS) for 1 h at room temperature.

(h) Incubate the exosomal samples with CD9 antibody, at dilution 1:500 overnight at 4 °C.

(i) Repeat the washing step with 1× PBS.

(j) Treat the exosomes with secondary antibody (rhodamine TRITC AffiniPure goat anti-mouse IgG H + L, at dilution 1:200) for 1 h at room temperature.

(k) To remove unbound antibodies, repeat the washing step with 1× PBS.

(l) Image the exosomes under fluorescence microscope for GFP and rhodamine at λEx/λEm 488/509 nm and 546/568 nm, respectively.

3. Extraction and quantification of exosomal proteins to assess exosomal suspension volume to be used for Click chemistry-mediated surface engineering.

(a) To 100 μL exosomal suspension of control, and the GFP-packing exosomes, add 67 μL of 20% PEG.

(b) Centrifuge at 14,000*g* for 5 min, at 4 °C to pellet exosomes.

   (c) Resuspend both, control, and the GFP-packing exosomes, in 100 μL RIPA buffer containing protease and phosphatase inhibitors.
   (d) Vortex the samples at 4 °C for 40 min.
   (e) Centrifuge the samples at 4 °C at 14,000*g* for 20 min.
   (f) Collect 100 μL supernatant containing protein fraction for quantification.
   (g) Using Pierce BCA Protein Assay Kit, and following the user guide, create a BSA standard curve to quantify the exosomal proteins.
   (h) Based on the standard curve, the concentrations of the exosome samples are calculated. [For example, we had concentrations of 733 μg/mL for the control exosomes and 955 μg/mL for the GFP-packed exosomes].

4. Modification of Exosomal Surface.
   (a) Add 70 mg, 0.6 mmol of tetrazine n-hydroxysuccinimide (NHS) ester and 58 mg, 0.6 mmol of 4-pentynoic acid to 2 mL of PBS.
   (b) Adjust the pH to 7.4 with sodium bicarbonate and stir the solution on ice for 1 h.
   (c) Add 92 mg, 0.6 mmol of 1-ethyl-3-(3-carbodiimide dimethylaminopropyl) to the solution and stir on ice for 1 h.
   (d) Add 4 μL of this reaction mixture to each, 160 μg of control, and 160 μg of GFP-packed exosome suspension in PBS.
   (e) Stirred the mixtures for 24 h at 4 °C.
   (f) Using a Sephadex G-25 column conditioned with 1× PBS, purify the exosomes and get rid of the excess reaction solution.
   (g) The final concentration for the control and the GFP-packed samples is 5 mg/mL exosomes, 2 mM tetrazine-NHS-ester.

5. Modification of anti-Il1RAP.
   (a) Add 70 mg, 0.6 mmol of TCO-PEG4 n-hydroxysuccinimide (NHS) ester and 58 mg, 0.6 mmol of 4-pentynoic acid to 2 mL of PBS.
   (b) Adjust the pH to 7.4 with sodium bicarbonate and stir the resulting solution on ice for 1 h.
   (c) Add 92 mg, 0.6 mmol of 1-ethyl-3-(3-carbodiimide dimethylaminopropyl) to the reaction mixture and stir the solution on ice for 1 h.

(d) Add 4 μL of this reaction mixture to 20 μL of anti-IL1RAP [rabbit polyclonal antibody to IL1RAP, isotype IgG, unconjugated, biorbyt orb 184435] and stir for 24 h at 4 °C.

(e) Purify the antibody and get rid of the excess reaction mix using a Sephadex G-25 column conditioned with 1× PBS.

(f) The final concentration of the modified antibody is 5 mg/mL, 2 mM TCO-NHS-Ester.

6. Click chemistry.

   (a) In a 1.5 mL microcentrifuge tube have conditioned control exosomes and GFP-packed exosomes, each in the suspension of 150 μL PBS.

   (b) To each exosome sample, add the following in a stepwise manner, at room temperature: 7 μL of 0.32 M copper (II) sulfate pentahydrate, 35.5 μL of 1.44 M L-ascorbic acid (pH buffered to 7.4 with sodium bicarbonate), 16.4 μL of 0.27 M bathophenanthrolinedisulfonic acid disodium salt trihydrate, and all of the conditioned antibody.

   (c) Add the conditioned antibodies and exosomes in 1:2 mol equivalents, respectively. Dissolve all the reactants in PBS before use. Stir the solutions at room temperature for 3 h.

   (d) Run the samples through a Sephadex G-25 column conditioned with PBS.

Dot Blot to Confirm Presence of Antibody on Exosomal Surface.

1. Membrane conditioning.

   (a) Activate the PVDF membrane by incubating in 100% methanol for 5 min at room temperature for optimal protein binding.

   (b) Wash the membrane in double-distilled water twice, for 5 min each time, to remove methanol traces.

2. Sample loading.

   (a) Apply the following samples to the PVDF membrane in distant spots: 20 μL of the control click chemistry final reaction sample, 20 μL of the GFP-packed click chemistry final reaction sample, 20 μL HEK 293 [no modification treatment] exosomes in RIPA buffer, and 10 μL anti-IL1RAP (nonmodified) antibody.

   (b) Allow the membrane to dry for 20 min at room temperature.

3. Blocking and washing.

   (a) Treat the membrane with blocking buffer containing 5% bovine serum albumin (BSA) in Tris-buffered saline with 0.1% Tween-20 (TBST).

(b) Incubate for 1 h at room temperature on an orbital shaker (80 rpm).

(c) Wash the membrane 3 times with 1× TBST. During each wash, keep the membrane on orbital shaker for 5 min.

4. Antibody incubation and wash.

(a) Incubate the membrane with 5 mL of primary antibody (1:10,000 dilution of anti-rabbit in blocking solution, horseradish peroxidase) for 1.5 h on orbital shaker, at room temperature.

(b) Repeat the washing step as before.

5. Detection and imaging

(a) Mix 2 mL SuperSignal West Femto Stable Peroxide Buffer and 2 mL SuperSignal West Femto Luminol/Enhancer Solution to prepare ECL reagent.

(b) Incubate the membrane ECL solution for 5 min on benchtop.

(c) We have imaged the membrane using a ChemiDocTM MP Imaging System Chemiluminescence application with Rapid Auto-exposure, no illumination, and 647SP Filter.

3.2 Enhanced and Targeted Localization of EVs/Exosomes to Specific Cells or Organs (Fig. 6)

3.2.1 Adding Peptide Tags BHP1 (Brain Homing Peptide 1) and NCAM (Neural Adhesion Molecule) to Target EV/Exosome Localization

1. Cell culture-HEK293 in T75 flasks [9].

2. Co-transfection of HEK293 cells using Lipofectamine 2000 and PEG-NaCl-mediated exosome precipitation [9].

(a) Use 2.5 μg of each, XPack CMV-XP-GFP-EF1α-Puro Expression Lentivector, and XStamp™-NCAM Lentivector for co-transfection for the first HEK293 flask.

(b) Use 2.5 μg of each, XPack CMV-XP-GFP-EF1α-Puro Expression Lentivector, and XStamp™-BHP1 Lentivector for co-transfecting HEK293 cultured in a separate flask.

(c) Keep one HEK293 culture flask without transfection to collect exosomes for control exosomal treatment.

(d) Collect the conditioned media in each flask separately and precipitate the exosomes [9].

3. Cell culture of NSC and ADSC.

(a) Thaw Matrigel on ice at 4 °C, overnight.

(b) Using pre-chilled pipette and tubes, mix 67.75 μL Matrigel stock with 6 mL of 1× PBS. The PBS should be at refrigeration temperature as well because Matrigel congeals above 10 °C.

(c) Add 45 μL of Matrigel mix to each well of the 8-chamber slide and incubate overnight at 37 °C and 5% CO_2.

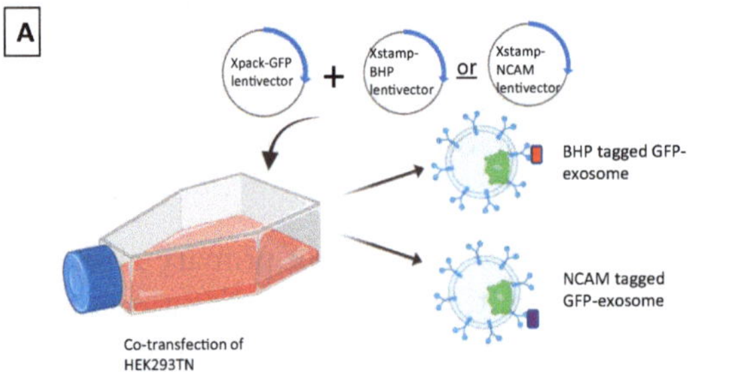

B

	NSC	ADSC
Control	-	-
GFP	+	+
GFP + BHP	++	++
GFP + NCAM	++	+

NSC and ADSC: Comparison of fluorescence intensity

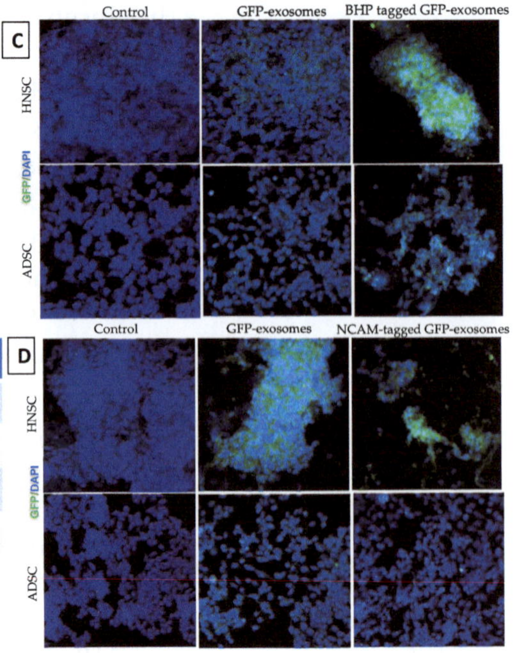

Fig. 6 Enhanced and targeted localization of EVs/exosomes. Adding peptide tags BHP and NCAM to target EV/exosome localization to NSC and ADSC. (**a**) Schematic of addition of brain and neuron localization tag to exosomes. (**b**) Comparison of fluorescence intensity between NSC and ADSC. (**c**) Addition of BHP tag to exosomes and their localization to NSC and ADSC. (**d**) Addition of NCAM tag to exosomes and their localization to NSC and ADSC

(d) The next day, remove the supernatant PBS, without puncturing the Matrigel layer at the bottom.

(e) Seed 10,000 cells of NSC and ADSC on the Matrigel layer with NESC and ADSC media, respectively, and incubate at 37 °C and 5% CO_2 until undefined differentiation.

4. Exosome treatment of NSC and ADSC

(a) Replace old culture media from each chamber of NSC and ADSC cultures with fresh media.

(b) Add 100 μL of exosomal suspension to each chamber. NSC and ADSC should receive exosomes of NCAM, BHP, and control in separate chambers. Do not mix the exosome types.

(c) Continue the incubation for the next 48 h.

5. Fixing cells

(a) Remove culture media from each chamber of the slide.

(b) Add 1× PBA to the chamber and gently shake the slide at room temperature for 5 min.

(c) Remove the PBS and cover the cells with 4% PFA.

(d) Incubate the cells in the chamber slide at room temperature for 10 min.

(e) Remove the PFA and carefully remove the chamber grid.

(f) Add a drop of DAPI-mounting media.

(g) Place a coverslip and seal with a clear microscope slide sealer.

6. Microscopy

(a) Visualize the slide under fluorescent microscope.

(b) The GFP has λEx/λEm of 488/509 nm, and the DAPI can be imaged at λEx/λEm 360/460 nm.

3.3 Visualization of EV/Exosome Uptake and Internalization (Fig. 7)

3.3.1 Staining the Cells Using Fluorescent Lipophilic Dyes

1. Cell culture.

(a) Culture GBM and NSC in suspension flasks [9].

2. Lipophilic staining (*see* **Note 10**).

(a) Transfer the NSCs and GBM cells from the stock flasks to 1 mL conical tubes and singularize the neurospheres with gentle, repeated aspiration.

(b) Centrifuge the cell suspensions at 0.3 rcf for 5 min at room temperature.

(c) Resuspend each cell pellet in 1.5 mL of fresh NSC media.

(d) Treat the cells with 1 uM dye. GBM are dyed red with DiI, and the NSC are dyed green with DiO so that the EVs secreted by each cell line take the color of the parent cell [14]. Reserve untreated cells as controls.

(e) Mix the dye with the cell suspension with brief and repeated pipetting.

(f) Seeded the cells along with the dye and the culture media in sterile 35 mm Petri dishes.

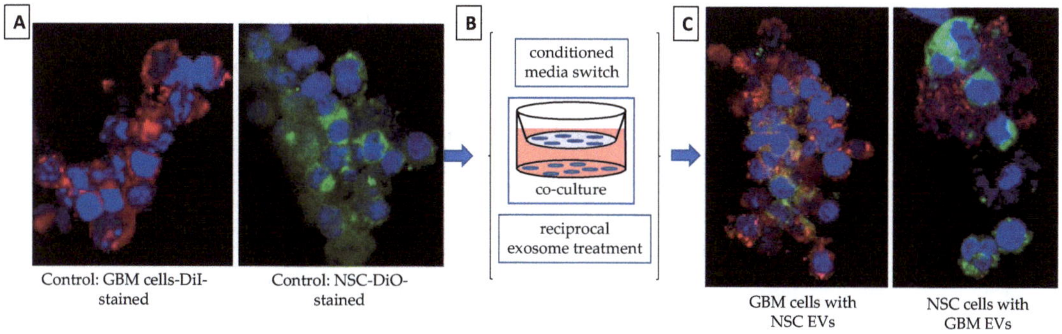

Fig. 7 Confocal micrographs of EV/exosome uptake and internalization. (**a**) Control: DiI-stained GBM (red) and DiO-stained NSC (green), with DAPI-stained nuclei (blue). (**b**) Methods of EV/exosome treatment. (**c**) Internalization of NSC-secreted green EVs by GBM. GBM-secreted red EVs internalized by green NSC

(g) Incubate the cells for 2 h at 37 °C on a rotating shaker at 30 rpm.

(h) Centrifuge the cells with a brief spin and discard the supernatant.

(i) Resuspended the pellet of dyed cells in 3 mL NSC media and divide NSC and GBM for 3 distinct methods of EV treatment.

 (i) Conditioned media switch.

 (ii) Coculture.

 (iii) Reciprocal exosome treatment.

(j) Culture the NSC and GBM separately in 6-well nontissue culture-treated plates with 3 mL fresh NSC media.

(k) Incubate the cells at 37 °C for 24 h with 5% CO_2.

3.3.2 Coculture, Conditioned Media Exchange, or a Direct EV/ Exosome Treatment

1. Conditioned media exchange.

 (a) Seed 300,000 GBM and NSC separately in each well of a 6-well, nontissue culture-treated plate.

 (b) After 24 h, collect the cells along with the conditioned media.

 (c) Centrifuge the cells briefly.

 (d) Collect the conditioned media from each cell type and keep it warm at 37 °C.

 (e) Resuspend the GBM cell pellet in the conditioned media of NSC.

 (f) Similarly, resuspend the NSC cell pellet in the conditioned media from GBM.

 (g) Continue the incubation for another 24 h.

2. Coculture.

 (a) In a 6-well nontreated plate, add 2 mL of fresh NSC media along with 150,000 DiO-stained NSC.

 (b) Place a cell insert of 0.4 micrometer pore size in each well, making sure that the bottom of the insert is completely immersed in the media.

 (c) Incubate the plate-insert assembly at 37 °C for 2–4 h with 5% CO_2.

 (d) Remove the plate and add 200,000 of DiI-stained GBM, resuspended in 1 mL of NSC media.

 (e) Continue the incubation for another 24 h.

3. Reciprocal EV/exosome treatment.

 (a) Seed 300,000 GBM and NSC separately in each well of a 6-well, nontissue culture-treated plate.

(b) After 24 h, briefly spin down the cells and collect the supernatant for EV precipitation.

(c) Resuspend the cell pellets in fresh NSC media and return the plate to incubator for continued incubation.

(d) Collect the EVs using PEG/NaCl precipitation reagent. If exosomes are to be used, further process the EV sample using CD63-coated magnetic beads and other exosome purification methods [9].

(e) Treat the dyed GBM and NSC with collected exosomal/EV suspension. Make sure that the GBM are treated with NSC exosomes and vice versa.

(f) Continue the incubation for another 24 h.

3.3.3 Confocal Microscopy for Confirmation of EV/Exosome Uptake and Internalization

1. Fixing cells.
   (a) Centrifuge the cells at 0.3 rcf for 5 min, at room temperature.
   (b) Resuspend the cell pellet in 400 μL of 4% PFA and incubated at room temperature for ~15 min.
   (c) With another round of centrifugation, discard PFA and resuspend the cell pellet in 500 μL ice-cold 1× PBS.
   (d) Repeat the washing process twice to get rid of traces of PFA.
   (e) Finally, resuspend the cell pellet in 50 μL of 1× PBS.
   (f) Drop 5 μL the fixed cell suspension on a clean microscope slide.
2. Slide preparation.
   (a) Add a drop of DAPI staining solution to the cell suspension on the slide (*see* **Note 11**).
   (b) Place a coverslip on the cells, carefully, avoiding air-bubble formation.
   (c) Seal the edges of the coverslip with a clear sealant.
   (d) Protect the slide from light and store at 4 °C until ready for fluorescent microscopy.

4 Notes

1. HEK293 and neural stem cell culture methods are described in detail in our previously published book chapter [9].
2. The PBS free from calcium and magnesium is recommended for mammalian cell culture.
3. The process of Lipofectamine® 2000-mediated transfection of HEK293 cells and precipitation as well as subsequent

purification of RFP-packed exosomes has been previously described in detail [9]. The chapter also details the immunoaffinity purification of pre-enriched exosomes as well as the recipes for NSC and HEK cell culture media.

4. Green buffers to be used with both, polymerase enzyme and digestion enzyme, have proprietary formulations containing blue and yellow dyes. Do not add DNA loading dye if green buffers are used in PCR or digestion reactions instead of colorless buffers.
5. The PCR product can either be purified directly after thermocycling <u>or</u> it can be eluted from the agarose gel after visual confirmation via gel electrophoresis. Either way, the PCR product must be cleaned before digestion with restriction enzymes.
6. Stbl3 is a preferred E. coli strain to host, maintain, and replicate lentiviral plasmids [15].
7. For an efficient transformation, the water bath must be at 42 °C.
8. Do not pour methanol down the drain. Collect it in a clearly labeled container and discard it as hazardous waste according to the local regulatory guideline.
9. The exosomes are not quantified before adding to the coverslip. 10 μL exosomal suspension from 300 μL stock should be sufficient for each coverslip. The exosomal stock is prepared from 10 mL of 24-h conditioned media collected from ~90% confluent T-75 flask of HEK293 cells.
10. For all the three methods of EV transfer, i.e., coculture, conditioned media switch as well as reciprocal EV treatment, the staining method is exactly the same.
11. Do not use a mounting medium. Organic solvents, including the polar organic solvent such as glycerol present in the mounting media, would solubilize the lipophilic dyes, resulting in a high background while imaging.

References

1. Doyle LM, Wang MZ (2019) Overview of extracellular vesicles, their origin, composition, purpose, and methods for exosome isolation and analysis. Cell 8(7):727
2. Kanada M, Bachmann MH, Hardy JW, Frimannson DO, Bronsart L, Wang A et al (2015) Differential fates of biomolecules delivered to target cells via extracellular vesicles. Proc Natl Acad Sci 112(12):E1433–E1442
3. Banks WA, Sharma P, Bullock KM, Hansen KM, Ludwig N, Whiteside TL (2020) Transport of extracellular vesicles across the blood-brain barrier: brain pharmacokinetics and effects of inflammation. Int J Mol Sci 21(12):4407
4. Vaidya M, Sreerama S, Gaviria M, Sugaya K (2022) Exposure to a pathological condition may be required for the cells to secrete exosomes containing mtDNA aberration. J Nucleic Acids 2022:1
5. Rajagopal C, Harikumar K (2018) The origin and functions of exosomes in cancer. Front Oncol 8:66

6. Edgar JR (2016) Q&A: what are exosomes, exactly? BMC Biol 14(1):1–7
7. Massaro C, Sgueglia G, Frattolillo V, Baglio SR, Altucci L, Dell'Aversana C (2020) Extracellular vesicle-based nucleic acid delivery: current advances and future perspectives in cancer therapeutic strategies. Pharmaceutics 12(10): 980
8. Busatto S, Iannotta D, Walker SA, Di Marzio L, Wolfram J (2021) A simple and quick method for loading proteins in extracellular vesicles. Pharmaceuticals 14(4):356
9. Vaidya M, Sugaya K (2021) Methods for the detection of circulating pseudogenes and their use as cancer biomarkers, Pseudogenes. Springer, pp 339–360
10. Ciccone L, Nencetti S, Socci S, Orlandini E (2021) Neuroglobin and neuroprotection: the role of natural and synthetic compounds in neuroglobin pharmacological induction. Neural Regen Res 16(12):2353
11. Fiocchetti M, Cracco P, Montalesi E, Fernandez VS, Stuart JA, Marino M (2021) Neuroglobin and mitochondria: the impact on neurodegenerative diseases. Arch Biochem Biophys 701:108823
12. Smyth T, Petrova K, Payton NM, Persaud I, Redzic JS, Graner MW et al (2014) Surface functionalization of exosomes using click chemistry. Bioconjug Chem 25(10): 1777–1784
13. Bandyopadhyay S (2021) Role of neuron and glia in Alzheimer's disease and associated vascular dysfunction. Front Aging Neurosci 13: 274
14. Dörsam B, Bösl T, Reiners KS, Barnert S, Schubert R, Shatnyeva O et al (2018) Hodgkin lymphoma-derived extracellular vesicles change the secretome of fibroblasts toward a CAF phenotype. Front Immunol 9:1358
15. Al-Allaf FA, Tolmachov OE, Zambetti LP, Tchetchelnitski V, Mehmet H (2013) Remarkable stability of an instability-prone lentiviral vector plasmid in Escherichia coli Stbl3. 3 Biotech 3(1):61–70

Chapter 14

Quantitative Analysis of Extracellular Vesicle Release Using Artificial MicroRNAs

Albert Lu

Abstract

Despite the widely used concept of extracellular vesicle (EV)-mediated intercellular communication, we are still far from understanding what is the exact role of such nanosized vesicles in human physiology and disease. Thus, development of new methods and tools that enable the study of fundamental EV biology is valuable for advancing the field. Typically, EV production and release are monitored using approaches that rely on either antibody-based FACS assays or genetically encoded fluorescent proteins. We previously devised artificially barcoded exosomal microRNAs (bEXOmiRs) that were used as high-throughput reporters of EV release. In the first part of this protocol, basic steps and considerations for the design and cloning of bEXOmiRs are explained in detail. Next, analysis of bEXOmiR expression and abundance in cells and isolated EVs is described.

Key words Artificial microRNAs, bEXOmiRs, Extracellular vesicles, Exosomes, Stem-loop RT-PCR

Abbreviations

bEXOmiRs	barcoded exosomal microRNAs
EV	extracellular vesicle
miRNAs	microRNAs
nt	Nucleotides
RT	Reverse transcription

1 Introduction

Extracellular vesicles (EVs), including exosomes, mediate transfer of biological molecules such as nucleic acids between neighboring and distant cells. Many recent reports suggest that EVs may play important roles in both normal physiology and the pathogenesis of multiple disease states. However, our understanding of the underlying molecular mechanisms by which the different and highly

Seppo Vainio (ed.), *Cell-Secreted Vesicles: Methods and Protocols*, Methods in Molecular Biology, vol. 2668,
https://doi.org/10.1007/978-1-0716-3203-1_14,

heterogeneous EVs are generated, released, and taken up by acceptor cells is still limited [1, 2]. Despite recent advances, progress in the field has been hampered due to the lack of isolation methods enabling homogeneous purification of a specific type of EV or the development of high-throughput approaches to study the different aspects of their biology [3].

EVs contain a wide range of bioactive cargoes either in their lumen or associated to their enclosing lipid bilayer, which appear to induce biological effects upon delivery in acceptor cells. EV cargoes include antigens, signaling proteins, lipids, and nucleic acids such as microRNAs (miRNAs) [1]. Of particular interest are the multiple roles that EV-miRNAs play in different disease contexts, including cancer [4, 5]. For example, miRNAs released in EVs derived from both tumor and mesenchymal cells are thought to promote metastasis by multiple mechanisms [4–7]. Moreover, EV-associated miRNAs secreted by hypothalamic stem cells have been proposed to regulate aging in mice [8]. Recent studies have identified a number of RNA-binding proteins that have been proposed to play a role in targeting specific subsets of miRNAs into EVs [5]. Some of these proteins, including hnRNPA2B1, SYNCRIP, or Lupus La protein or FMR1 among others, have been proposed to regulate EV sorting of miRNAs by binding to specific motifs found in the mature miRNA sequence [9–13]. On the other hand, alternative EV-independent secretory pathways may also participate in miRNA export [14]. Nevertheless, how miRNAs and other EV cargoes are specifically packaged and released by EVs remains poorly understood.

In an effort to further our understanding of EV biology, we recently developed a genome-wide CRISPR/Cas9 screening approach to systematically identify factors involved in the regulation of fundamental aspects of EVs, including biogenesis, cargo packaging, or release [15]. This strategy used unprecedented molecular barcoding technology to monitor EV release at high-throughput level. Taking advantage of current knowledge of EV-mediated miRNA export, we engineered artificially barcoded exosomal miRNAs (bEXOmiRs) that could be expressed, targeted, and exported in EVs. bEXOmiRs were designed to contain a 15-nucleotide (nt) random sequence (barcode) at the base of the miRNA stem region (*see* Fig. 1), followed by a constant 7 nt motif (EXO motif) previously reported to promote hnRNPA2B1-mediated sorting of miRNAs into exosomes [9]. bEXOmiRs are transcribed, processed into 22 nt single-stranded RNAs (like other endogenous miRNAs), and released in EVs (*see* Fig. 1). This chapter describes the basic steps needed for the design and expression of bEXOmiRs using cultured cells followed by quantification of reporter abundance in EV fractions. These artificial miRNA reporters can be used in many applications ranging from monitoring EV release or cargo sorting, to discovery of motifs driving selective miRNA export.

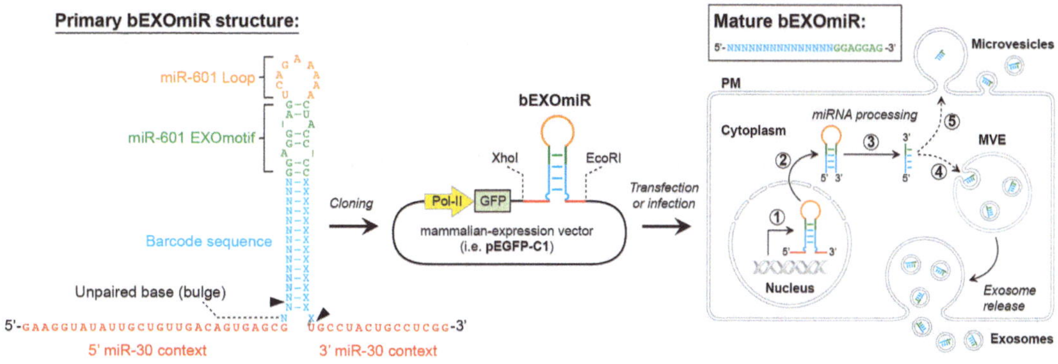

Fig. 1 Design and expression of artificially barcoded exosomal microRNAs (bEXOmiRs). (Left, panel) detailed design of bEXOmiR reporters. The double-stranded stem region contains a 15 nt-random sequence (barcode; highlighted in blue "N" and "X" letters) with perfect base complementarity except for the first nucleotide pair at the base (bulge). The barcode sequence is followed by a constant 7 nt exosome-targeting motif (EXOmotif, green) that contains two asymmetric bulges (one in each strand). The EXOmotif and loop sequence (orange) were derived from endogenous miR-601. Finally, miR-30 context sequences (red) are flanking the base of the stem region. "N" or "X" indicates any ribonucleotide base. (Middle panel) schematic of the pEGFP-C1 mammalian expression plasmid used for cloning bEXOmiRs. XhoI and EcoRI restriction sites are indicated. (Right panel) diagram of a bEXOmiR-expressing cell showing the canonical pathway of miRNA processing and EV sorting before release in EVs (either exosomes or microvesicles). **Step 1** shows transcription of primary bEXOmiRs in the nucleus where they are processed by the microprocessor machinery (Drosha and cofactors) before export (**step 2**) into the cytoplasm where maturation continues (**step 3**) by Dicer and Argonaute proteins. Single-stranded mature bEXOmiRs can be sorted into exosomes (**step 4**) or microvesicles (**step 5**) and subsequently released by cells. MVE, multivesicular endosomes; PM, plasma membrane

2 Materials

2.1 Design of an Artificially Barcoded Exosomal microRNAs (bEXOmiRs)

1. Custom DNA oligonucleotides (standard desalted) for bEXOmiR constructions (*see* **Note 1**).
2. Ultrapure water (Nuclease-free).

2.2 Cloning of bEXOmiRs in a Mammalian Expression Vector

1. bEXOmiR expression plasmid: pEGFP-C1 (widely available in most basic research laboratories).
2. Cloning enzymes: T4 Polynucleotide kinase (PNK), XhoI, EcoRI, T4 DNA ligase.
3. T4 PNK buffer.
4. Restriction enzyme buffers.
5. T4 DNA ligase buffer.
6. Ultrapure water (nuclease-free).
7. LB agar Miller plates for bacterial colony selection: 100 μg/mL carbenicillin, 25 μg/mL chloramphenicol, 10 g/L peptone, 5 g/L yeast extract, 10 g/L NaCl, 12 g/L agar.

8. LB Miller broth for bacterial cultures: 100 μg/mL carbenicillin, 25 μg/ mL chloramphenicol, 10 g/L peptone, 5 g/L yeast extract, 10 g/L NaCl.
9. NanoDrop 2000 UV-vis spectrophotometer.
10. Polyethylenimine (PEI; PolySciences Inc.).
11. QIAprep spin miniprep kit (Qiagen).
12. QIAquick PCR Purification Kit (Qiagen).

2.3 Analysis of bEXOmiR Expression and Abundance in Cells and Isolated EVs

1. Stem-loop primer (5′ > 3′): GTCGTATCCAGTGCAGGG TCCGAGGTATTCGCACTGGATACGACCTCCTC.
2. PCR bEXOmiR-specific primer (see protocol below for details).
3. PCR universal reverse primer (5′ > 3′): CCAGTGCAGGG TCCGAGGTA.
4. RT-PCR enzymes: SuperScript™ III reverse transcriptase (Invitrogen), Vent® DNA Polymerase (New England BioLabs), Phusion® High-Fidelity DNA Polymerase (New England BioLabs).
5. O'RangeRuler® 20 bp DNA Ladder (Thermo Fisher).
6. Ultrapure water (Nuclease-free).

2.4 EV Isolation Materials

1. Millex-GP Syringe Filter Unit, 0.22 μm (Millipore).
2. 70 mL, Polycarbonate Bottle Assembly, 38 × 102 mm (Beckman).
3. 1 mL, Open-Top Thickwall Polycarbonate Tube, 11 × 34 mm (Beckman).
4. Type 45 Ti Fixed-Angle Titanium Rotor (Beckman).
5. TLA-120.2 Fixed-Angle Rotor (Beckman).
6. Phosphate-buffered saline (PBS): 137 mM NaCl, 2.7 mM KCl, 10 mM Na2HPO4, 2 mM KH2PO4.

2.5 Cell Culture

1. Cells of interest: HEK293T and K562 cells.
2. Dulbecco's modified Eagle's medium (DMEM) for HEK293T cells (K562 cells use RPMI medium).
3. Fetal bovine serum (FBS).
4. Penicillin/streptomycin, 10,000 U/mL.
5. 0.05% (w/v) trypsin, 5 mM EDTA diluted in PBS.
6. Phosphate-buffered saline (PBS).

3 Methods

All procedures are carried out at room temperature under aseptic conditions using filter tips, unless otherwise specified. We recommend to set up PCRs under a hood properly equipped.

3.1 Design of an Artificially Barcoded Exosomal microRNAs (bEXOmiRs)

The design of bEXOmiRs uses miR-30-based sequences and features [15], together with the loop sequence and exosome-targeting motif (EXOmotif) of the EV-enriched miR-601 [9]. A random stretch of 15 nucleotides (nt), so-called barcode sequence, is placed at the 5′ end of the stem region (*see* Fig. 1). After intracellular processing, the resulting 22 nt-long mature bEXOmiR is predicted to comprise the barcode sequence followed by the EXOmotif (*see* **Note 1**).

1. Order the following pair of custom DNA oligonucleotides:

   TOP strand oligonucleotide (5′ > 3′):

   [TCGA**G**]**AAGAAGGTATATTGCTGTTGACAGTGAGCG** NNNNN
NNNNNNNNNNGGAGGAGTCAGAAAAACTACCCCXXXXXXXXXX
XXXXX**TGCCTACTGCCTCGGACTTCAAGGGG**TCAGTCA[G].

   BOTTOM strand oligonucleotide (5′ > 3′):

   [AATTC]TGACTGA**CCCCTTGAAGTCCGAGGCAGTAGGCA**
NNNNNNNNNNNNNNNGGGGTAGTTTTTCTGACTCCTCCXXXX
XXXXXXXXXXX**CGCTCACTGTCAACAGCAATATACCTTCTT**[**C**]

   (a) miR-30 context sequences are indicated in bold letters.

   (b) Insert a barcode sequence (15 nt random stretch) in positions denoted by "N" in the Top and Bottom strand oligonucleotides (*see* **Notes 2** and **3**), followed by inserting the corresponding reverse complement sequence in positions denoted by "X." After this, change the first letter of the "N" stretch of both TOP and BOTTOM strand oligonucleotides, as follows: If the first base of the barcode is A or T, change it to C. If the first base is C or G, convert it to A. This unpaired base between the complementary "N" and "X" stretches will generate a structural bulge in the final, folded miRNA necessary for initial processing (*see* Fig. 1 and **Note 4**).

   (c) Letters between brackets correspond to digested XhoI (**CTCGAG**) and EcoRI (**GAATTC**) restriction site sequences (used later for cloning, see below) resulting after digestion (restriction sites can vary depending on the target mammalian expression vector used for cloning). Note that a G of the digested XhoI site overlaps with the 5′-miR-30 context sequence.

Table 1
5' Phosphorylation of Oligonucleotides

Components	Amount
100 μM oligonucleotide stock	2 μL
10x T4 DNA ligase buffer (*see* **Note 6**)	2 μL
T4 polynucleotide kinase (10 U/μL)	1 μL
Sterile nuclease-free water	15 μL
Total reaction volume	20 μL

Table 2
Annealing of Phosphorylated Oligonucleotides

Components	Amount
Phosphorylated TOP strand oligonucleotide	5 μL
Phosphorylated BOTTOM strand oligonucleotide	5 μL
Sterile nuclease-free water	90 μL
Total reaction volume	100 μL

3.2 Cloning of bEXOmiRs in a Mammalian Expression Vector

1. Dissolve lyophilized oligonucleotides with sterile nuclease-free water to a 100 μM stock.
2. Perform oligonucleotide 5′ phosphorylation (*see* Table 1 for reaction components; *see* **Note 5**).
3. Incubate at 37 °C for 1 hour followed by 20 min at 65 °C to heat-inactivate T4 PNK. Phosphorylated oligonucleotides can be stored at −20 °C until further use.
4. Perform annealing of phosphorylated oligonucleotides (*see* Table 2).
5. Heat a block up to 95 °C. Insert tubes in the hot block and incubate for 3 min (*see* **Note 7**). Immediately after this, sit the block on a room temperature bench and it cools down for 1 h. Then, chill tubes on ice. Annealed products can be stored at −20 °C until further use.
6. While performing oligonucleotide annealing, digest target mammalian expression vector (pEGFP-C1) with XhoI and EcoRI restriction enzymes for 2 h at 37 °C (*see* **Note 8**). Set up a reaction as shown in Table 3.
7. Purify digested plasmid vector using a PCR purification column and elute in 30 μL of sterile nuclease-free water.
8. Quantify by nanodrop DNA concentrations of both annealed oligonucleotides and digested pEGFP-C1 plasmid.

Table 3
Target Plasmid Digestion

Components	Amount
1 μg of purified plasmid vector (pEGFP-C1)	x μL
10X NEB 2.1 buffer	2 μL
EcoRI restriction enzyme (20 U/μL)	1 μL
XhoI restriction enzyme (20 U/μL)	1 μL
Sterile nuclease-free water	16-x μL
Total reaction volume	20 μL

Table 4
Ligation of Oligonucleotides into XhoI/ECoRI-Digested pEGFP-C1

Components	Amount
Annealed oligonucleotides stock	X μL
Digested pEGFP-C1	Y μL
10x T4 DNA ligase buffer	1 μL
T4 DNA ligase (10 U/μL)	1 μL
Sterile nuclease-free water	8-(X + Y) μL
Total reaction volume	10 μL

9. Ligate annealed oligonucleotides into XhoI/ECoRI-digested pEGFP-C1. Setup a positive control reaction using a molar ratio of 1:3 (vector:insert; *see* **Note 9**) and a negative control reaction containing only digested plasmid DNA (*see* Table 4). Incubate at room temperature for 1 h.
10. Transform chemically competent DH5α *E. coli* cells with 5 μL of positive/negative control ligation reaction mixture. Spread cells onto LB agar plates containing 100 μg/mL carbenicillin, and incubate the plates at 37 °C overnight.
11. Check the plates after overnight incubation. Pick 3–5 individual colonies from the positive ligation plate (*see* **Note 10**) and inoculate them into 5 mL LB broth containing 100 μg/mL of carbenicillin. Grow bacterial cultures at 37 °C overnight in an incubator shaker set at 180 rpm.
12. Purify plasmid DNA using QIAprep spin miniprep kit. Next, quantify DNA concentration by Nanodrop.
13. Verify positive clones by sequencing with a forward primer (5'- CATGGTCCTGCTGGAGTTCGTG-3′) located within the GFP ORF. Alternatively, use a reverse primer (5'-TAAG CTGCAATAAACAAG-3′) located in the plasmid's poly (A) signal.

3.3 Analysis of bEXOmiR Expression in HEK293T Cells

1. Seed HEK293T cells in DMEM medium supplemented with 10% (v/v) FBS, 100 U/mL of potassium penicillin, and 100 U/mL of streptomycin sulfate, and maintained in a 37 ° C incubator with 5% CO_2 supply. Split cells every 2–3 days when ~100% confluence is reached.
2. Transfect one 10-cm dish of HEK293T cells seeded at a 40–50% confluency using PEI transfection method (*see* **Note 11**):
   (a) Dissolve 3 μg of pEGFP-C1-bEXOmiR plasmid (*see* Subheading 3.2) in 970 μL of serum-free and antibiotic-free cell culture medium.
   (b) Add 30 μL of PEI into this solution (*see* **Note 12**), without touching the walls of the tube. Mix by inversion.
   (c) Incubate for 15–30 min at room temperature. During this time, add 9 mL of fresh medium in the plate to be transfected.
   (d) Add the PEI/DNA complex mixture dropwise, swirl the plate gently, and grow cells at 37 °C for 24–48 h.
3. Gently wash cells twice in cold PBS. Then proceed to detach cells by vigorously pipetting cold PBS directly onto the plate. Transfer volume of detached cells into a fresh 15 mL conical tube.
4. Sediment cells by centrifuging at 300 g for 5 min. Aspirate supernatant carefully without disrupting the pellet.
5. Add 1 mL of Trizol® solution and isolate total cell RNA (*see* **Note 13**).
6. Proceed to reverse transcribe 0.5–1 μg of total cell RNA. First-strand cDNA synthesis is performed using a highly stable stem-loop (S-L) primer, which increases specificity of the reaction and lengthens the size of the target miRNA (22 nt) up to >60 nt (*see* Fig. 2).
7. In a PCR tube, add 1 μL of a 100 μM stock of S-L primer and dilute it into 99 μL of sterile nuclease-free water. Denature the 1 μM S-L primer solution at 65 °C for 5 min. Immediately after this, chill tube on ice.
8. Assemble RT reactions (*see* **Note 14** and Table 5).
9. Insert tubes in a thermal cycler and incubate for 30 min at 16 ° C. This step is followed by pulsed RT (*see* **Note 15**) of 60 cycles at 30 °C for 30 s, 42 °C for 30 s and 50 °C for 1 s. Finally, inactivate reverse transcriptase at 85 °C for 5 min. Hold tubes at 4 °C. cDNA can be stored at −20 °C until further use.
10. After this, perform a PCR using the previously obtained cDNA (*see* Table 6). This reaction utilizes a universal reverse primer (complementary to an internal sequence of the initial S-L

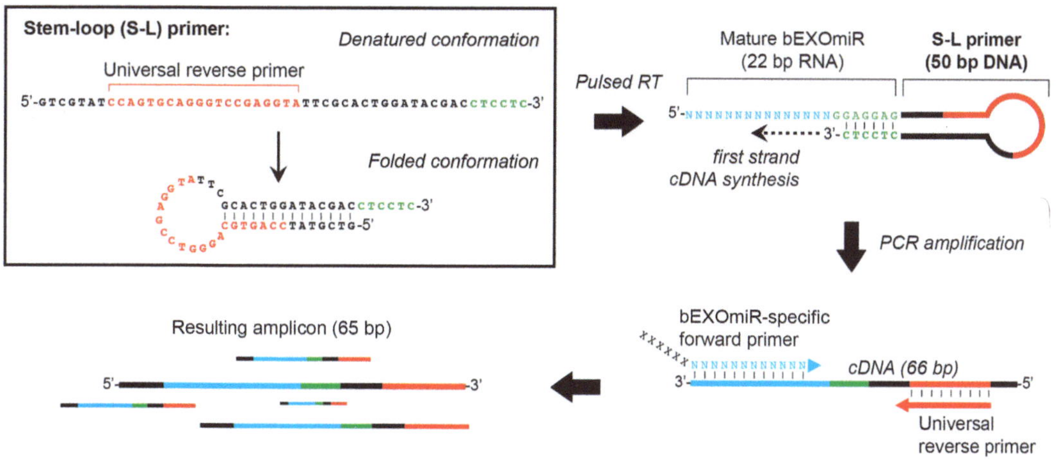

Fig. 2 Overall workflow of the stem-loop RT-PCR protocol. Diagram in the boxed shows stem-loop primer (S-L primer) in denatured (linear) and folded (hairpin) conformations. During reverse transcription, S-L primer anneals at the 3′ end as depicted. The resulting cDNA (66 bp) is amplified by PCR with a forward bEXOmiR-specific primer and a reverse universal primer as indicated. The resulting amplicon is 65 bp

Table 5
First-Strand cDNA Synthesis Reaction

Components	Amount
cDNA template	X μL
0.1 μM denatured S-L primer	1 μL
10 mM dNTP mix	0.5 μL
0.1 M DTT	2 μL
5x first-Strand buffer	4 μL
RNAse OUT (40 U/μL)	0.1 μL
SuperScript III reverse transcriptase (200 U/μL).	0.25 μL
Sterile nuclease-free water	12.15-X μL
Total reaction volume	20 μL

primer; *see* Fig. 2), together with a bEXOmiR-specific forward primer (*see* **Note 16**). It is important to include negative control reactions (*see* Fig. 3a and b). As a start, use from 2 to 5 μL of cDNA.

11. Insert tubes in a thermal cycler and use the following PCR protocol: 95 °C for 3 min, 25–28 cycles of 95 °C 30 sec/60 °C 30 sec/72 °C 10 sec, 72 °C 2 min, Hold at 4 °C.
12. Cast a 3.5% agarose-TAE gel (3.5 g agarose, 100 mL TAE buffer) at 100 V until well-resolved (40–45 mins). Save a lane

Table 6
PCR Amplification Reaction

Components	Amount
cDNA	X μL
100 μM bEXOmiR-specific forward S-L primer	0.25 μL
100 μM universal reverse primer	0.25 μL
10 mM dNTP mix	0.5 μL
10X ThermoPol® reaction buffer	2.5 μL
Vent® DNA polymerase (2 U/μL; *see* **Note 17**)	0.5 μL
Sterile nuclease-free water	16-X μL
Total reaction volume	20 μL

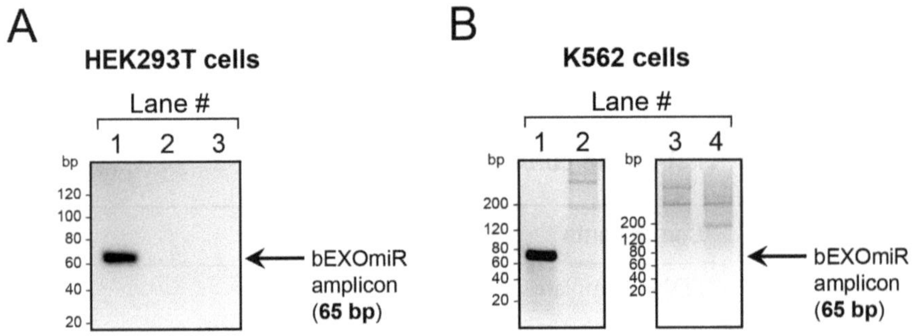

Fig. 3 Confirmation of bEXOmiR expression in mammalian cells. (**a**) Lane 1 shows an RT-PCR amplification band (65 bp) corresponding to a bEXOmiR (5'-CGGGCUAAAGGUUUCGGAGGAG-3′) expressed in HEK293T cells (total cell RNA was used in the RT reaction). Lanes 2 and 3 correspond to negative control reactions in which the reverse transcription (RT) reaction was performed in the absence of stem-loop primer (lane 2) or reverse transcriptase enzyme (lane 3). Molecular mass marker mobility is shown at left in base pairs (bp). (**b**) Lane 1 shows an RT-PCR amplification band (65 bp) corresponding to a bEXOmiR (5′- CGACAUGUCUGCCAAGGAG-GAG-3′) stably expressed in K562 cells (total cell RNA was used in the RT reaction). Lane 2 corresponds to a negative control reaction in which a cDNA obtained from an uninfected cell (not expressing a bEXOmiR) was used. Lanes 3 and 4 correspond to negative control reactions in which the PCR used a non-specific forward primer (lane 3) or in which the RT reaction was performed in the absence of stem-loop primer (lane 4). Molecular mass marker mobility is shown at left in base pairs (bp)

for running 4 μL of O'RangeRuler® 20 bp DNA Ladder. bEXOmiR amplicon (65 bp) should migrate between the 60 bp and 80 bp ladder markers (*see* Fig. 3a and b).

3.4 Detection and Quantification of bEXOmiRs in Isolated EVs

This section of the protocol is divided into two sub-sections: (i) EV isolation and EV-RNA purification followed by (ii) quantification of bEXOmiRs by RT-qPCR.

1. Transfect one 15-cm dish of HEK293T cells seeded at a ~ 40% confluency using the PEI transfection method (as detailed above). As a starting point, use 9 μg of pEGFP-C1-bEXOmiR plasmid. Add 27 mL of fresh EV-depleted media (*see* **Note 18**).
2. Add the PEI/DNA complex mixture dropwise, swirl the plate gently, and grow cells at 37 °C for 48 h. As an alternative to transfection, a cell line stably expressing bEXOmiRs can also be used (*see* **Notes 8** and **19**).
3. Carefully transfer supernatant to a fresh 50 mL conical tube. Collect cells as described above (*see* Subheading 3.3, **steps 3** and **4**).
4. Proceed to isolate EVs using a standard ultracentrifugation protocol (REF), as follows:
   (a) Centrifuge supernatant at 300 *g* for 5 min at 4 °C (use a tabletop centrifuge). This centrifugation step pellets live cells.
   (b) After this, collect supernatant without disrupting the pellet and centrifuge it at 2100 *g* for 20 min at 4 °C (use a tabletop centrifuge). This centrifugation step pellets dead cells.
   (c) Carefully collect supernatant and filter it through a 0.2 μm filter unit.
   (d) Using the appropriate tubes and rotor, centrifuge samples at 100,000 *g* for at least 70 min at 4 °C (*see* **Note 20**).
   (e) Decant tubes to remove supernatants and wipe the walls of the tubes carefully (without touching the pellet) until remaining media is absorbed.
   (f) Immediately after this, chill tubes on ice and resuspended pellets by pipetting 20 times up and down 800 μL of cold PBS (do not generate bubbles from pipetting). Then, transfer volume to 1 mL micro-ultracentrifuge tubes.
   (g) Repeat the previous step with 300 μL of cold PBS (to collect all remaining EV pellet material).
   (h) Centrifuge samples using the appropriate rotor at 100,000 g for 70 min in a micro-ultracentrifuge.
   (i) Discard supernatant and either freeze exosome pellets or proceed with downstream applications (*see* **Note 21**). If purifying EVs for the first time, an initial immunoblot analysis is recommended to confirm purity of the fraction (*see* Fig. 4a).
5. Add 500 μL of Trizol® solution directly into the micro-ultracentrifuge tube and gently pipette up and down to dissolve the EV pellet. Transfer the resuspended pellet into a fresh RNase-free tube.

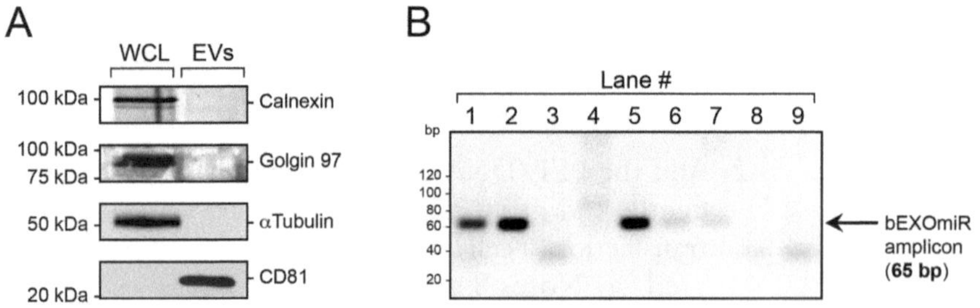

Fig. 4 Analysis of bEXOmiR abundance in isolated EV fractions. (**a**) Biochemical characterization of whole cell lysate (WCL) and EV fractions derived from HEK293T cells is shown in immunoblot panels using antibodies for the EV marker CD81, Calnexin (ER), Golgin 97 (Golgi), and α-Tubulin (Cytoskeleton). Note that ER, Golgi, and cytoskeleton markers are excluded from isolated EVs. Molecular mass marker mobility is shown at left in kilodaltons. (**b**) Lanes 1 and 2 show RT-PCR amplification bands (65 bp) corresponding to bEXOmiRs (lane 1: 5′- AUACGUAGUACGGGAGGAGGAG-3′; lane 2: 5'-CGGGCUAAAGGUUUCGGAGGAG-3′) expressed in HEK293T cells (RNA extracted from EVs was used in RT reactions). Lanes 3 and 4 correspond to negative control reactions in which the PCR used non-specific forward primers (lane 3 used the same cDNA as in lane 1 but amplified with primers specific for bEXOmiR amplified in lane 2; lane 4 used the same cDNA as in lane 2 but amplified with primers specific for bEXOmiR amplified in lane 1). Lanes 5–7 correspond to PCRs in which EV-RNA was treated with PBS (lane 5) or RNase (lanes 6 and 7) before reverse transcription (RT). Lane 5 (PBS control) shows an RT-PCR amplification band (65 bp) corresponding to the same bEXOmiR (5-'-CGGGCUAAAGGUUUCGGAGGAG-3′) amplified in lane 7 (note that RNase treatment significantly reduced amplification). Lane 6 shows the same results as in lane 7 but for a different bEXOmiR (5′- AUACGUA-GUACGGGAGGAGGAG-3′). Lanes 8–9 correspond to negative control reactions in which the RT reaction was performed in the absence of stem-loop primer (lane 8) or EV-RNA (lane 9). Molecular mass marker mobility is shown at left in base pairs (bp)

6. Repeat the previous step to collect all remaining material in the micro-ultracentrifuge tube (*see* **Note 22**). Samples can be stored at −20 °C until further use.
7. Isolate Total cell- and EV-RNA (*see* **Note 23**).
8. Perform a stem-loop RT-PCR using the previously isolated RNAs (follow the same protocol described above in Subheading 3.3). Use 200–500 ng of EV-RNA (*see* **Note 14**).
9. An initial non-quantitative PCR amplification is recommended to confirm bEXOmiR expression in both cellular and EV fractions (*see* **Note 24**). Add in the reaction from 2 to 5 μL of cDNA generated in the previous step. This initial PCR will also allow to optimize cDNA amounts needed for quantitative PCR (qPCR) experiments.
10. Assemble PCRs and run them as detailed in Subheading 3.3, **step 10**. It is important to include negative control reactions (*see* Fig. 4b).
11. Run the whole PCR volume reaction in a 3.5% agarose gel and run PCR products at 100 V until well-resolved. Save a lane for running 4 μL of O'RangeRuler® 20 bp DNA Ladder. bEXO-miR amplicon (65 bp) should migrate between the 60 bp and 80 bp ladder markers (*see* Fig. 4b).

Table 7
Assembly of qPCRs

Components	Amount
cDNA template	X μL
20 μM bEXOmiR-specific forward S-L primer	0.4 μL
20 μM universal reverse primer	0.4 μL
1X SYBR® green I (Invitrogen S-7563)	0.4 μL
1X ROX reference dye (life technologies)	0.4 μL
10 mM dNTP	0.4 μL
5X Phusion® HF buffer	4 μL
Phusion® high-Fidelity DNA polymerase (2 U/μL)	0.5 μL
Sterile nuclease-free water	13.5-X μL
Total reaction volume	20 μL

12. Proceed to perform a qPCR analysis of miRNA expression using a SYBR® green qPCR assay (TaqMan probe-based assays can also be applied).
13. Set up qPCRs in a multi-well plate according to Table 7 (it is recommended to prepare a master mix with all components except for cDNA).
14. Seal the multi-well plate, centrifuge it (this brings all reaction volume down and eliminates bubbles generated from pipetting), and insert it into a real-time thermocycler.
15. Run reactions with the following settings: 95 °C for 3 min, 40 cycles of 95 °C 15 sec/60 °C 20 sec/72 °C 3 sec.
16. Perform dissociation curve analysis by denaturing samples at 95 °C for 15 sec, followed by cooling to 65 °C, then collect fluorescence from 65 °C to 95 °C at 0.2 °C/sec.
17. Analyze qPCR results using the delta-delta Ct method [16]. Quantitative measurement of bEXOmiR abundance will serve as a proxy of EV release [15]. As we previously showed, thousands of bEXOmiRs can be transduced at a low multiplicity of infection (MOI), individually expressed in pooled format and quantified by next-generation sequencing [15].

4 Notes

1. Since processing of endogenous miR-30 and miR-601 yields 22 nt-long sequences, it is recommended to maintain this length for the predicted mature bEXOmiR.

2. Random sequences can be generated online (https://molbiotools.com/randomsequencegenerator.php). Nevertheless, supplementary file 5 ("Sequencing counts of bEXOmiRs in EV fraction sub-library screens") of our previous study [15], contains thousands of bEXOmiR sequences from which 15 nt-long barcodes can be obtained. It is recommended to use those barcodes with the highest number of sequencing counts from either WT or Cas9 ACOC screens data sheets, which we confirmed to be efficiently exported in EVs [15]. The first column in the data sheets contains the id (example "ENSG00000101144_BMP7_ACOC_10342.10") of the sgRNA-bEXOmiR associations used in our study. Next use this id to find the corresponding bEXOmiR sequence in WT or Cas9 ACOC screens data sheets of supplementary file 4 [15]. These data were obtained using K562 cells.
3. Important rules regarding the barcode sequence should be considered:
   - GC content should be between 40% and 60% (6 to 9 GC pairs out of 15).
   - Avoid the following patterns AAAA, UUUU, CCCC, GGGG, GAAUUC, and CCUAGG.
   - The first base pair should be altered to create a mismatch at the base of the hairpin to ensure proper miRNA processing.
   - Perform BLAST search using the 22 nt-long bEXOmiR sequence to make sure there is no high similarity to any known gene, miRNA, etc.
4. The native pri-miR-30 transcript is predicted to contain a bulge at the base of the stem which is necessary for proper cleavage by the RNase III enzyme Drosha [17].
5. Oligonucleotide phosphorylation can be omitted if phosphorylated oligonucleotides are ordered.
6. T4 PNK buffer comes with no ATP. T4 DNA ligase buffer comes with the enough ATP to perform phosphorylation reactions.
7. For convenience, this step can be performed in a thermocycler if handling a large number of reactions. In that case, choose the following program: 95 °C 3 min / cool the reaction to 25 °C over 45 min. Then chill tubes on ice.
8. Cloning between XhoI and EcoRI sites places the artificial miR downstream of the GFP ORF in the pEGFP-C1 plasmid. Co-expression of a fluorescent reporter may be useful as a control of transfection efficiency and for selection of transfected/infected cells. Alternative mammalian expression vectors can also be used. We have previously cloned and expressed bEXOmiRs downstream a GFP ORF in a lentiviral expression vector (Addgene #89360) [15].

9. Higher molar ratios can also be used to ensure successful ligation.
10. The number of colonies picked (from the positive control ligation plate) varies depending on how many colonies appear in the negative control ligation plate. For example, if the negative and positive control plates contain 2 and 4 colonies respectively, there is theoretically a 50% chance that each of the colonies in the positive control plate is a false positive. Thus, pick a save number of colonies accordingly.
11. This transfection protocol is optimal for HEK293T cells plated on a 10-cm plate (scale up or down accordingly). Up to 6 μg of DNA can be transfected in a 10 cm plate. Change the medium on day 1 post-transfection if toxicity is observed. Transfection efficiency is 80–90%.
12. To prepare PEI transfection reagent, dissolve 50 mg of polyethylenimine (Linear, MW 25000) in 40 mL of sterile water. Mix solution using a magnetic stirrer at 50–60 °C until it turns transparent. Adjust pH to 7.4 by adding NaOH solution. Adjust final volume of 50 mL. Filter-sterilize using a 0.2 μm filter unit. Store at −20 °C. Make 1 mL aliquots and store them at −20 °C (Once thawed aliquots can be used within 2 weeks).
13. Right after pipetting out the water phase (containing RNA) as described in the Trizol® RNA extraction protocol, it is recommended to add 2 μL of glycogen as a carrier molecule to increase RNA isolation yields (especially when isolating RNA from EVs).
14. It is advised to include the following negative control reactions: no S-L primer reaction and no reverse transcriptase reaction.
15. A pulsed RT reaction is recommended to increase the efficiency of first-strand cDNA synthesis.
16. For the design of bEXOmiR-specific forward primers include 6-7 nt 5′ overhang sequence to adjust for an appropriate Tm (*see* Fig. 2). The overhang is followed by 12 nt of the barcode sequence.
17. Alternative polymerases may also be used.
18. Cell culture media depleted of bovine EVs (EV-depleted media) is prepared by centrifuging FBS-containing media at 100,000 g for 24 h at 4 °C. Carefully collect supernatant (EV-depleted media) and filter-sterilize it before use. Non-depleted regular media can also be used since bEXOmiRs are artificial and should not match with any existing microRNA in nature.
19. Using a lentivirus expression system, we previously generated K562 cells stably expressing bEXOmiRs [15]. Setup 60 mL cultures of bEXOmiR-expressing K562 cells at ~0.250*10^6 cells/mL. Grow cells for 48 h before EV isolation.

20. Different ultracentrifuge tubes and rotors can be used. When using a fixed-angle rotor, the pellet will be on the side at the bottom. On the other hand, in a swinging-bucket rotor the pellet will be located on the center at the bottom.
21. This centrifugation step pellets exosomes. An additional last step consisting of a floatation through a 30% sucrose (diluted in deuterated water) cushion can be included to further purify small EVs. Protein aggregates and other non-specific sedimented material will not float.
22. To normalize RNA extraction and reverse transcription variability, a 22 nt-long spike-in RNA can be exogenously added right after dissolving the EV pellet. Make sure to include a GGAGGA sequence in the 3′ end of the spike RNA (spike-in RNA will use the same stem-loop and PCR universal reverse primers as bEXOmiRs, but different PCR forward primer).
23. Alternatively, to increase reproducibility among multiple replicates/conditions, after collection of the aqueous phase obtained in the Trizol® protocol, RNA extraction can be continued using the miRNeasy Micro Kit (Qiagen).
24. Our previous work revealed highly heterogenous EV targeting efficiencies among thousands of barcodes [15]. It is thus recommend performing an initial confirmation of the bEXOmiR reporter expression in both cellular and EV fractions. Here below are 3 examples of bEXOmiR sequences that showed good expression levels in either HEK293T or K562 cells and derived EVs (EXOmotif in bold letters):

bEXOmiRs efficiently targeted to HEK293T-derived EVs (5′ > 3′):
CGGGCUAAAGGUUUC**GGAGGAG.**
UUGUAGGUAUCCCAG**GGAGGAG.**
AUACGUAGUACGGGA**GGAGGAG.**

bEXOmiRs efficiently targeted to K562-derived EVs (5′ > 3′):
CGACAUGUCUGCCAA**GGAGGAG.**
AGAGGAAUCACUGGU**GGAGGAG.**
CGGAUCCUUCUCCAC**GGAGGAG.**

References

1. van Niel G, D'Angelo G, Raposo G (2018) Shedding light on the cell biology of extracellular vesicles. Nat Rev Mol Cell Biol 19(4): 213–228. https://doi.org/10.1038/nrm.2017.125
2. Mathieu M, Martin-Jaular L, Lavieu G, Théry C (2019) Specificities of secretion and uptake of exosomes and other extracellular vesicles for cell-to-cell communication. Nat Cell Biol 21(1):9–17. https://doi.org/10.1038/s41556-018-0250-9
3. van Niel G, Carter D, Clayton A, Lambert DW, Raposo G, Vader P (2022) Challenges and directions in studying cell-cell communication by extracellular vesicles. Nat Rev Mol Cell Biol 23(5):369–382. https://doi.org/10.1038/s41580-022-00460-3
4. Becker A, Thakur BK, Weiss JM, Kim HS, Peinado H, Lyden D (2016) Extracellular vesicles in cancer: cell-to-cell mediators of metastasis. Cancer Cell 30(6):836–848. https://doi.org/10.1016/j.ccell.2016.10.009

5. Bhome R, Del Vecchio F, Lee GH, Bullock MD, Primrose JN, Sayan AE, Mirnezami AH (2018) Exosomal microRNAs (exomiRs): small molecules with a big role in cancer. Cancer Lett 420:228–235. https://doi.org/10.1016/j.canlet.2018.02.002
6. Zhou W, Fong MY, Min Y, Somlo G, Liu L, Palomares MR, Yu Y, Chow A, O'Connor ST, Chin AR, Yen Y, Wang Y, Marcusson EG, Chu P, Wu J, Wu X, Li AX, Li Z, Gao H, Ren X et al (2014) Cancer-secreted miR-105 destroys vascular endothelial barriers to promote metastasis. Cancer Cell 25(4):501–515. https://doi.org/10.1016/j.ccr.2014.03.007
7. Zhang L, Zhang S, Yao J, Lowery FJ, Zhang Q, Huang WC, Li P, Li M, Wang X, Zhang C, Wang H, Ellis K, Cheerathodi M, McCarty JH, Palmieri D, Saunus J, Lakhani S, Huang S, Sahin AA, Aldape KD et al (2015) Microenvironment-induced PTEN loss by exosomal microRNA primes brain metastasis outgrowth. Nature 527(7576):100–104. https://doi.org/10.1038/nature15376
8. Zhang Y, Kim MS, Jia B, Yan J, Zuniga-Hertz JP, Han C, Cai D (2017) Hypothalamic stem cells control ageing speed partly through exosomal miRNAs. Nature 548(7665):52–57. https://doi.org/10.1038/nature23282
9. Villarroya-Beltri C, Gutiérrez-Vázquez C, Sánchez-Cabo F, Pérez-Hernández D, Vázquez J, Martin-Cofreces N, Martinez-Herrera DJ, Pascual-Montano A, Mittelbrunn M, Sánchez-Madrid F (2013) Sumoylated hnRNPA2B1 controls the sorting of miRNAs into exosomes through binding to specific motifs. Nat Commun 4:2980. https://doi.org/10.1038/ncomms3980
10. Santangelo L, Giurato G, Cicchini C, Montaldo C, Mancone C, Tarallo R, Battistelli C, Alonzi T, Weisz A, Tripodi M (2016) The RNA-binding protein SYNCRIP is a component of the hepatocyte Exosomal machinery controlling MicroRNA sorting. Cell Rep 17(3):799–808. https://doi.org/10.1016/j.celrep.2016.09.031
11. Temoche-Diaz MM, Shurtleff MJ, Nottingham RM, Yao J, Fadadu RP, Lambowitz AM, Schekman R (2019) Distinct mechanisms of microRNA sorting into cancer cell-derived extracellular vesicle subtypes. eLife 8:e47544. https://doi.org/10.7554/eLife.47544
12. Wozniak AL, Adams A, King KE, Dunn W, Christenson LK, Hung WT, Weinman SA (2020) The RNA binding protein FMR1 controls selective exosomal miRNA cargo loading during inflammation. J Cell Biol 219(10): e201912074. https://doi.org/10.1083/jcb.201912074
13. Garcia-Martin R, Wang G, Brandão BB, Zanotto TM, Shah S, Kumar Patel S, Schilling B, Kahn CR (2022) MicroRNA sequence codes for small extracellular vesicle release and cellular retention. Nature 601(7893):446–451. https://doi.org/10.1038/s41586-021-04234-3
14. Jeppesen DK, Fenix AM, Franklin JL, Higginbotham JN, Zhang Q, Zimmerman LJ, Liebler DC, Ping J, Liu Q, Evans R, Fissell WH, Patton JG, Rome LH, Burnette DT, Coffey RJ (2019) Reassessment of exosome composition. Cell 177(2):428–445.e18. https://doi.org/10.1016/j.cell.2019.02.029
15. Lu A, Wawro P, Morgens DW, Portela F, Bassik MC, Pfeffer SR (2018) Genome-wide interrogation of extracellular vesicle biology using barcoded miRNAs. eLife 7:e41460. https://doi.org/10.7554/eLife.41460
16. Schmittgen TD, Livak KJ (2008) Analyzing real-time PCR data by the comparative C(T) method. Nat Protoc 3(6):1101–1108. https://doi.org/10.1038/nprot.2008.73
17. Lee Y, Ahn C, Han J, Choi H, Kim J, Yim J, Lee J, Provost P, Rådmark O, Kim S, Kim VN (2003) The nuclear RNase III Drosha initiates microRNA processing. Nature 425(6956): 415–419. https://doi.org/10.1038/nature01957

Part III

Systems and Models to Study EVs

Chapter 15

Purification of Bacterial-Enriched Extracellular Vesicle Samples from Feces by Density Gradient Ultracentrifugation

Nadiya Byts, Olha Makieieva, Artem Zhyvolozhnyi, Genevieve Bart, Johanna Korvala, Jenni Hekkala, Sonja Salmi, Anatoliy Samoylenko, and Justus Reunanen

Abstract

Commensal microbiota has huge impact on the maintenance of human health, its dysregulation being associated with the development of a plethora of diseases. Release of bacterial extracellular vesicles (BEVs) is a fundamental mechanism of systemic microbiome influence on the host organism. Nevertheless, due to the technical challenges of isolation methods, BEV composition and functions remain poorly characterized. Hereby, we describe the up-to-date protocol for isolation of BEV-enriched samples from human feces. Fecal extracellular vesicles (EVs) are purified through the orthogonal implementation of filtration, size-exclusion chromatography (SEC), and density gradient ultracentrifugation. EVs are first separated from bacteria, flagella, and cell debris by size. In the next steps, BEVs are separated from host-derived EVs by density. The quality of vesicle preparation is estimated via immuno-TEM (transmission electron microscopy) for the presence of vesicle-like structures expressing EV markers and via NTA (nanoparticle tracking analysis) for assaying particle concentration and size. Distribution of EVs of human origin in gradient fractions is estimated using antibodies against human exosomal markers with Western blot and ExoView R100 imaging platform. The enrichment for BEVs in vesicle preparation is estimated by Western blot for the presence of bacterial OMVs (outer membrane vesicles) marker and OmpA (outer membrane protein A). Taken together, our study describes a detailed protocol for EV preparation with enrichment for BEVs from feces with a purity level suitable for bioactivity functional assays.

Key words Bacterial extracellular vesicles, Human feces, Commensal bacteria, Density gradient ultracentrifugation

---

Nadiya Byts and Olha Makieieva contributed equally.

Seppo Vainio (ed.), *Cell-Secreted Vesicles: Methods and Protocols*, Methods in Molecular Biology, vol. 2668, https://doi.org/10.1007/978-1-0716-3203-1_15,

1 Introduction

There is an exponentially growing interest toward EVs, acknowledging biological relevance of EVs in health and disease [1–5]. A huge diversity between the methods of isolation, purification, and characterization of EVs through emerging studies complicates progress in this new field of research. Therefore, there is a strong demand on method characterization and standardization [1, 6–8]. This is especially relevant when studying BEVs, extracellular vesicles of bacterial origin, due to the additional challenge to separate BEVs from the host-derived EVs [9, 10]. Hereby, we describe in detail the protocol for isolation of BEV-enriched EV samples from human feces, modified from Tulkens et al. [9].

Fecal crude extract in addition to EVs of both host and bacterial origin contains such components as lipoproteins, chylomicrons, protein aggregates, low-molecular weight proteins, and fibers [11]. In the procedure we describe here (Fig. 1), the EVs are separated first from bacteria, flagella, and cell debris by size [9]. In the next steps, BEVs are separated by density from host EVs of human origin [9, 12]. Submitting the crude extract to density gradient centrifugation allows the enrichment for BEVs with minimal contamination by other biological components, due to differences in buoyant density [9, 11]. Ten fractions of 1 mL each are collected after centrifugation. BEVs (1.133–1.201 g/mL) are mostly allocated to fractions 6 and 7 (numbering from top to bottom of gradient), while human EVs (1.083–1.111 g/mL) are more abundant in fractions 4 and 5. Size, concentration, and marker expression in EVs were characterized as described elsewhere [5, 13]. The resulting BEV-enriched EV suspension has concentration yield in the range of 10^{9-11} particles per milliliter, containing bilayered membrane nanostructures. The diameter range reaches up to 500 nm, with abundant presence of sub-200 nm particles, which is consistent with the presence of small EVs. BEV preparation is positive for gram-negative bacterial marker outer membrane protein A (OmpA) by Western blot and immuno-TEM, while presence of human tetraspanin CD63 is detected with immuno-TEM, Western blot, and ExoView R100 imaging platform.

2 Materials

Make all solutions using ultrapure water and analytical-grade reagents. Prepare and store all reagents at room temperature (unless indicated otherwise). PBS (phosphate-buffered saline) used for homogenization, in washing steps and in NTA, should be prepared from commercial source and be sterile-filtered (0.45 μm, 50 mL aliquots stored at +4 °C) to remove nanoparticle contaminants of different origins in the buffer.

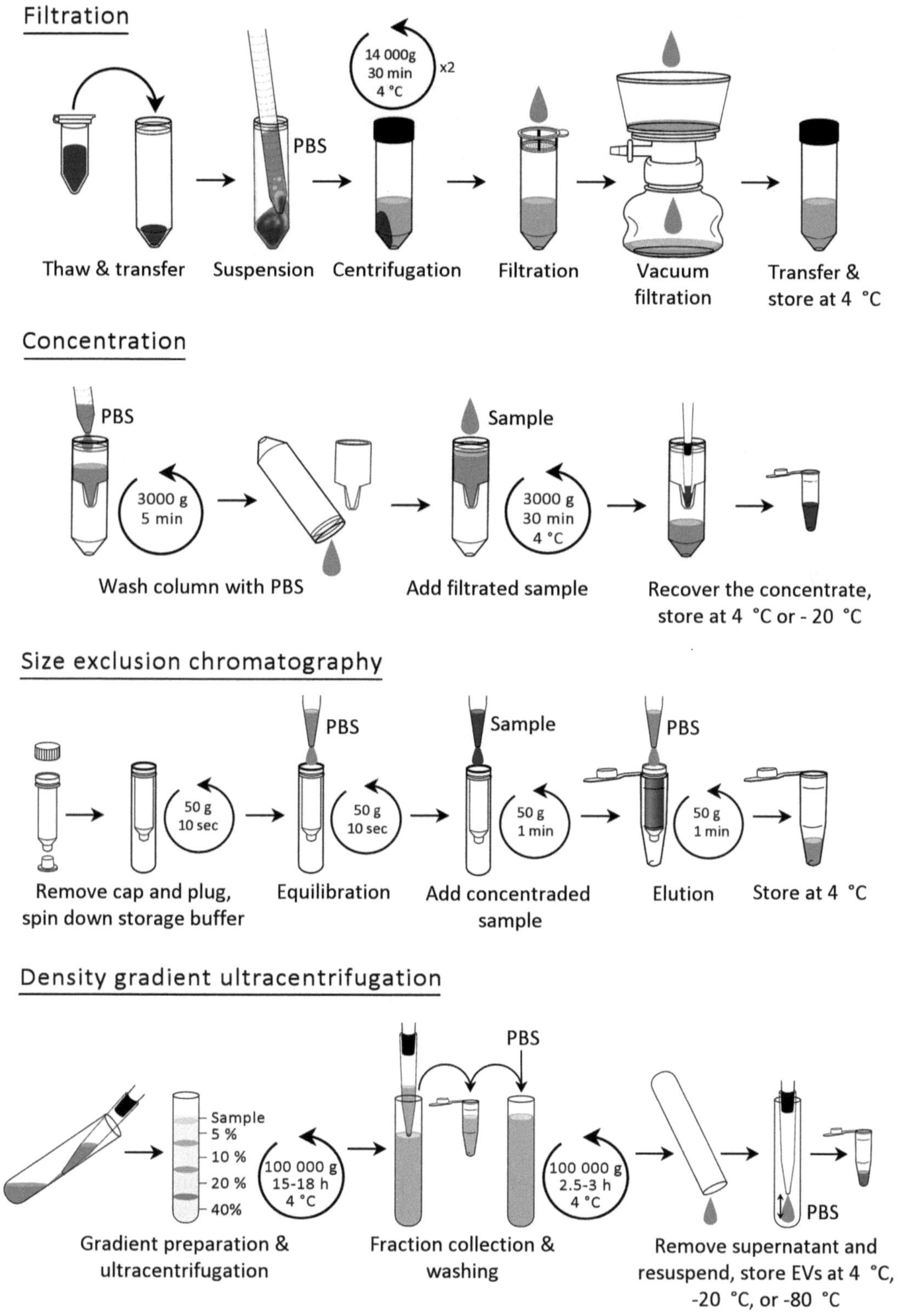

Fig. 1 Scheme of EV isolation from fecal material

2.1 Materials and Reagents Required

1. Human (*see* **Note 1**) feces stored at −80 °C.
2. 50 mL falcon tubes.
3. Cell strainer, 40 μm pore.
4. Vac Filter PES Membrane 150 mL Pore 0.45 μm.
5. Amicon® Ultra-15 Centrifugal Filter Unit, 15 mL (Millipore, #UFC910024; *see* **Note 2**).
6. Exo-Spin™ Mini-Columns (Cell Guidance Systems, #EX03).
7. Optiprep™ (#115778535; STEMCELL Technologies), stored protected from light.
8. 13.2 mL, open-top thin-wall ultra- clear tube, 14 × 89 mm (Beckman Coulter, #344059).
9. 13.2 mL, 14 × 89 mm polypropylene centrifuge tubes (Beckman Coulter, #331372).
10. Ultracentrifuge (we use ultracentrifuge Optima™ L-100/ L90K preparative ultracentrifuge with SW 41 Ti swinging-bucket rotor (Beckmann)).

2.2 Buffers

1. 1 M sucrose for density cushion. Dissolve 171.145 g ultrapure sucrose in sterile ddH_2O to 500 mL of total volume. Sterile filter with 0.45 μm pore size.
2. 1 M Tris–HCl. Dissolve 31.52 g Tris–HCl in sterile ddH2O to 200 mL of total volume. Sterile filter with 0.45 μm pore size.
3. Homogenization buffer (HB): 0.25 M sucrose, 10 mM Tris–HCl, 1 mM EDTA, pH 7.4). Mix 125 mL 1 M sucrose with 5 mL 1 M Tris–HCl, add ddH_2O to total volume of 500 mL dissolve 0.1861 g EDTA. Sterile filter with 0.45 μm pore size.
4. Working solution buffer (WSB): 0.25 M sucrose, 60 mM Tris–HCl, 6 mM EDTA, pH 7.4). Mix 125 mL 1 M sucrose with 30 mL 1 M Tris–HCl, add ddH_2O to total volume of 500 mL, dissolve 1.1166 g EDTA. Sterile filter with 0.45 μm pore size.

   These stock solutions can be stored at +4 °C in dark for about a year. Sterile filtration should remove almost all nanoparticle contaminants coming from the reagents.
5. Working solution (WS, 1× WSB: 5× Optiprep™, v/v). Example for two gradients: 2 mL WSB + 10 mL OptiPrep™. Mix vigorously by vortexing upon pipetting.

2.3 Density Gradient Preparation

Four layers of different percentages of iodixanol (component of OptiPrep medium, 5%, 10%, 20%, and 40%) are used to form a discontinuous iodixanol gradient [12]. The gradient is formed during **step 1** in Subheading 3.4 by layering 2.5 mL of 40%, 2.5 mL of 20%, 2.5 mL of 10%, and 2.2 mL of 5% solutions on top of each other in a thin-wall ultra-clear tube.

Table 1
Pipetting maps for four gradient layers for variable number of samples

% of iodixanol	WS, mL	HB, mL
2 samples WS = 2 mL WSB + 10 mL OptiPrep™		
5	0.6	5.4
10	1.2	4.8
20	2.4	3.6
40	4.8	1.2
4 samples WS = 3.33 mL WSB + 16.67 mL OptiPrep		
5	1.2	10.8
10	2.4	9.6
20	4.8	7.2
40	9.6	2.4
6 samples WS = 4.5 mL WSB + 22.5 mL OptiPrep		
5	1.5	13.5
10	3.4	13.6
20	6.8	10.2
40	13.6	3.4
10 samples WS: 7.5 mL WSB + 37.5 mL OptiPrep		
5	2.5	22.5
10	5.6	22.4
20	11.2	16.8
40	22.4	5.6

To prepare separate layers of gradient, mix appropriate amounts of working solution (from **step 5** in Subheading 2.2) and homogenization buffer (from **step 3** in Subheading 2.2) and vortex vigorously (Table 1). We recommend to prepare gradient layers fresh.

2.4 NanoSight Measurements

1. Milli-Q water.
2. 1 mL syringes.
3. Cleaning tissues.
4. 2 mL tubes.

2.5 Immuno-Electron Microscopy

1. Formvar-coated copper grids (400 mesh).
2. Grid coated with carbon and glow-discharged.
3. 1% glutaraldehyde in PBS (pH 7.4).
4. 2% uranyl acetate (UA) in H2O.
5. 1% BSA in PBS (pH 7.4).
6. 2% methylcellulose-UA (0.4%).
7. Protein A-gold complex (10 nm size).
8. Anti-CD63 antibody (1:100, Santa Cruz).
9. Anti-OmpA (1:100, ARC).

2.6 Protein Isolation and Western Blotting

1. Running buffer (Tris–glycine/SDS) (25 mM Tris base; 190 mM glycine; 0.1% SDS; pH 8.3).
2. Transfer buffer (25 mM Tris base; 190 mM glycine; 20% methanol; pH 8.3).
3. RIPA buffer (Cell Signaling Technology).
4. Inhibitor cocktail cOmplete ULTRA (Roche).
5. Phosphatase inhibitor cocktail 2 (Sigma-Aldrich).
6. BCA Protein Assay Kit (Pierce).
7. 4× Laemmli loading buffer (Thermo Scientific).
8. 4% and 10% SDS–PAGE.
9. Nitrocellulose membrane (Macherey-Nagel).
10. TotalStain kit (Azure biosystems).
11. Anti-CD63 (1:1000, Santa Cruz).
12. Anti-OmpA (1:1000, ARC).
13. Peroxidase-conjugated IgG antibodies (1:5000, Agilent Technologies).
14. Lumi-Light Western Blotting Substrate (Roche Diagnostics).
15. PageRuler Plus Prestained Protein Ladder (Thermo Scientific).

2.7 Analysis of EVs Using ExoView

1. ExoView Human Tetraspanin Kit (NanoView Biosciences).
2. Milli-Q water.
3. Aluminum foil.
4. PCR plate seals.
5. Tweezers.
6. 24-well cell culture plate (Corning).
7. 10-cm Petri dish.
8. 50 mL tubes.

3 Methods

The brief scheme of the isolation method is shown in Fig. 1.

3.1 Stool Sample Filtration

During this step, the undigested parts, cell debris, fibers, pili, and flagella are removed from the stool sample [9]. Keep the samples on ice between all the steps. Carry out **steps 1** and **2** under laminar flow cabinet.

1. Melt feces raw material on ice for 10 min.
2. Transfer app. 1 g (*see* **Note 3**) of feces into a 50 mL falcon tube and resuspend by pipetting in 15 mL PBS (*see* **Note 4**).
3. Centrifuge sample at 14,000*g* for 30 min at 4 °C.
4. Transfer supernatant into a new falcon tube and centrifuge again (14,000*g*, 30 min, +4 °C).
5. Filter supernatant through 40 μm cell strainer into a new 50 mL falcon tube.
6. Filtrate the sample again using Vac Filter PES Membrane 150 mL, 0.45 μm pore. Transfer the filtrate into 50 mL falcon tube and store at +4 °C. Process the filtrate within a few days.

3.2 Concentration of the Fecal Filtrate

Concentrate the fecal filtrate using Amicon® Ultra-15 Centrifugal Filter Unit (*see* **Note 2**).

1. Recover the membrane from glycerin according to manufacturer instructions: add 10 mL PBS to the top part of the tube, centrifuge 3000 g 5 min, remove PBS flow-through by pouring off from the main tube.
2. Add the fecal filtrate (from **step 6** in Subheading 3.1, max. 15 mL) to the top part of the tube and centrifuge (3000*g*, +4 °C for 30 min) using swinging-bucket rotor (*see* **Note 5**).
3. Collect all of obtained concentrate from the top part of the device to a 1.5 mL Eppendorf cup; pipetting is performed by side-to-side sweeping motion from the bottom of the filter. Discard the flow-through.
4. The obtained volume varies from 200 μL to 1000 μL. If the volume is more than 200 μL (which is used in the next steps), we recommend to repeat the concentration procedure reusing Amicon® Filter Unit. This will insure normalization of concentrate per amount of primary stool material when further applied on gradient. Store the concentrate at 4 °C, if you are planning to use it soon and at −20 °C for long-term storage.

3.3 Isolation of Extracellular Vesicles by Size-Exclusion Chromatography

Exo-Spin™ Mini-Columns are used to isolate extracellular vesicles from the sample according to manufacturer instructions.

1. Prepare the Exo-Spin™ Mini-Column by removing the outlet plug and screw cap and place the spin column into a waste collection tube. Briefly spin down at 50 g for 10 s to remove preservative buffer (*see* **Note 6**).
2. Equilibrate the column by adding 200 μL 1× PBS and spin down again at 50 g for 10 s, and discard the eluate. If any PBS remains above the top filter, remove it carefully with a micropipette or spin again.
3. Carefully apply 200 μL of concentrated filtrate (obtained during **step 4** in Subheading 3.2) to the top of the column and place the column into the waste tube. Centrifuge at 50*g* for 60 s and discard the eluate.
4. Transfer the column into a fresh 1.5 mL Eppendorf cup and add 200 μL 1xPBS on the top of the column. Centrifuge at 50*g* for 60 s and collect the eluate containing the purified EVs. Secure the lid with parafilm so that the sample will not evaporate upon short-term storage at +4 °C.

3.4 BEV Enrichment by Density Gradient Ultracentrifugation

1. The gradient was formed by layering 2.5 mL of 40%, 2.5 mL of 20%, 2.5 mL of 10%, and 2.2 mL of 5% solutions on top of each other in a 13.2 mL thin-wall ultra-clear tube (from bottom to the top, solution composition described in 2.3) (*see* **Note 7**).
2. Overlay 200 μL of EVs containing sample (from **step 4** in Subheading 3.3) onto the top of the gradient (*see* **Note 8**).
3. Place tubes into adaptors (*see* **Notes 9** and **10**) and equilibrate (*see* **Note 11**).
4. Centrifuge 15–18 h, 100,000*g*, +4 °C. Centrifugation is done using ultracentrifuge, such as Optima L-100/L90 ultracentrifuge with SW 41 Ti swinging-bucket rotor. Collect gradient fractions of 1 mL from the top of the gradient to fresh thin-wall (clear-wall is not necessary here) polypropylene tubes (*see* **Notes 12** and **13**). This results in 10 fractions.
5. Wash fractions: Add PBS to 10 mL and centrifuge for 2.5–3 h at 100,000*g* and 4 °C (*see* **Note 14**) to pellet EVs.
6. Remove supernatant (by pulling supernatant to waste container very carefully) and suspend the "invisible" pellet containing EVs into 100 μL of PBS (*see* **Note 15**).
7. At this step, BEVs are allocated with a high specificity to dissolved pellets from gradient fractions 6 and 7 as described in introduction [9]. Store the BEVs / PBS solution at +4 °C for short-time storage and at −20 °C for long-time storage. Storage over months should be done at −80 °C; this preserves

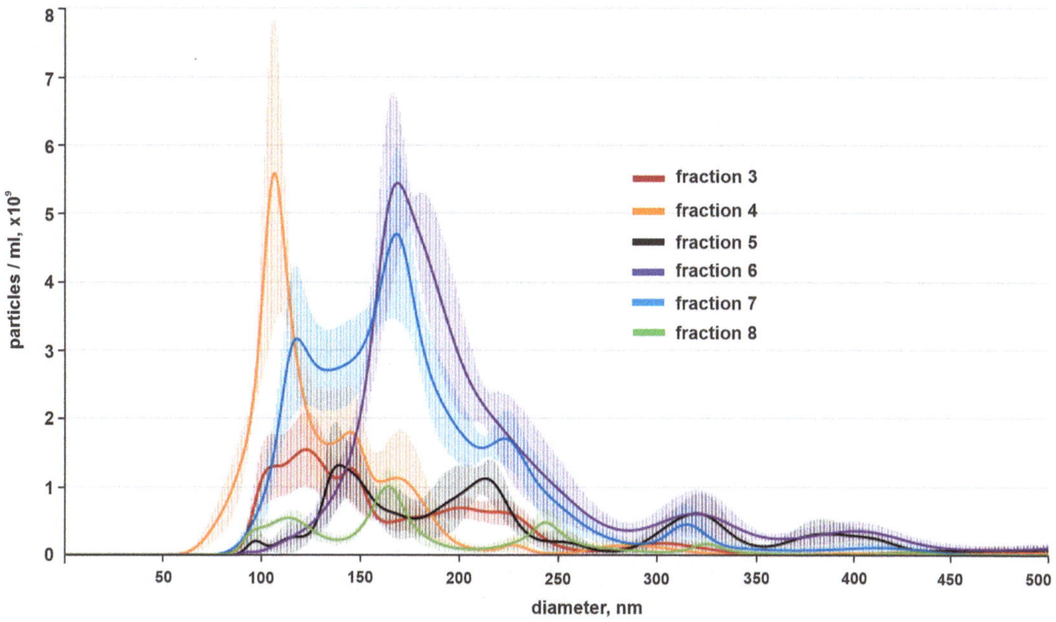

Fig. 2 Concentration and size distribution of EVs in different gradient fractions by nanoparticle tracking analysis (NTA). Representative graph is shown. Data represent mean ± SEM from $n = 4$ measurements

RNA and protein content [14]. Aliquot to avoid multiple freeze–thaw cycles.

8. We recommend to use BEVs in 24 h after being thawed for functional studies [14].
9. Characterize resulting vesicle suspension according to a position statement of the International Society for Extracellular Vesicles on Minimal information for studies of extracellular vesicles [1]. To illustrate here the efficiency of isolation method (Fig. 1), we show the data from Nanosight analysis (Fig. 2), immuno-transmission electron microscopy (Fig. 3), ExoView (Fig. 4), and Western blot analysis (Fig. 5) on EVs from different gradient fractions.

Nanosight results illustrate that particles are not homogeneously distributed through gradient fractions, but are located in higher concentrations in fractions 4, 6, and 7 (Fig. 2). As expected, the size of particles increases with fraction number and is higher in fractions 6 and 7 comparing to fractions 4 sand 5 (Fig. 2).

Expression of human and bacterial EV biomarkers was further demonstrated using immuno-TEM with antibodies against CD63 or OmpA (Fig. 3). ExoView analysis was used to characterize the fractions on expression of human EV biomarkers. In ExoView, the anti-tetraspanin antibodies are immobilized on the chips and bind the EVs present in sample. At the next step of the assay, the chip-

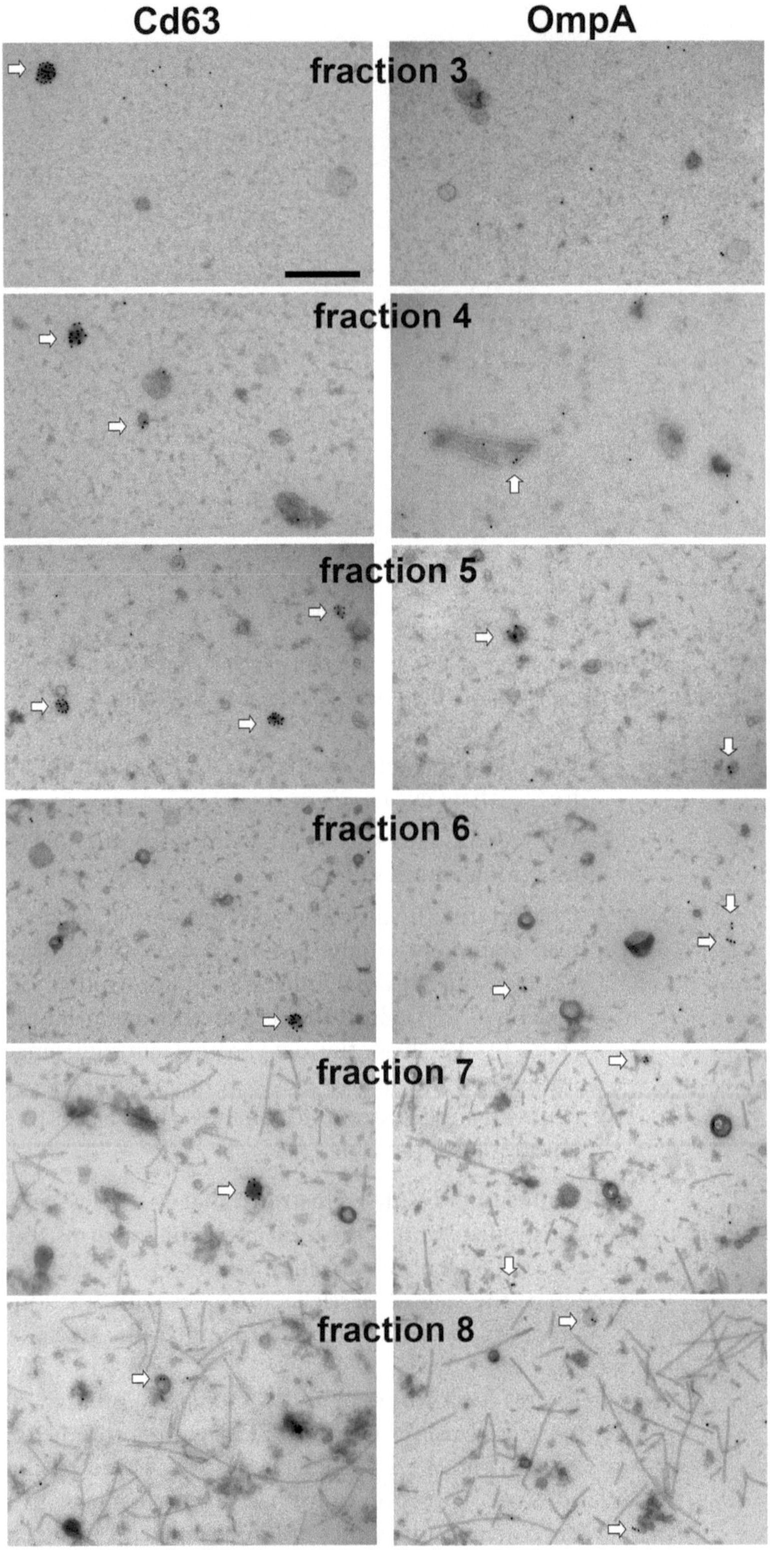

Fig. 3 Immuno-TEM on EVs with anti-CD63 and anti-OmpA antibody. Representative micrographs of vesicle preparations corresponding to different gradient fractions. Scale bar is 400 nm. Arrows indicate co-localization of at least 2 positive dots, which presumably represent EVs positive for markers analyzed

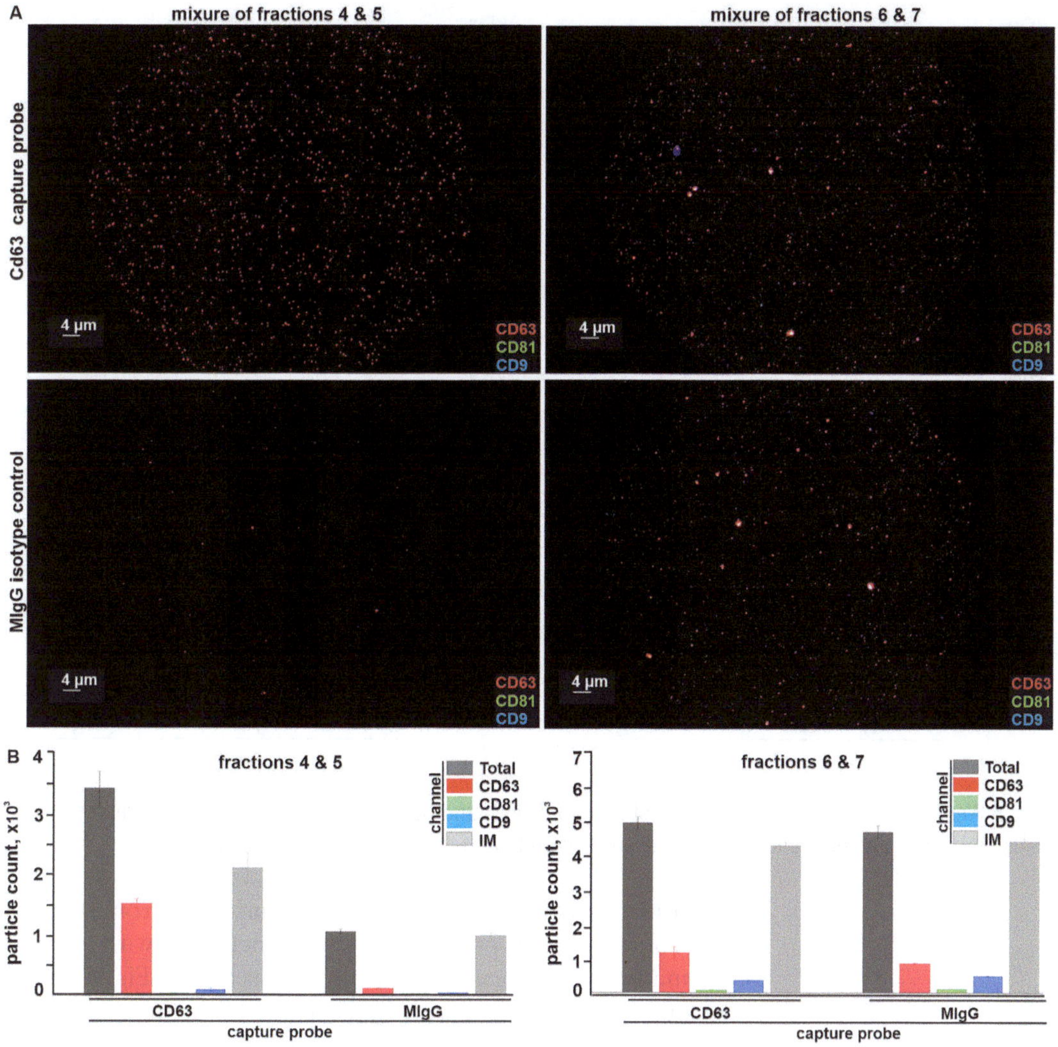

Fig. 4 Expression of human tetraspanins in EVs from different fractions by ExoView. Data from slots with either CD63 or MIgG control capture probe are shown from the chips treated with either mixture of fractions 4 & 5 or mixture of fractions 6 & 7. (**a**) Representative images of individual slots on the chips. Colored dots represent single vesicles captured by antibodies (CD63 or MIgG) anchored on corresponding slots. Dot's color depends on the fluorescently labeled detection antibodies (red for CD63, green for CD81, and blue for CD9). (**b**) Quantification of images. Total number of detected EVs in a sample is shown by dark gray bars; number of EVs expressing CD63, CD81, and CD9 is depicted by red, green, and blue bars, correspondingly. Number of EVs detected by interference microscopy (IM) is shown by light gray bars and indicates number of particles of 50–200 nm size. Data represent mean ± SEM from $n = 3$ slots on a chip

fixed EVs are stained with fluorescently labeled antibodies against specific human EV proteins. We found that the ratio of CD63-positive EVs to all EVs in a sample representing mix of fractions 4 & 5 was about 1:2 and much lower in fractions 6 & 7, about 1:5 (Fig. 4). These differences may reflect higher amount of EVs of non-human origin in the fractions 6 and 7. The ratio between

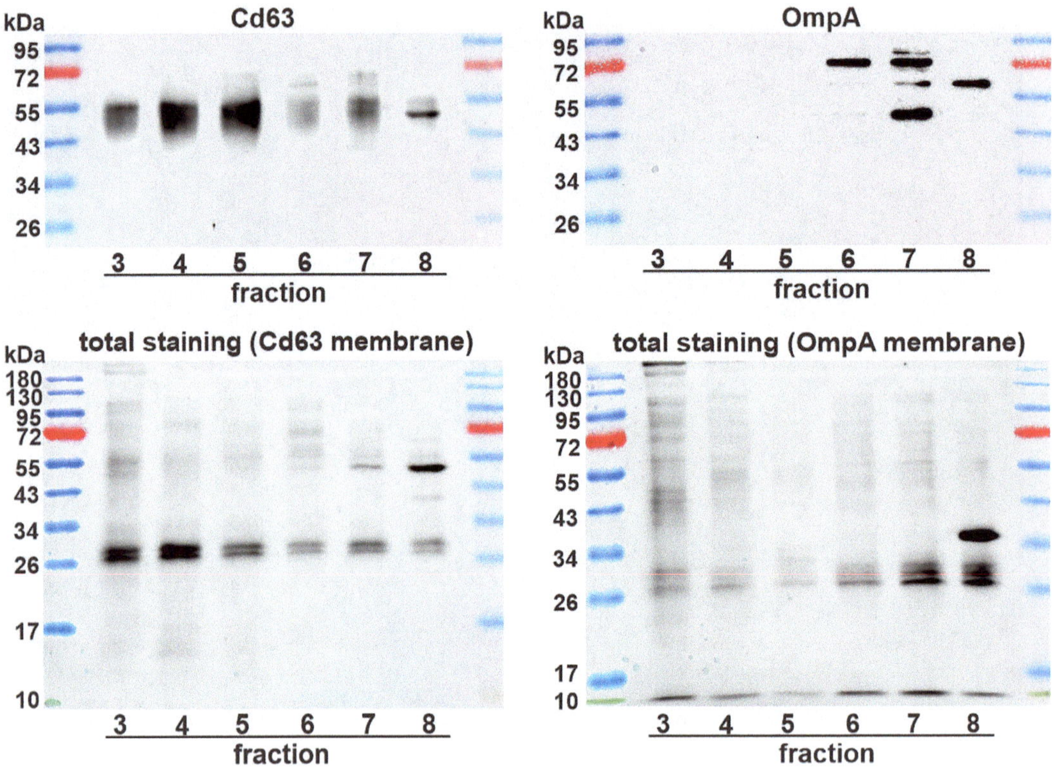

Fig. 5 Western blot with antibodies against human (CD63) and bacterial (OmpA) EV markers. Representative blots. Total staining serves to confirm presence of protein in different gradient fractions

signal from slot with CD63 capture probe and signal from slot with isotype control capture probe is high when incubated with fractions 4 & 5 and low when incubated with fractions 6 & 7 (Fig. 4). High background signal in slots with IgG capture probe incubated with fractions 6 & 7 (Fig. 4) may be explained by the presence of significant amounts of debris, fibers, and flagella in fractions 7 and 8, as seen in Immuno-TEM (Fig. 3), and presumably reflects their unspecific binding. Only minority of fecal EVs expressed CD81 and CD9 markers. In our samples, the signal from syntenin-1 capture probe is under the limit of detection (data not shown). Taken together, our ExoView results show relatively higher amount of CD63-positive particles in mixture of fractions 4 & 5 versus mixture of fractions 6 & 7.

In accordance with ExoView results, the Western blot data on EV lysates confirm high levels of CD63 expression in fractions 3 and 4 with decreased expression in fractions 6 and 7 (Fig. 5). In the opposite pattern, levels of OmpA expression are low or undetectable in fractions 4 and 5 and high in fractions 6 and 7 (Fig. 5). The levels of CD9, CD81, and syntenin-1 expression by Western blot were under the limit of detection (data not shown). In summary, the Exoview and Western blot results together confirm partial separation of human and bacterial vesicles to different gradient fractions.

3.5 Methods Used for EV Characterization

3.5.1 NanoSight Measurements

The concentrations and size distributions of the EV samples were characterized by NTA using the NanoSight NM300 instrument (Malvern Panalytical) equipped with a 405 nm laser and syringe pump and supplemented with NTA software version 3.4.4 (Malvern Panalytical). Samples were diluted with sterile-filtrated PBS before measurements in 1:1000 ratio. Four 60 s videos were recorded of each sample with screen gain 7, camera level 14, and detection threshold set up to 3.

3.5.2 Immuno-Electron Microscopy

EV samples were analyzed by transmission electron microscopy (TEM). 2 μL of each sample was deposited on a Formvar carbonated grid and after negative staining with 2% uranyl acetate and immunostaining with anti-CD63 antibody (1:100, sc-15363, Santa Cruz) or anti-OmpA (1:100, #111228, ARC) examined using the Tecnai G2 Spirit transmission electron microscope (FEI Europe). Protein A-gold complex (10 nm) served to detect the primary antibodies. Images were captured with a charge-coupled device camera (Quemesa, Olympus Soft Imaging Solutions GMBH) at magnification 1:49,000.

3.5.3 Protein Isolation and Western Blotting

Isolated EVs were lysed for 20 min on ice in RIPA buffer (#9806, Cell Signaling Technology) containing protease inhibitor cocktail cOmplete ULTRA (Roche) and phosphatase inhibitor cocktail 2 (Sigma-Aldrich). Samples were centrifuged for 10 min at 20,000 g, and supernatants were collected and stored at −20 °C. The protein concentration was measured via BCA Protein Assay Kit (#23227, Pierce). 8 μg of EV proteins was mixed with 4× Laemmli loading buffer (#J60015-AD, Thermo Scientific) and boiled at 95 ° C for 5 min (ddH2O was used to normalize to equal sample volume if needed prior to adding Laemmli buffer). Proteins (8 μg per slot) were concentrated on 4% SDS–PAGE (about 15 min at 10 mA) and separated on 10% SDS–PAGE (about 60 min at 30 mA) and then transferred to nitrocellulose membrane (#741280, Macherey-Nagel) (blotting conditions: 80 min, 300 V/gel). TotalStain kit (AC 2227, Azure Biosystems) was used according to manufacturer protocol for the whole protein labeling. After being blocked for 30 min, membranes were probed at +4 °C overnight with the following antibodies: CD63 (1:1000, sc-15,363, Santa Cruz) and OmpA (1:1000, #111228, ARC) in 5% milk. The respective secondary peroxidase-conjugated IgG antibodies (1:5000, P0448, Agilent Technologies) were then applied to the membranes. The Lumi-Light Western Blotting Substrate (Roche Diagnostics) was used to visualize signal. PageRuler Plus Prestained Protein Ladder (#26616, Thermo Scientific) was used as the molecular weight marker.

3.5.4 Analysis of EVs Using ExoView

The ExoView R100 platform (NanoView Biosciences) was used to detect expression of commonly used EV biomarkers. In ExoView, the anti-tetraspanins (CD63, CD81, CD9) or isotype control (mouse-IgG) antibodies (capture probes) are immobilized on the chips to bind the EVs for analysis. At the next step of the assay, the chip-fixed EVs are stained with fluorescently labeled anti-tetraspanins antibodies and imaged in different channels. For each individual capture probe, 3 separate replicates (slots) are performed. The human ExoView Tetraspanin kit (EV-TETRA-C) was used according to the manufacturer's instructions for cargo and surface membrane immuno-fluorescence staining. Briefly, isolated fecal EVs (5×10^4 particles per slot) were carefully loaded on individual slots and incubated for the 24 h. After that, the chips were washed three times on an orbital shaker to remove unbound particles. Afterward, the chips were incubated for 1 h with a cocktail of human anti-CD81 (#555675, BD Pharmingen), human anti-CD63 (#556019, BD Pharmingen), and human anti-CD9 (V P018, BioLegend) fluorescently labeled antibodies. The immunostained chips were washed three times in PBS, once in deionized water and dried. Image and data acquisition of the stained chips were performed with the ExoView R100 (NanoView Biosciences) and the data analysis with the ExoViewer 3 (NanoView Biosciences).

4 Notes

1. The same preparation protocol can be adopted for isolation of BEVs from mouse feces. It successfully worked for 8-week-old mice using feces material taken directly from cecum during surgical intervention. All material obtained was used per isolation. For murine exosome depletion, Exosome Isolation Kit Pan, mouse (Miltenyi Biotec, #130-117-039) was used following the manufacturer's instructions [10].
2. Instead Centricon® Plus-70 centrifugal filter devices (#UFC710008, Merck Millipore) can be used according to manufacturer instructions.
3. To increase the yield, 4–5 g of feces raw material can be taken, triturated in 50 mL PBS, filtered using Vac Filter PES Membrane 1000 mL, 0.45 μm pore, and subsequently concentrated using Centricon® Plus-70 centrifugal filter devices (#UFC710008, Merck Millipore). The average obtained volume of concentrate is about 4 mL. From those 800 μL can be loaded on top of the gradient (2.4 mL per fraction). After gradient ultracentrifugation, first 600 μL from top is thrown away, counted as "0" fraction.

4. Preincubation of feces in PBS on ice makes suspension process easier. Be careful, fecal matter clogs pipets easily and this can cause splashing!
5. If you plan to start overnight ultracentrifugation during the same day, you can prepare gradient solutions at this point to save some time.
6. Do not spin at too high speed or for too long as this may desiccate or compress the resin.
7. To create a gradient, lean the ultracentrifuge tube onto the table at an angle. Pipette the gradient solutions against the tube wall steadily and very slowly without disturbing the interface, maximum 1 mL at a time.
8. The interfaces of the two lowest gradient layers (with the highest density) are usually very evident, while interfaces between upper fractions are less distinguishable.
9. Make sure the adaptors are completely dry. Always use all adaptors for centrifugation, placed in the rotor in their specific order.
10. Provide regular maintenance to adaptors according to manufacture description. For Beckman adaptors: Wash adaptors gently with mild detergent, rinse with both tap water and milli-Q water, and finally rinse inner parts of the adaptors with 70% ethanol. Air-dry adaptors completely, preferably overnight at room temperature. Add lubricant to adaptor cap screw threads. Add grease to both sides of gaskets. If needed, replace worn out gaskets with new ones. Place gaskets and screw caps back into the adaptors.
11. Check first whether empty adaptor pairs are in balance. Then, balance adaptors together with tubes containing gradient (the acceptable difference in weigh is ≤0.05 g for Optima ultracentrifuge), using either PBS or 5% gradient solution.
12. Do the pipetting from the middle of each fraction because the interfaces between the fractions can mix. Carefully keep the pipette tip close to the surface.
13. Alternatively, the fractions can be stored unwashed at +4 °C for short-term storage or at −20 °C for long-time storage.
14. We recommend to pipette fraction directly to PBS: for example, first 2 mL of PBS on the bottom of the tube, then add fraction on top, then add 7 mL of PBS, and mix well by pipetting. Sixteen-hour centrifugation can be used if more convenient for researcher.
15. Preferably make suspension in two steps: Add first 50 μL, pipette thoroughly back and forth, occasionally scratch the round bottom, transfer to collection e-cup, and then add

another 50 μL and repeat the procedure. Do not overpipette, it can lead to partial lysis of membrane and DNA release, visible when TEM is performed.

Acknowledgments

The authors acknowledge Ilkka Miinalainen and Electron microscopy core facility (BioCenter Oulu) for their help with TEM. JR thanks Academy of Finland for funding (project number 243032491).

References

1. Théry C et al (2018) Minimal information for studies of extracellular vesicles 2018 (MISEV2018): a position statement of the International Society for Extracellular Vesicles and update of the MISEV2014 guidelines. J Extracell Vesicles 7(1). https://doi.org/10.1080/20013078.2018.1535750
2. Park KS et al (2018) Sepsis-like systemic inflammation induced by nano-sized extracellular vesicles from feces. Front Microbiol 9 (Aug). https://doi.org/10.3389/fmicb.2018.01735
3. Amatya SB, Salmi S, Kainulainen V, Karihtala P, Reunanen J (2021) Bacterial extracellular vesicles in gastrointestinal tract cancer: an unexplored territory. Cancers 13(21). https://doi.org/10.3390/cancers13215450
4. Kameli N, Borman R, López-Iglesias C, Savelkoul P, Stassen FRM (2021) Characterization of feces-derived bacterial membrane vesicles and the impact of their origin on the inflammatory response. Front Cell Infect Microbiol 11. https://doi.org/10.3389/fcimb.2021.667987
5. Samoylenko A et al (2021) Time-gated Raman spectroscopy and proteomics analyses of hypoxic and normoxic renal carcinoma extracellular vesicles. Sci Rep 11(1):19594. https://doi.org/10.1038/s41598-021-99004-6
6. Royo F, Théry C, Falcón-Pérez JM, Nieuwland R, Witwer KW (2020) Methods for separation and characterization of extracellular vesicles: results of a worldwide survey performed by the ISEV rigor and standardization subcommittee. Cell 9(9). https://doi.org/10.3390/cells9091955
7. Editorial (2021) Updating MISEV: evolving the minimal requirements for studies of extracellular vesicles. J Extracell Vesicles 10:e12182
8. Witwer KW et al (2017) Updating the MISEV minimal requirements for extracellular vesicle studies: building bridges to reproducibility. J Extracell Vesicles 6(1):1396823. https://doi.org/10.1080/20013078.2017.1396823
9. Tulkens J, De Wever O, Hendrix A (2020) Analyzing bacterial extracellular vesicles in human body fluids by orthogonal biophysical separation and biochemical characterization. Nat Protoc 15(1):40. https://doi.org/10.1038/s41596-019-0236-5
10. Mosby CA, Bhar S, Phillips MB, Edelmann MJ, Jones MK (2022) Interaction with mammalian enteric viruses alters outer membrane vesicle production and content by commensal bacteria, vol 11. J Extracell, Vesicles, p e12172
11. Simonsen JB (2017) What are we looking at? Extracellular vesicles, lipoproteins, or both? Circ Res 121(8):920. https://doi.org/10.1161/CIRCRESAHA.117.311767
12. Van Deun J et al (2014) The impact of disparate isolation methods for extracellular vesicles on downstream RNA profiling. J Extracell Vesicles 3(1). https://doi.org/10.3402/jev.v3.24858
13. Bart G et al (2021) Characterization of nucleic acids from extracellular vesicle-enriched human sweat. BMC Genomics 22(1):425. https://doi.org/10.1186/s12864-021-07733-9
14. Mendt M et al (2018) Generation and testing of clinical-grade exosomes for pancreatic cancer. JCI Insight 3(8). https://doi.org/10.1172/jci.insight.99263

Chapter 16

Isolation and Characterization of Extracellular Vesicles from Lymphocytes

Lujain Al-Ghadir and Zhi Chen

Abstract

Extracellular vesicles (EVs) are nanosized particles secreted by all cells as a means of communication. When it comes to the immune system, most of the studies have focused on the regulation of T cells by EVs derived from other cells, such as dendritic cells, tumor cells, and mesenchymal stem cells. Nevertheless, the communication between T cells, and from T cells to other cells via EVs must also exist and influence various physiological and pathological functions. Here, we describe sequential filtration; a new method for the physical isolation of vesicles based on their size. Furthermore, we describe several methods that can be applied to characterize both size and markers of the isolated EVs derived from T cells. This protocol overcomes the limitations of some of the current methods and offers a high yield of EVs from a low number of T cells.

Key words Extracellular vesicles, Exosomes, Microvesicles, Primary T cells, Sequential filtration

1 Introduction

Extracellular vesicles (EVs) are highly "systemically heterogeneous" in their size and cargo composition, which impedes our understanding of their function as intercellular communication agents. Integrating the understanding of physical and molecular EV heterogeneity is a crucial step towards accomplishing many of the EV's promises [1].

Even with the increasing interest of exosomes and microvesicles due to their role in intercellular communication, a few studies have focused on T lymphocytes derived EVs, especially primary T cells. This is due to the difficulty of attaining a mass of high-purity cells, varied responses to different stimulations, a great diversity of cell subsets, and the difficulty of obtaining both high and pure EV isolates. At present, most of the studies have focused on the regulation of T cells by other EVs derived from nonlymphoid cells, such as dendritic cells, mesenchymal stem cells, and tumor cells.

Seppo Vainio (ed.), *Cell-Secreted Vesicles: Methods and Protocols*, Methods in Molecular Biology, vol. 2668,
https://doi.org/10.1007/978-1-0716-3203-1_16,

Nevertheless, the communication between the different subsets of T lymphocytes via EVs must also exist, and the effects of these EVs may be more precise and potent [2].

Several isolations, purifications, detections, and characterization techniques are used to obtain different subsets of EVs from body fluids and cell culture media, each claiming to yield the ideal fraction. The purification method defines the nature of EV subtypes mixture in the isolate, which in turn will determine its biological function and biochemical properties. An exemplary EV preparation would have a high-recovery yield, no contamination with free proteins, no cross-contamination with other EV subgroups, and minor physical or chemical damage following the isolation procedure. So far, there are no technologies to achieve these golden standards for EV preparations, but some strategies, such as the coupling of existing methods, are currently used to attain the "optimal EV preparation" [1].

Only recently, researchers in this field acknowledge the experimental limitations and the difficulty of separating the different subpopulations of EVs, which precludes a clear attribution of a certain function to the different types of released vesicles. In addition, most of the developed methods are intended for purifying EVs from large volumes and rich sources of EVs, such as physiological fluids and a high number of cells. Therefore, the coupling of techniques or an additional washing step of the sample, in the case of a low number of cells, such as primary cells, can yield a high-purity fraction, yet a very low yield that is challenging for any further functional analysis. One asset would certainly be the development of more accurate methods for optimal purification and analysis of different subpopulations of vesicles to which an origin and function can be attributed [3].

To better define the relationships among cargo and EV size and provenance and to separate EVs, we employed a differential filtration approach adapted from [4] with some alterations to isolate intact EVs derived from a low number of T cells. This physical separation method isolates EVs based on their size and therefore deconstructs their heterogeneity. The sequential filtration (SF) protocol is based on decreasing pore size filters to separate medium extracellular vesicles (mEVs; ≥0.22 μm, including microvesicles) and small extracellular vesicles (sEVs; ≥0.02 μm, including exosomes). This protocol offers several advantages over current methods, including a very high yield of concentrated EVs, less impact on EV structure, better separation between EVs, higher recovery of the total RNA content of the EVs, and scalability. Additionally, this method was less time-consuming than differential centrifugation (DC). Therefore, EVs isolated with this method maintained their functional and physical integrity in further assays and characterization with different methods for their purity, yield, and morphology.

2 Materials

Prepare all solutions with Milli-Q water. Store all reagents at room temperature (unless indicated otherwise). Diligently follow all waste disposal regulations when disposing of waste materials.

2.1 Cell Culture

1. Jurkat cells, an immortalized line of human T lymphocytes (American Type Culture Collection, Manassas, Va).
2. RPMI 1640 medium with L-glutamine and sodium bicarbonate (Sigma-Aldrich, St Louis, MO, USA).
3. Fetal calf serum (Gibco, Waltham, MA, USA).
4. Penicillin–streptomycin–glutamine 100× solution (P/S/G) (Gibco, Waltham, MA, USA).
5. 25 cm^2 and 75 cm^2 Nunclon cell culture flask (Thermo Fisher Scientific, Waltham, MA, USA).
6. Trypan blue (Thermo Fisher Scientific, Waltham, MA, USA).

2.2 Activation of Cells

1. X-VIVO 15 medium (Lonza, Walkersville, MD, USA).
2. PMA (Sigma-Aldrich, St Louis, MO, USA).
3. Ionomycin (Sigma-Aldrich, St Louis, MO, USA).
4. 75 cm^2 Nunclon cell culture flask (Thermo Fisher Scientific, Waltham, MA, USA).

2.3 Isolation of EVs

1. Phosphate-buffered saline (PBS).
2. 1 mL and 20 mL syringe (Henke-Sass, Wolf, Tuttlingen, DE).
3. 2 μm syringe filter (Buckinghamshire, UK).
4. 0.8 μm (Millipore, Burlington, MA, USA).
5. 0.22 μm filter (Sartorius Stedim Biotech, Aubagne, Fr).
6. Nalgene high-speed round-bottom PPCO centrifuge tubes (Thermo Fisher Scientific, Waltham, MA, USA).
7. 1.5 mL and 2 mL low protein-binding Eppendorf tubes (Eppendorf SE, Hamburg, DE).

2.4 Nanoparticle Tracking Analysis (NTA)

1. 1 mL disposable syringe.
2. Fiber-optic Cleaning Tissue for Fiber End.
3. 1.5 mL low protein-binding Eppendorf tubes (Eppendorf SE, Hamburg, DE).
4. Milli-Q water.
5. Syringe pump.

2.5 Transmission Electron Microscopy (TEM)

1. Formvar-coated copper TEM grids (300 mesh, Science Services, München, Germany); coat the grids with carbon and glow-discharged for 1 min before sample application (*see* **Note 1**).
2. Humidity chamber; place a petri dish or plastic chamber with wet paper at the bottom on a bench free of vibrations.
3. PBS.
4. Straight strong point Boley-style tweezers.
5. Platinum loop.
6. 1% glutaraldehyde in PBS (pH 7.4); store it at 4 ° C. Filter it before you use it (*see* **Note 2**).
7. Neutral 2% UA in Milli-Q water; store it at −20 ° C. Spin it down at 13.3 × *g*, 4 ° C for 45 min and filter it right before you use it.
8. 2% methylcellulose-UA (0.4%); stored it at 4 ° C. Spin it down at 13.3 × *g*, 4 ° C for 45 min right before you use it.
9. Icebox.
10. 10 cm glass dish.
11. Cover to protect from light.
12. Filter papers.

2.6 ExoView

1. ExoView Human Tetraspanin Kit (NanoView Biosciences, Boston, USA).
2. Milli-Q water.
3. Orbital shaker or microplate shaker with digital settings capable of shaking at 500 RPM.
4. EMS-style tweezers with Carbon Fiber tips, Fine Tip.
5. Straight strong point Boley-style tweezers.
6. Square/flat tip tweezer.
7. 24-well flat-bottom, cell culture plate.
8. 15 mL and 50 mL falcon tubes.
9. Adhesive PCR plate seals.
10. Plate seal roller.
11. 10 cm sterile Petri dish.
12. Aluminum foil.

2.7 Equipment

1. AH-629/36 rotor for Sorvall WX-90 ultracentrifuge (Thermo Fisher Scientific, Waltham, MA, USA).
2. NanoSight NS300 NTA (Malvern, UK).
3. TECNAI-spirit G 2 TEM (FEI, Eindhoven, Netherlands).
4. ExoView (NanoView Biosciences, Boston, USA).

3 Methods

Carry out all procedures at room temperature unless otherwise specified.

3.1 Cell Line and Culture Conditions

1. Culture $10x10^6$ Jurkat cells in a complete RPMI 1640 medium supplemented with 10% fetal calf serum and 1% P/S/G 100× solution and maintain the cells in a 25 cm^2 Nunclon cell culture flask with a filter cap to allow continuous venting.
2. Incubate the cells in a humidified atmosphere at 37 °C with 5% CO_2.
3. Determine the number and assess the viability of the cells, before adding more or changing the medium by using the trypan blue exclusion assay.
4. Change the medium every 48 h or whenever needed and keep the cell density between 0.3–0.5 × 10^6 cells/ mL of growth media.
5. Passage the cells to a 75 cm^2 cell culture flask once they reached a density of 0.8 × 10^6 cells/ mL.

3.2 In Vitro Activation of Jurkat T Cells

1. Determine the number and assess the viability of the cells, before changing the medium and starting the activation of cells by using the trypan blue exclusion assay.
2. Wash $20x10^6$ of Jurkat cells twice with X-VIVO 15 medium to remove any residue of serum and reduce the contamination of the EV fractions with exogenous FBS-derived EVs.
3. To stimulate the cells and induce the secretion of EVs, culture the cells in X-VIVO 15 medium supplemented with 1% P/S/G 100× solution, 25 ng/mL PMA, and 500 ng/mL ionomycin for 6 h and maintain it in a density of 1.0–3.0 × 10^6 cells/ mL in a 75 cm^2 Nunclon cell culture flask with a filter cap, as reported [5].
4. Centrifuge Jurkat cell suspension for 10 min at 300 × g, 4 °C to pellet the cells and collect the supernatant for the isolation of EVs.
5. Count the cells and assess the viability to give the denominator for the vesicles/cell calculation and to make sure the medium is not heavily contaminated with membrane fragments that are not EVs.
6. Resuspended the cell-pellet either with lysis buffer for RNA and protein isolation, or directly use the pellet for intracellular staining.

3.3 Isolation of T Cell-Derived EVs: Sequential Filtration

1. Deplete the conditioned media of dead cells by low-speed centrifugation at 2000 × *g* for 15 min, 4 °C.
2. Following centrifugation, transfer the supernatant into a 20 mL syringe carefully and discard the pellet.
3. Filter the cleared conditioned media sequentially through a series of syringe filters with decreasing pore size, 2 μm, and 0.8 μm by applying minimal pressure.
4. Split the filtrate to 15 mL per filter and further filter it through the 0.22 μm filter, this filtrate contains the sEVs.
5. Wash the filter with 10 mL of PBS to elute the rest of the sEVs.
6. Recover the captured medium-sized vesicles on the filters by reverse flow in approximately 25 mL of PBS.
7. For RNA and protein isolation, further concentrate the flow-through media by pelleting at 100,000 × *g* UC for 120 min, 4 °C in AH-629/36 rotor for Sorvall WX-90 ultracentrifuge and resuspended in sterile 1× PBS (pH 7.4) as illustrated in (Fig. 1) (*see* **Note 3**).

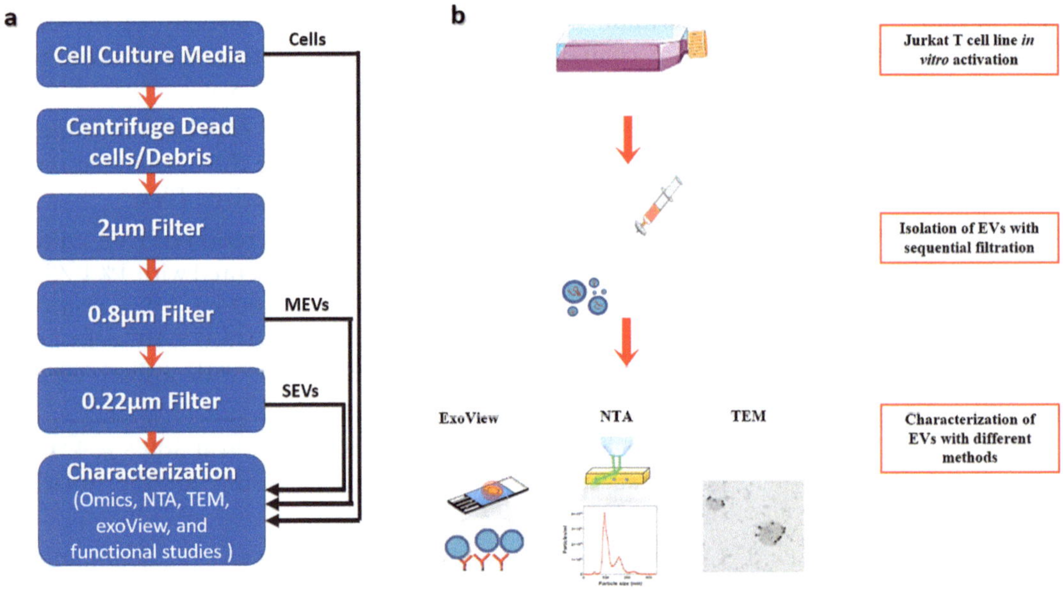

Fig. 1 Schematic representation of experiments workflow. (**a**) The pipeline of the sequential filtration-based vesicle isolation procedure. Cell culture supernatants from primary T cells were cleared of cellular debris and dead cells by low-speed centrifugation, and vesicles were partitioned by filtering the supernatants through consecutively reduced pore sizes (2 μm, 0.8 μm, 0.22 μm). (**b**) Sequential filtration is used to isolate EVs from Jurkat cells. The efficiency and purity of vesicles isolated are determined, and vesicles are characterized using a number of methods: ExoView, nanoparticle tracking analysis (NTA), and transmission electron microscopy (TEM). (The figure was produced partially using art pieces provided by Servier Medical Art. Servier Medical Art by Servier is licensed under Creative Commons Attribution 3.0 Unported License [6])

3.4 Characterization of EVs Isolated from T Cells

3.4.1 Extracellular Vesicle Characterization with NTA System

NTA is a technique that allows real-time visualization and a rapid sizing and counting of cellular vesicles. It is comprised of a laser light scattering microscope, charge-coupled device camera (CCD), and proprietary analytical software. The EVs are visualized by a light scattering using a light microscope and their Brownian motion is then determined by a highly sensitive CCD camera and the mean velocity of each vesicle is computed with image processing software. EVs from 30 to 1000 nm in diameter at a total concentration range of 10^8–10^9 can be counted with relatively high sensitivity. The NTA software is then able to identify and track individual EVs moving under Brownian motion and relates the rate of movement to the particle size [7].

Purified Jurkat cells derived EVs can be analyzed through NanoSight NS300 NTA for quantitation and size distribution.

The following settings in the protocol were set according to the manufacturer's software manual (NanoSight NS300 User Manual, MAN0541- 01-EN-00, 2017).

1. Once you set the top plate in its place and connect any tubing, the system is ready to load a sample. You can load the samples into the chamber with a syringe manually or using a syringe pump accessory connected to the inlet tubing (*see* **Note 4**).
2. Dilute all samples as a start to 1:100 in filtered sterile Milli-Q water to a final volume of 1 mL (*see* **Note 5**).
3. Fill a syringe with the sample after vortexing.
4. Remove any air bubbles from the syringe.
5. Place the syringe into the luer port on the end of the inlet tubing, ensuring liquid-to-liquid contact if the system has already been loaded with buffer solution.
6. Start injecting the sample slowly into the chamber. The top plate should not be loaded at speeds exceeding 1 mL/ 10 s.
7. In case of loading the sample with a syringe pump, once the liquid is seen emerging in the waste tubing, place the syringe into the syringe pump holder as described in the NanoSight Syringe Pump Operating Manual.
8. In the NTA software, once an image can be seen on the capture screen, finetune the focus as appropriate and increase camera level until all particles are distinctly visible, but when the image starts to show colored pixels, which means the light from the image is saturating, then reduce the camera level until approximately 10% of particles displaying colored pixels.
9. Do not exceed a particle signal saturation over 20% (level 14).
10. Determine the ideal detection threshold to include as many particles as possible with the restrictions that 10–100 red

crosses are counted while only <10% are not associated with distinct particles and the blue cross count is limited to 5. Usually, a detection threshold of 3 is ideal for EVs.

11. Set eight videos of 1 min to capture and then are averaged for each measurement, under the following conditions: cell temperature: 25 °C; syringe speed: 50 μL/s.
12. Start the capture using "Create and Run Script." The recording can be canceled at any time by using the red "Abort" button. After capture, the videos will be analyzed by the in-build NanoSight Software NTA 3.3 Build 3.3.104. Hardware: embedded blue laser: 45 milliwatts at 405 nm (Violate); camera: scientific complementary metal-oxide-semiconductor.
13. Make sure that the particle concentration is in the range of 5×10^7–9.9×10^9 particles/mL, which is approximately 20–100 particles in the field of the view (*see* **Note 6**).
14. Make sure that the number of completed tracks in NTA measurements is always greater than the proposed minimum of 1000 in order to minimize data skewing based on single large particles (Fig. 2).

3.4.2 TEM Imaging of EVs

Given their small size, EVs isolated from human Jurkat T cells can only be directly observed under TEM. This will allow you to visually inspect the vesicle population and observe the effect of the isolation method on their morphology and evaluate their purity. Isolated EVs with different methods can be negatively stained for TEM as shown in (Fig. 3) to micrograph the EVs.

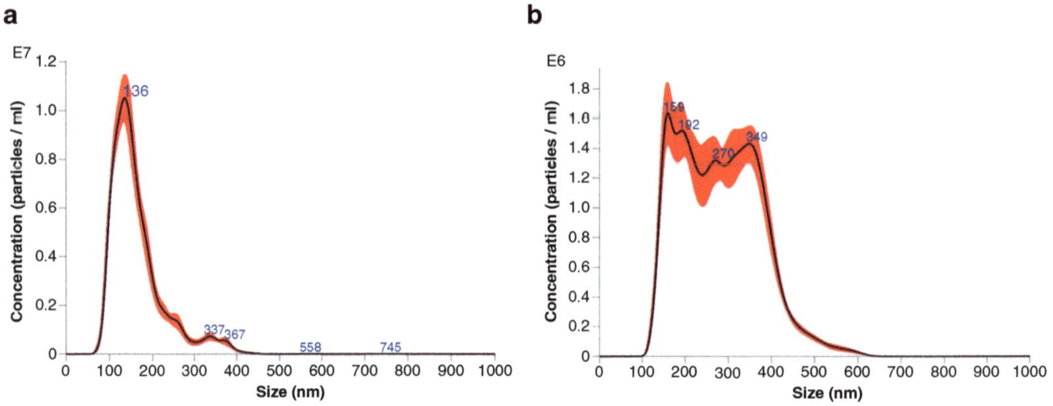

Fig. 2 Results from NTA analysis of EVs purified from Jurkat T cells. (**a**) Small EVs derived from Jurkat cells (≤200 nm, including exosomes) are filtered from a 0.22 μm filter and (**b**) microsize (≤800 nm, including microvesicles) captured on the filter. The curve describes the relationship between particle number distribution (left *y*-axis) and particle size (*x*-axis)

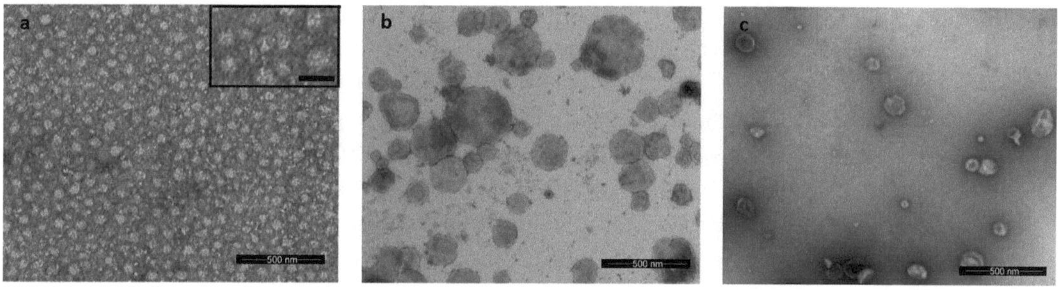

Fig. 3 Quality control of the fraction separation. Representative photomicrographs of negative stain transmission electron microscopy (TEM; 30,000 × magnification) of EVs derived from Jurkat T cells isolated with different methods: sequential filtration (SF) indicating that intact vesicles of (**a**) small size (≤200 nm, including exosomes) are filtered from 0.22 μm filter and (**b**) medium size (≤800 nm, including microvesicles) captured on the filter compared to (**c**) differential centrifugation (DC) (scale bars 500 nm)

1. Perform all the incubation and washing steps by floating the grid on top of droplets of the different solutions used in this protocol, place the droplets on parafilm in the humidity chamber except for **step 10** (*see* **Note** 7).
2. Mix gently the sample solution with a pipette and use at least 3 μL of sample/grid.
3. Place the coated grids on the parafilm using the tweezers, and pipette 3 μL of resuspended EVs with 1× PBS (pH 7.4) on top of the grid (*see* **Notes** 7 and **8**).
4. Incubate the grids and sample solution for 20 min to allow the membrane to absorb (*see* **Note 9**).
5. Transfer the grids using the tweezers to 100 μL droplets of 1× PBS (pH 7.4) to wash the grids twice each in a fresh droplet for 1 min (*see* **Notes** 7 and **10**).
6. Fix the samples by incubating grids on a drop of filtered 1% glutaraldehyde in 1× PBS (pH 7.4) for 5 min in the fume hood (*see* **Notes** 7 and **11**).
7. Wash the grids 8 times for 1 min each time in a fresh drop of Milli-Q water to make sure that the membrane of the grid is never dried out but remained wet (*see* **Notes** 7 and **12**).
8. Contrast the grids with neutral 2% uranyl acetate in Milli-Q water for 5 min in the fume hood. Do this step in the dark as neutral 2% uranyl acetate is sensitive to light.
9. Place 10 cm glass dish with wet parafilm on it on ice.
10. Put 4 fresh drops on the wet parafilm vertically: 1 of Milli-Q water and 3 of 2% methylcellulose- uranyl acetate (0.4%) solution for each grid. Ensure that the last drop of 2% methylcellulose- uranyl acetate (0.4%) is large in size.

11. Transfer the grid using the tweezers to the drop of Milli-Q water to wash quickly and dry the edges of the grid with wet tissues to remove any residue of Milli-Q water.
12. Dip the grid in the first and second drops of 2% methylcellulose-uranyl acetate (0.4%) solution quickly using the loop and then place on the third to coat on ice for 5 min. Protect samples from light by using a cover (*see* **Note 13**).
13. Repeat **steps 10–12** for the rest of the samples.
14. Remove each grid with the help of the loop and drag the loop in angel on filter paper to remove most of the solution.
15. Place the edge of a small piece of filter paper cut into a triangle shape to the middle of the loop to remove the rest of the solution. Let the grid air dry at 25 °C for at least 10 min while keeping the grid on the loop.
16. Repeat **steps 14** and **16** for the rest of the samples.
17. The girds are ready to be visualized or transferred to a grid storage box using the tweezers (*see* **Note 14**).

The grids were visualized at 80 kV, and images were acquired using TECNAI-spirit G 2 TEM, which has 30,000 × magnification for EV visualization. iTEM 5.2 software (QUEMESA, Olympus Soft Imaging Solutions) was used for imaging. Image pixel sizes that were obtained from TEM were 0.462 nm. Vesicle preparations were captured on the 0.22 μm-sized filter and its flow-through revealed size and morphology consistent with those reported for sEVs (exosomes) and mEVs (microvesicles), respectively, as seen on the micrographs (Fig. 3a, b). TEM also showed a more uniform distribution in size in the filtrates compared with the complex, heterogeneous mixture observed in 100,000 × *g* of DC (Fig. 3b). Interestingly, the EVs isolated with SF appeared to have less granular background staining than the other DC, presumably because this approach pellets both EVs and other particles. In comparison with SF, the morphology of EVs was altered mostly in the sample isolated by DC due to the high pressure that ruptures the membrane of EVs and could further affect functional characterization. On the other hand, SF allowed for intact regular-shaped vesicles to be filtrated.

3.4.3 Extracellular Vesicles Analyses with ExoView

ExoView is a newly developed technique that provides a single vesicle characterization by using a tetraspanin chip labeled with distinct antibodies to immunocapture EVs through their surface markers. The size and concentration of EVs are also automatically measured. This technique helps also in understanding whether specific surface markers are colocalized on individual vesicles [8]. For this part, use the ExoView Human Tetraspanin Kit by following the manufacturer's instructions.

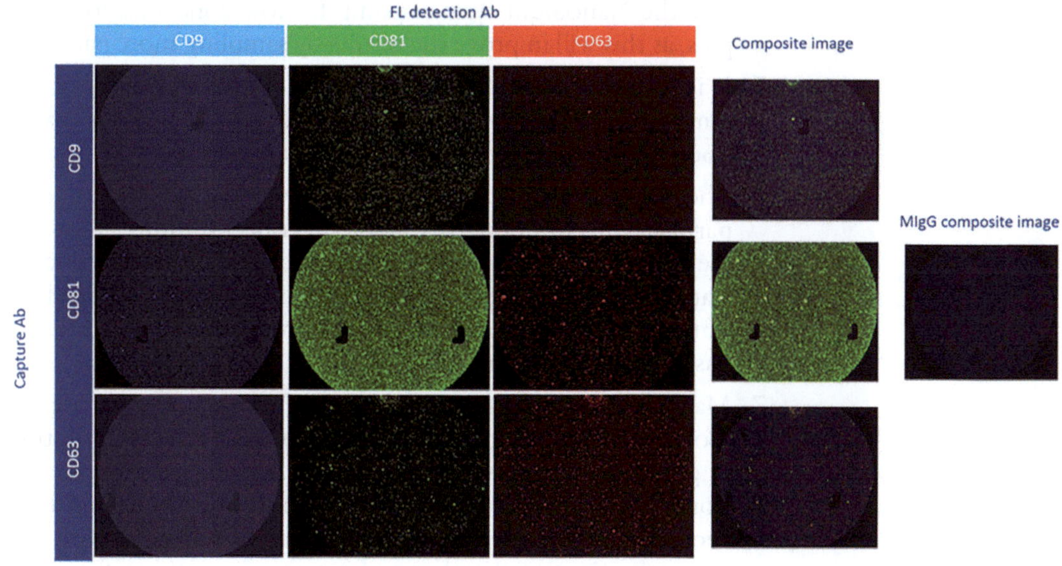

Fig. 4 Characterization of single extracellular vesicle surface markers. Representative photomicrographs of immunofluorescence labeling of exosomal markers from small extracellular vesicles filtered fractions (≤200 nm) isolated from Jurkat cells. Images were acquired on the three different fluorescence channels: blue particles are vesicles positive for CD63; green particles are vesicles positive for CD81 whereas red particles are vesicles positive for CD9

The whole spot images from the analyses show a brilliant display of the canonical exosomal markers (CD9, CD81, and CD63) stained individual vesicles from the 0.02 μm fractions, revealing different phenotypes (Fig. 4). The highest single was observed in CD81-positive EVs. However, single-positive CD63 and CD9 vesicles exhibit low signals, in comparison with CD81.

4 Notes

1. Use the coated grids within 1 h after coating them with glow discharge, if not used, then the coating has to be redone. Always coat more grids than what you need to have extra in case they were damaged during handling and staining.
2. Once the color of 1% glutaraldehyde in PBS (pH 7.4) turns yellow, then it is no longer useable.
3. Running the samples at this high speed with any slight differences in the uneven distribution of weight around the rotor can cause the centrifuge to explode. Thus, do not forget to balance your sample tubes in the rotor buckets by weighing them using a digital precise scale. Place the buckets of the same weight in the opposite direction in the rotor. Run all the buckets in the rotor even if you are not using them because it will help with maintaining the balance.

4. Use the NanoSight syringe pump for low-concentration samples, as this will improve the results by sampling more particles.
5. It is recommended that before proceeding with the analysis, a sample of any buffer or diluent is checked to confirm that it does not contain any contaminating nanoparticles.
6. The Nanosight instrument can only work with this range of particle concentration. An exorbitant concentration of the sample may prevent accurate particle tracking and a lower concentration requires longer capture and analysis time to produce statistically significant results. Therefore, adjust your dilution based on this concentration.
7. Make sure that the size of the droplets is big enough to use the loop to transfer the grids between droplets of the same solution (e.g., washes), and tweezers between droplets of different solutions. Using the tweezers to transfer the grids to the next droplet will also help to dry the edges of the grids with the wet paper in the chamber to remove any residue from the previous droplet. Always hold the grid from the edges to avoid breaking it.
8. The shiny side of the grid should be at the bottom facing the parafilm.
9. Centrifuge neutral 2% UA and 2% methylcellulose-UA (0.4%) right after the incubation time is over so they will be ready to use later in the protocol.
10. After incubation is done and before washing the grids, turn the gird so that the shiny side is facing upwards and keep it like that for the rest of the process.
11. Exposure to glutaraldehyde has been commonly associated with symptoms of respiratory tract irritation; thus, once the 5 min is over, use a paster pipette to take the droplets and discard the solution in its specific waste bottle. Keep the paster pipette with the syringe you used in the fume hood for at least 10 min before discarding them. Wash the filter with 50 mL of water that is added to the solution-specific waste bottle before you discard the filter in the regular waste bin.
12. You can keep the grids in Milli-Q water for a longer time.
13. You can use the rest of 2% methylcellulose-UA (0.4%) for future staining so store it at 4 °C. However, pipette back the used droplets into an Eppendorf tube and rinse it with 50 mL of water before discarding it into its specific waste.
14. Discard all the solutions you used under the fume hood in their specific waste bottles.

Acknowledgments

This work was supported by the Academy of Finland grant No. 325965 (Z.C.) and the University of Oulu.

References

1. Colombo M, Raposo G, Théry C (2014) Biogenesis, secretion, and intercellular interactions of exosomes and other extracellular vesicles. Annu Rev Cell Dev Biol 30:255–289
2. Lu J, Wu J, Tian J, Wang S (2018) Role of T cell-derived exosomes in immunoregulation. Immunol Res 66(3):313–322
3. Gould SJ, Raposo G (2013) As we wait: coping with an imperfect nomenclature for extracellular vesicles. J Extracell Vesicles 2. https://doi.org/10.3402/jev.v2i0.20389
4. Wei Z, Batagov AO, Schinelli S, Wang J, Wang Y, El Fatimy R, Rabinovsky R, Balaj L, Chen CC, Hochberg F, Carter B, Breakefield XO, Krichevsky AM (2017) Coding and non-coding landscape of extracellular RNA released by human glioma stem cells. Nat Commun 8: 1145
5. Németh A et al (2017) Antibiotic-induced release of small extracellular vesicles (exosomes) with surface-associated DNA. Sci Rep 7(1):8202
6. SMART. Servier Medical Art. http://smart.servier.com/
7. Dragovic RA, Gardiner C, Brooks AS, Tannetta DS, Ferguson DJP, Hole P, Carr B, Redman CWG, Harris AL, Dobson PJ, Harrison P, Sargent IL (2011) Sizing and phenotyping of cellular vesicles using nanoparticle tracking analysis. Nanomed Nanotechnol Biol Med 7(6):780–788
8. Holliday LS, Truzman E, Zuo J, Han G, Torres-Medina R, Rody WJ (2019) Extracellular vesicle identification in tooth movement models. Orthod Craniofac Res 22(S1):101–106

Chapter 17

Biogenesis of Mesoporous Silica Nanoparticles Enclosed in Extracellular Vesicles by Mouse Renal Adenocarcinoma Cells

Feby Wijaya Pratiwi, Keerthanaa Balasubramanian Shanthi, Olha Makieieva, Zih An Chen, Artem Zhyvolozhnyi, Ilkka Miinalainen, Genevieve Bart, Anatoliy Samoylenko, and Si-Han Wu

Abstract

Integrating the versatility of synthetic nanoparticles to natural biomaterials, such as cells or cell membranes, has gained considerable attention as promising alternative cargo delivery platforms in recent years. Extracellular vesicles (EVs), natural nanomaterials composed of a protein-rich lipid bilayer secreted by cells, have also shown advantages and great potential as a nano delivery platform in combination with synthetic particles due to their specific natural properties in overcoming several biology hurdles possessed in the recipient cell. Therefore, the preservation of EV's origin properties is critical for their application as nanocarriers. This chapter will describe the encapsulation procedure of MSN encapsulated in EV membrane derived from mouse renal adenocarcinoma (Renca) cells through biogenesis. The FMSN-enclosed EVs produced through this approach still contain preserved EV's natural membrane properties.

Key words Mesoporous silica nanoparticle, Extracellular vesicle, Biogenesis, Hybrid nanoparticles

1 Introduction

Integrating the versatility of synthetic nanoparticles to natural biomaterials, such as cells or cell membranes, has gained considerable attention as promising alternative cargo delivery platforms in recent years. Nanoparticles coated with various cell membranes from red blood cells (RBCs), cancer cells [2, 7], platelets [3], white blood cells [15], and macrophages [6] could significantly suppress non-specific cytotoxicity, improve circulation, and targeting capacity. Extracellular vesicles (EVs), as key players in intercellular communication, are natural nanomaterials composed of a protein-rich lipid bilayer secreted by mammalian cells. EVs have shown advantages in immune-evasiveness when used autologously, overcoming

Seppo Vainio (ed.), *Cell-Secreted Vesicles: Methods and Protocols*, Methods in Molecular Biology, vol. 2668,
https://doi.org/10.1007/978-1-0716-3203-1_17,

biological barriers, enhancing cellular uptake, and target-homing characteristics to distinctive cells depending on the EV's membrane proteins [4, 18]. Given that the specific molecules on the EV surface membrane may be essential to their role, preservation of EV membrane integrity and stability is critical for their application as nanocarriers.

Construction of EV-biomimetic nanoparticles through iterative physical extrusion or freeze/thaw process to fuse EVs and nanoparticles may affect the protein integrity on vesicle membranes, thereby compromising their biofunctions [5, 12, 14]. Therefore, an efficient approach to construct EV-biomimetic nanoparticles without interfering with the membrane integrity is vital to develop a targeted nanodelivery platform. Mesoporous silica nanoparticles (MSN) have been widely used as drug carriers due to their excellent drug-loading capacity, facile surface functionality, and biocompatibility [1, 8, 10, 19]. This chapter will describe MSN encapsulated in EV membrane derived from mouse renal adenocarcinoma (Renca) cells. A step-by-step procedure of preparing the fluorescence MSN (FMSN), Renca cell culture, producing the FMSN enclosed EVs, and several characterization and analysis methods (Fig. 1) will be discussed in the following section.

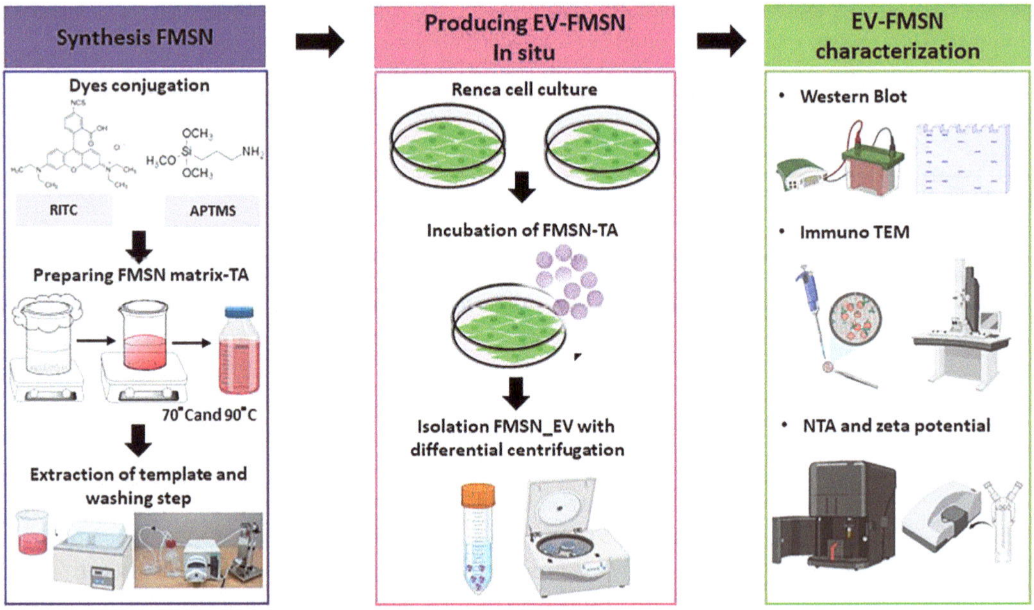

Fig. 1 Schematic of EV-FMSN production and characterization

2 Materials

2.1 Fluorescence Mesoporous Silica Nanoparticle 25 nm TA

- Tetraethyl orthosilicate (TEOS, 98%).
- Silane polyethylene glycol (Silane-PEG) and trimethylammonium chloride (Silane-PEG-TA).
- Hydrochloric acid (37%).
- Cetyltrimethylammonium bromide (CTAB, >99%).
- Aminopropyltrimethoxysilane (APTMS, 95%).
- Rhodamine B isothiocyanate (RITC).
- Ethyl alcohol (EtOH, 99%).
- Ammonia hydroxide (35 wt%) and ammonium nitrate (99%).

2.2 Producing Mouse Renal Adenocarcinoma Cell-Derived EV-FMSN

- Mouse renal adenocarcinoma-derived cells (Renca cell line) purchased from ATCC® product numbers # CRL-294.
- EV-depleted cell culture medium is used in the experiment.

  Centrifugate the glutamine-containing Dulbecco's modified Eagle's medium (DMEM)/F-12, supplemented with 10% fetal bovine serum or free FBS, and 1% penicillin–streptomycin the medium contains 10% FBS at 100,000*g* overnight to remove the existing EV.
- EDTA Trypsin.
- T75 culture flasks, 15 cm dish.
- Centricon Plus-70 filter units (Merck Millipore, cut-off 100 K).
- Centrifuge tube (11 mL).

2.3 Western Blot

All the buffer and solution required in the western blots experiment are listed in Table 1.

2.4 Negative Staining TEM and Immuno-TEM

- Formvar-coated copper grids (400 mesh).
- Grid coated with carbon and glow-discharged.

  Before sample application, place the Formvar-coated copper grids (400 mesh) into a GloQube plus Glow Discharge system using tweezers to perform the glow discharge process (one grid is used for each sample) for 1 min.
- Humidity chamber (petri dish or plastic chamber with wet paper towel at the bottom).
- 1% glutaraldehyde in PBS (pH 7.4)
- neutral 2% uranyl acetate (UA) in H_2O.
- 1% BSA in PBS (pH 7.4).
- 2% methylcellulose-UA (0.4%).

Table 1
List of the Buffer and solution required in western blot

No	Solution	Component
1	1× SDS PAGE running buffer	25 mM Tris 192 mM glycine 0.1% SDS
2	1× transfer buffer (store at 4 °C)	25 mM Tris 192 mM glycine 20% ethanol
3	1× tris-buffered saline, 0.1% tween 20 (1× TBST)	20 mM Tris 150 mM sodium chloride pH 7.6 0.1% tween 20
4	Blocking solution	5% BSA in 1× TBST
5	Sample loading buffer (4×):	8% SDS 20% β mercaptoethanol 40% glycerol 0.008% bromophenol blue 0.125 M Tris-HCl pH 6.8

- Primary rat anti-mouse CD81 antibody Anti-CD81 Antibody (166029, Santa Cruz1:100 dilution).
- protein A-gold complex (10 nm size).

3 Methods

3.1 Preparing Fluorescence Mesoporous Silica Nanoparticle 25 nm TA

Synthesis of 25 nm polyethylene glycol (PEG)-TA MSNs containing rhodamine B isothiocyanate (RITC) based on a Stöber-like process developed by Mou et al. [11, 16] consists of the following steps: (1) preparation of rhodamine B isothiocyanate (RITC) conjugated with 3-aminopropyltrimethoxysilane (APTMS), (2) synthesis of fluorescence-labeled nanoparticles, and (3) extraction the CTAB template to form porous nanoparticle and washing step (Fig. 2).

Preparing RITC conjugation with APTMS

RITC should be placed at room temperature for 30 min before use. Use a brown sample bottle for dye preparation. Be careful when opening APTMS. It is easy to hydrolyze when exposed to air.

1. Dissolve 8 mg RITC in 5 mL 99.5% ethanol and stir for 15 min in the dark
2. Add 10 μL of APTMS in the dye solution and continuously stir for 1 h

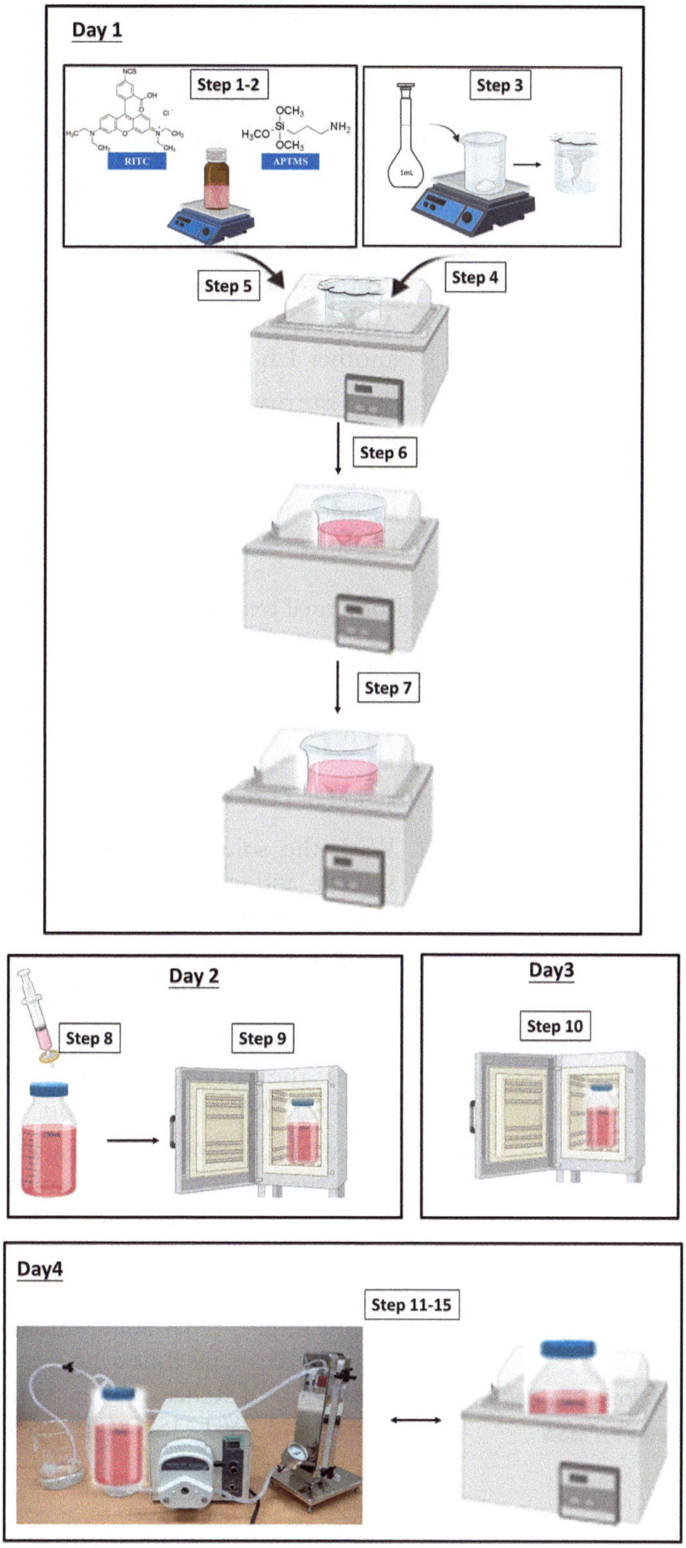

Fig. 2 Synthesis 25 nm fluorescence MSN-TA based on a Stöber-like process

Synthesis FMSN decorated with TA

3. Dissolve 7.0 g NH_3 28–30 wt% in 1 L ddH_2O (prepare it in a volumetric flask)
4. Place 150 mL of the prepared NH_3 solution in the beaker glass and add 0.29 g CTAB in the mixture solution and stir for 15 min in a water bath at temperature 60 °C
5. Add 2.5 mL RITC-APTMS and add 2 mL of TEOS drop by drop and stir for another 1 h at 60 °C
6. Add Silane-PEG or Silane-PEG TA drop by drop and stir for another 1 h at 60 °C
7. Stop stirring and keep the solution at 60 °C until volume decrease to one-third of the initial volume (overnight)
8. Filter the solution through a filter syringe (0.22 μm) and place it into another bottle with a cap for the hydrothermal treatment

Hydrothermal treatment

9. Place the synthesis solution in the oven at 70 °C for 16 h
10. Keep the synthesis solution in the oven at 90 °C for another 16 h

Extraction of pore template and washing step

11. Wash the solution through crossflow system using ethanol solvent and continue with water to collect the MSN and remove the other excess chemical in the solution
12. Extract the CTAB template using 339 μL HCl solution (36.5–38%)/in 20 mL 99.5% ethanol at 60 °C, stir 1 h
13. Repeat the washing step using the crossflow system in **step 10**
14. Repeat the extraction using 39 μL HCl solution (36.5–38%)/ in 20 mL 99.5% ethanol at 60 °C, stir 1 h
15. Repeat the washing step using crossflow system in **step 10** and finish the last step washing using a solvent which is intended to be used, for example, ethanol for store purposes or water when we want to use it immediately for a biology experiment

3.2 Producing Mouse Renal Adenocarcinoma Cell-Derived EV-FMSN

Producing mouse renal adenocarcinoma cells-derived EV-FMSN consists of three mains following steps: (1) culturing Renca cell line, (2) incubation of Renca cell line with FMSN, and (3) isolation of Renca EVs and EV-FMSN (Fig. 3).

Culturing the Renca cell line

1. Culture and maintain under standard conditions (37 °C, 5% CO_2, 95% humidity) in EV-depleted cell culture medium

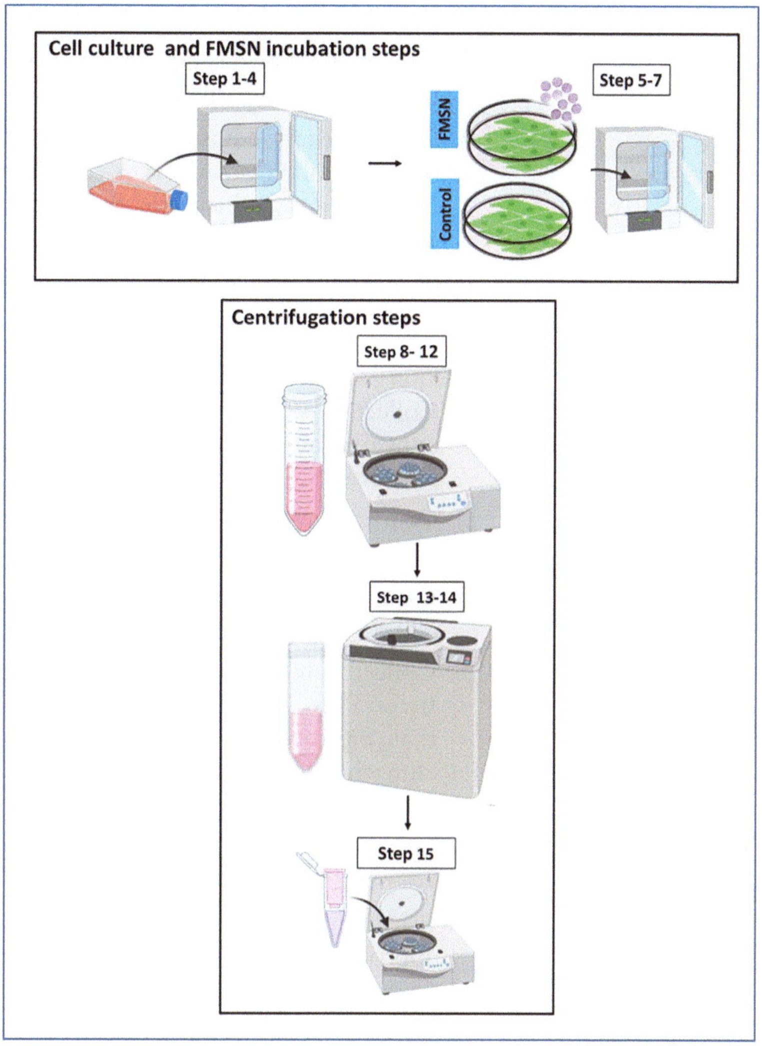

Fig. 3 Producing FMSN enclosed membrane vesicle via EV biogenesis from Renca cell

2. Prior to the initiation of experiments, maintain the cells at ≤75% confluency for at least 72 h by routine sub-culturing involving trypsinization, centrifugation (1000 × *g*, 5 min), and re-seeding in T75 culture flasks.
3. Determine the number of cells: mix the 20 μL of cell suspension with 20 μL of Trypan blue, measure 10 μL of the cell suspension in the TC20 Automated Cell Counter, and calculate the number of cells per mL.
4. Plate approximately 5×10^6 cells per 15 cm dish and incubate it overnight at 37 °C incubator. Check the cultured cell and replace the medium every 3 days until it reaches 80% confluency.

Incubation of Renca cell line with FMSN (Time 16 h)

5. Prepare FMSN dispersions (0.125 mg/mL) in fresh culture medium without FBS supplement
6. Aspirate the medium from the cell, wash with PBS 1×, and replace with fresh medium containing the MSN or without MSN (as control) and incubate for 6 h
7. After this point, discharge the medium which contained non-uptake MSN, wash with PBS1× for 2×, and replace it with fresh medium without FBS

Isolation of Renca EVs and EV-FMSN (Time 1 h)

8. Collect the culture medium which contains EV-FMSN and control culture medium (cells without treatment) after 16-h incubation
9. Centrifuge the culture medium from **step 8** at 5000 rpm for 15 min at 4 °C to remove all cell debris
10. Carefully collect the supernatant to a new 50 mL polypropylene tube (*see* **Note 1**).
11. Concentrate the supernatant collected at **step 10** using Centricon Plus-70 filter units at 4 °C for 10–20 min at 2500*g*. After that, flip the tube and centrifuge again at 1000*g* for 5 min to elute the solution from the filter.
12. Transfer the eluted concentrated samples to a polypropylene centrifuge tube (11 mL) and dilute with PBS to the total volume of 7 mL
13. Place the tube in the Type Sorvall TH-641 rotor with an appropriate balance and centrifuge at 40,000*g* for 15 h at 4 ° C. Set the acceleration and deceleration at the maximum rate to pellet out the EV-FMSN
14. Discard the supernatant carefully. Re-suspend the pellet (EVs or EV-FMSNs) in 200–500 μL of PBS (*see* **Note 2**)
15. Purify the obtained suspensions using Exo spin kit (EXO3, Cell Guidance Systems Ltd) according to the manufacturer's protocol. Each sample was eluted in 200 μL of PBS (*see* **Note 2**)

3.3 Characterization of Renca EV-FMSN

The following section has included several approaches to characterizing EVs and EV-FMSN isolated as suggested in the MISEV guidelines [13, 17]. The characterization includes TEM, NTA, and zeta potential to observe the morphology, size, and surface charge, respectively. At the same time, immune-TEM and Western blot are used to characterize the surface protein marker. For some EV characterization experiments, it is required to measure the EV protein and concentration. For EV protein measurement, we use a Pierce™ BCA Protein Assay Kit (Thermo Fisher Scientific, 23225) following the manufacturer's protocol.

3.3.1 Nanotracking Analysis (Time 1–2 h)

The concentrations and size distributions of the EV samples are measured using nano tracking analysis using the Malvern P analytical Nano Sight NM300 instrument equipped with a 405 nm laser. At the same time, the data were analyzed by NTA software version 3.4. Analysis of variance was performed using GraphPad Prism 8 software, with $p \leq 0.05$ considered statistically significant.

1. Wash the system before starting measurement by loading a 1 mL syringe of DI water without any air bubbles present in the syringe. The flow velocity should not exceed 0.05 mL per second.

   Confirm the cleanliness of the chamber by checking for any particles present in the NTA software image. Repeat the flush if necessary to remove remaining particles.
2. Prepare 1:1000 dilutions of EV, hybrid EV-MSN, and MSN using Milli-Q (MQ) water before measurements. The optimum concentration of the sample measurement is within the linear concentration range of the NTA software (5×10^7 to 8×10^8 particles/mL).
3. Fill 1 mL disposable syringe with sample and remove air bubble in the syringe. Load the sample syringe into the system up to 500 μL.
4. Place the 1 mL syringe, which contains the sample, in the syringe pump holder, then continue the recording of sample measurement.
5. Record each sample with four or eight of 60 s videos with camera level 14 and detection threshold set up at three.

3.3.2 Zeta Potential Measurement (Time 1–1.5 h)

Zeta potential is a general method to measure the surface potential of EVs, while used as an indicator of surface charge and colloidal stability influenced by surface chemistry or bioconjugation [9]. In this experiment (Fig. 4), the surface charge of FMSN, Renca EVs, and EV-FMSN are measured using zeta sizer lab Malvern, equipped with a laser source (wavelength 633 nm) at a scattered angle of 13°. The calibrating of the instrument follows the manufacturer's operating instructions by measuring the known surface charge (40 ± 5.8 mV) of 100 nm polystyrene nanoparticles.

1. Fill the measurement folded capillary cell (cuvette: cell DTS 1070) with Milli-Q water using syringe without introducing the bubble. To do so, fill half of the cell in an upside-down direction then turn the cell in the right position and continue to fill until the liquid level reaches the maximum height marked by the line "FILL MAX" without introducing bubbles.

   Air bubbles trapped in the could disturb the quality of the electrical field. Tap carefully to remove air bubble if it presents in the cell.

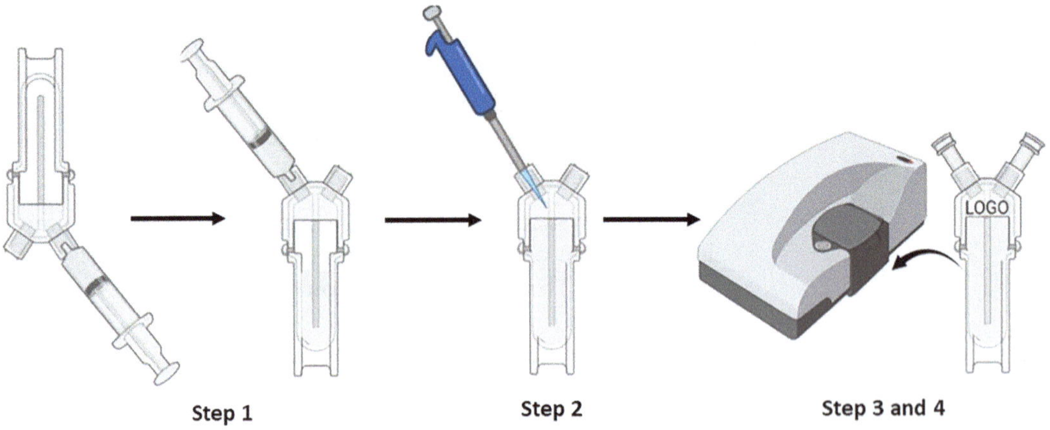

Fig. 4 Measuring the particle's surface potential using Zetasizer

2. Carefully introduce the EV sample (10 μL) in the bottom of the cuvette using PCR tips
3. Place the cap sequentially on the top holes of the filled cell. (One port is tightly capped, and the second cap is then placed on the second port with precaution to avoid causing cell pressure)
4. Place it in the apparatus in the right direction to run measurements. (For cells DTS 1070, the logo is oriented toward the front of the instrument).
5. Measure the zeta potential of sample for thrice at 25 °C, with automatic number of runs and voltage selection at equilibration time 120 s.
6. Analyze the collected data using Zetasizer software following a general-purpose setting as analysis mode.

3.3.3 Western Blot (Time 13 h)

Western blot is the most common method to detect the protein of interest in the samples. Western blot includes four different steps as follows: (1) SDS–PAGE electrophoresis, (2) protein transferring, (3) blocking membrane, and (4) detecting protein of interest. Extracellular vesicles (EVs) are very stable nanoparticles. The success of the western blot lies in most part with the purity of the EV sample. The western blot protocol works even for impure EV samples with 1×10^7 particles/μg of protein. This protocol can be modified according to the purity of your sample by increasing or decreasing the amount of protein used for the experiment. This protocol aims at breaking down the EVs and detection of EV markers.

1. Measure protein concentration of the EV sample. (Using Qubit protein assay or highly sensitive BCA assay). For best results, using 5–10 μg of EV protein is recommended. If the volume for 10 μg is over 25 μL, reduce the volume using a vacuum concentrator.

2. Add the required amount of loading dye to the sample.
3. Heat the samples for 5 min at 95 °C.
4. Prepare the polyacrylamide gel using clean glass plates and combs (12.5% separating gel and 4% stacking gel have given the best results for most EV markers).
5. Load the samples and preferred protein molecular weight visible marker on the gel.
6. Electrophorese at 80 V for 10 min until the samples are in the middle of stacking and separating gel.
7. Electrophorese at 120 V until all the bands of the visible marker are separated.
8. You can keep the gel running at 5–10 V if your blotting setup is not yet ready (*see* **Note 4**).
9. Equilibrate the supported nitrocellulose membrane with the ice-cold transfer buffer for 5 min.
10. Carefully remove the gel from the cast.
11. Arrange the Western blot setup.
12. Blot the membrane for 2–3 h at 200 V at 4 °C.

    As the blotting is done at a very high voltage, it is essential to have a good cooling setup. For example, place the blotting setup inside a cold room at 4 °C, along with the icepack inside the blotting chamber (*see* **Note 5**).
13. Take the membrane out of the blotting chamber.
14. Wash the membrane with 1× TBST.
15. Block the membrane with blocking solution for 1 h at RT.
16. Wash the membrane with 1× TBST 3 times for 5 min (*see* **Note 6**).
17. Incubate the membrane with primary antibody (1:500–1: 1000) diluted with blocking solution overnight at 4 °C.
18. Wash the membrane with 1× TBST 3 times for 5 min.
19. Incubate the membrane with a secondary antibody (1:5000) diluted with blocking solution (HRP conjugated, or fluorophore-conjugated) at RT for 1 h.
20. Wash the membrane with 1× TBST 3 times for 5 min.
21. Use the appropriate imaging method according to the secondary antibody.

3.3.4 Negative Staining TEM and Immuno-TEM

In principle, immunolabeling at the electron microscope (EM) level follows the same precepts as immunolabeling at the light microscope level; in tissues or cells, the location of an antigen of interest is identified by a specific antibody and must be visualized appropriately for investigation. Here, the detailed steps of sample preparation and imaging of EVs using both negative and immunostaining using transmission electron microscopy are explained.

1. Prepare the EV sample in 5–10 μL.
2. Mix the sample solution gently with pipette. Add 3 μL droplet on grid and incubate for20 min.
3. Wash the grid for 2 × 1 min in 100 μL drop of PBS and 1 × 10 min in 100 μL drop of 1% BSA in PBS.
4. Incubate the grid with a 5 μL drop of primary antibody (diluted with 0.1% BSA in PBS) for 20 min and wash the grid 3 × 1 min in a drop of 0.1% BSA in PBS.
5. Incubate the grid on a 5 μL drop of secondary antibody (diluted with 0.1% BSA in PBS) for 20 min and wash the grid 3 × 1 min in a drop of 0.1% BSA in PBS. This step is needed if primary antibody is monoclonal or produced in mouse, rat, or goat.
6. Incubate the grid on a 5 μL drop of gold-conjugated Protein A (dilution depends on product patch) for 20 min and wash the Grid 3 × 1 min in a drop of PBS.
7. Fix the sample by incubating grid on a drop of 1% glutaraldehyde in PBS for 5 min (*see* **Note 7**).
8. Wash the grid 8 × 1 min in a drop of dH_2O (*see* **Note 8**).
9. Stain the grid with neutral 2% UA in H_2O for 5 min (*see* **Note 9**).
10. Embed the grid with 2% methylcellulose-UA (0.4%) solution for 10 min and dried at RT.
11. Quickly remove the grid from the excess UA drop with loop and blot the excess fluid by pushing the loop sideways on Whatman no.1 filter paper so that thin film is left on the EV side of grid.
12. Air-dry the grids at room temperature for about 10 min (cover the grid partially with a culture dish to dry) and store in a grid storage box.
13. Acquire EM images of the sample using a Tecnai G2 Spirit transmission electron microscope (FEI, Eindhoven, The Netherlands). The micrographs collected in this manuscript were captured a charge-coupled device camera (Quemesa, Olympus Soft Imaging Solutions GMBH, Münster, Germany) at low magnification (20,000–40,000×) and high magnification (60,000–80,000×).

3.4 Anticipated Results

The 25 nm of MSN decorated with polyethylene glycol (PEG)-TA containing rhodamine B isothiocyanate (RITC), synthesized based on Stöber-like process, is used as payload carried by Renca EVs [11, 16]. MSN is chosen due to its excellent drug/dyes loading capacity, facile surface functionality, low toxicity, and high cell uptake efficiency. To produce EV-FMSN, the Renca cell line is cultured in an EV-depleted medium with the treatment of FMSN

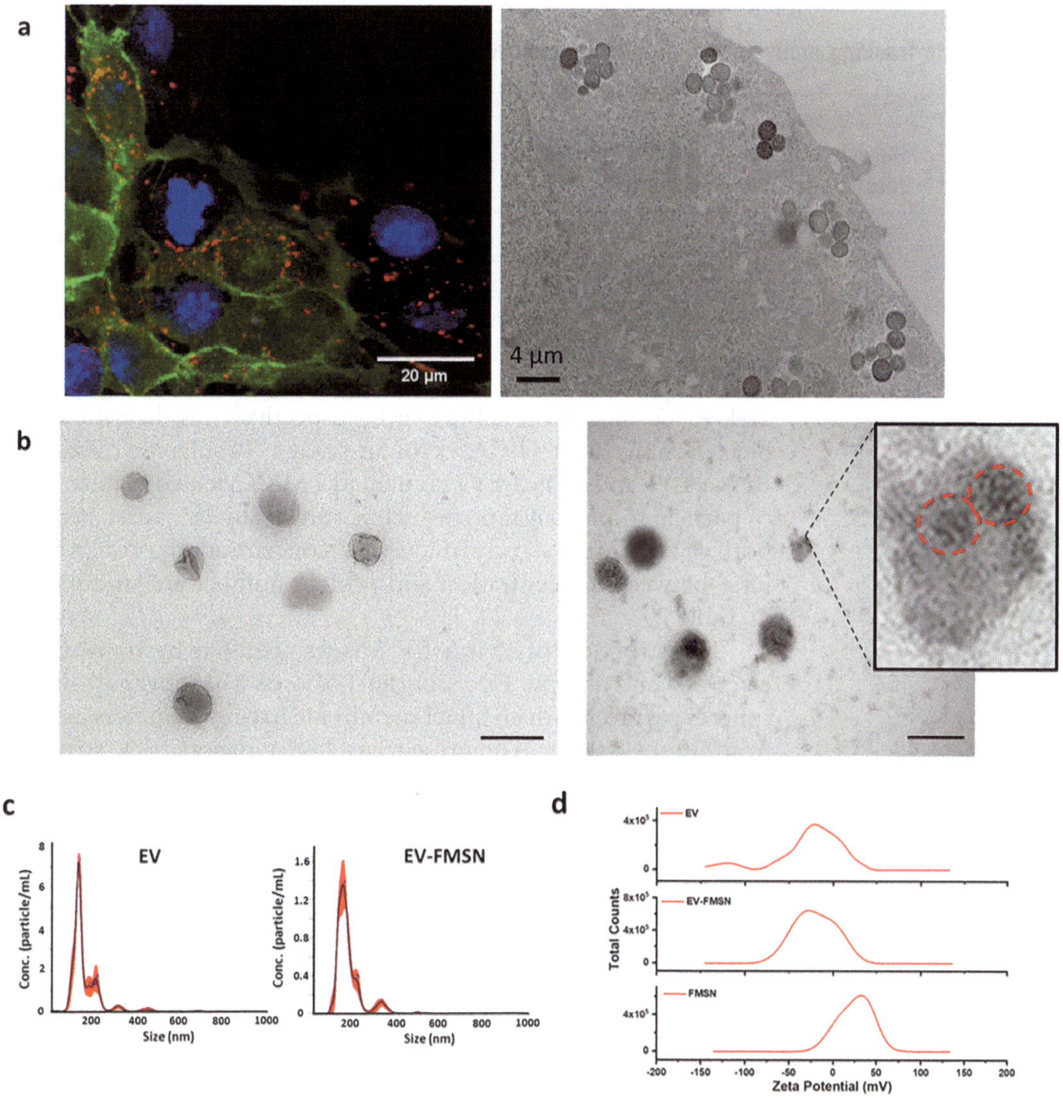

Fig. 5 (**a**) Confocal (left) and TEM (right) image of F-MSN uptake in Renca cell (**b**) TEM images of Renca EVs (left) and EV-FMSN (right). Scale bar, 200 nm. (**c**) NTA-based hydrodynamic size distribution of EVs and EV-FMSN and (**d**) zeta potential of EV, EV-FMSN, and FMSN

for 16 h. Figure 5a shows the uptake of FMSN in punctuate membrane-like vesicle within the cell. The secreted vesicle containing FMSN is isolated from the conditioned medium of Renca by differential centrifugation, ultracentrifugation, and additional purification from other protein and free FMSN presented in the solution by using Exo-spin kit.

The number of particles from 40 mL of cell culture medium determined via nanoparticle tracking analysis (NTA) is $1.38 \times 10^{10} \pm 1.95 \times 10^{7}$ and $1.82 \times 10^{11} \pm 4.58 \times 10^{7}$ particles/mL for Renka EV and EV- FMSN, respectively. The protein content of Renca EVs and Renca EVs contained FMSN is about

Table 2
Nanoparticle tracking analysis and protein concentration of Renca EVs and EV-FMSN

	Particle number (particle/mL)	Protein number (μg/mL)	Ratio particle/ protein (particle/μg)	zeta potential (mV)	Size (nm)
Renca EV	$1.38 \times 10^{11} \pm 1.95 \times 10^{7}$	264 ± 3.7	5.23×10^{8}	−20.13 ± 0.8	151.3 ± 3.2
Renca EV-MSN	$1.82 \times 10^{11} \pm 4.58 \times 10^{7}$	255 ± 2.0	7.14×10^{8}	−22.14 ± 1.2	127.7 ± 3.3

264 ± 3.7 and 255 ± 2.0 μg/mL, respectively (as determined by bicinchoninic acid (BCA) protein assay). Results revealed that Renca EVs and Renca EVs contained FMSN yielded similar value of particle to protein ratios which is shown similar purity between both of vesicle. All data related to EV concentration, protein yield, ratio between concentration and protein number are summarized in Table 2.

The structure of Renca EV images acquired by transmission electron microscopy Fig. 5b (left) shows a donut/cup-shaped spheres particles with an intact membrane having diameters ranging from 30 to 120 nm. The presence of lattice nanoparticle structure surrounding by membrane indicates the encapsulation of FMSN in EVs (Fig. 5b (right)). Moreover, EV-FMSN has an average diameter of 127.7 ± 3.3 nm (EV = 151.3 ± 3.2 nm) and zeta potential of −22.14 ± 1.2 mV (EV = −20.13 ± 0.8 mV) shown in Table 2 consistent with TEM morphology (Fig. 5a–d). Thus, the size and surface potential of EV- FMSN are similar to EVs indicate that the FMSN encapsulation did not significantly change the EV morphology.

Moreover, the Western blot analysis (Fig. 6a) showed the presence of EV-specific surface markers, such as CD63, and CD81, both in Renca EVs and in Renca EV-FMSN, indicating that Renca EV successfully encapsulates the FMSNs. This result is also supported by the Immuno-TEM image (Fig. 6b, c), which shows the CD-81 antibody binding on the surface of EV.

4 Notes

1. This step can be performed at room temperature. Be careful not to disturb the pellet or transfer any part of the pellet. Do not leave the samples for too long after the rotor has stopped to prevent part of the pellet from diffusing into the supernatant. Do not touch the pellet during the supernatant transfer. It is also preferable to leave 5 mL of supernatant. The pellet (cell) re-suspended in PBS can be used immediately or stored at − 80 °C.

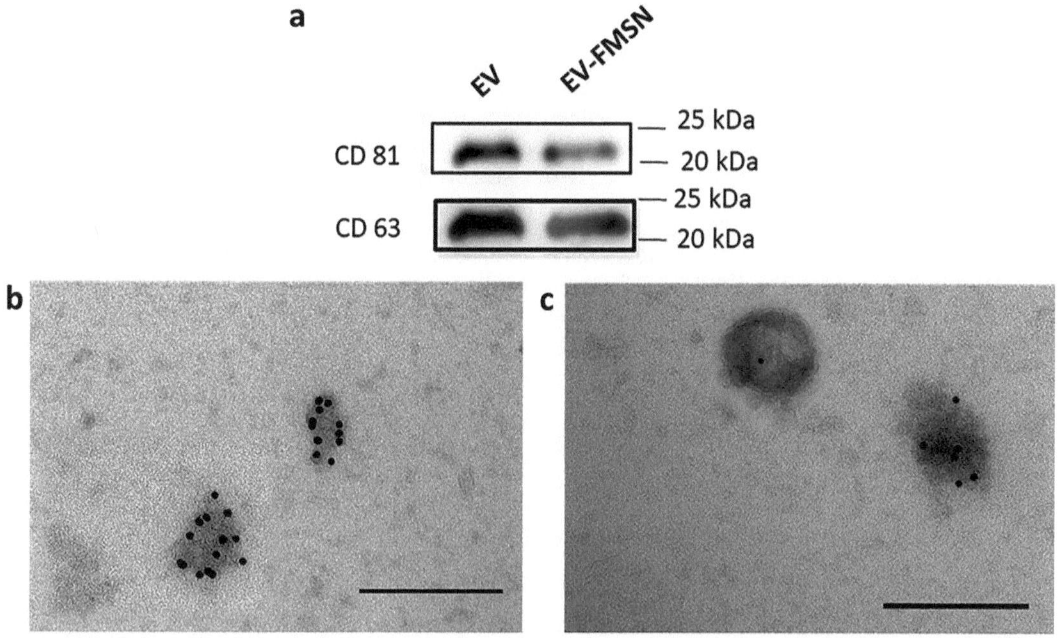

Fig. 6 (**a**) Western blot analysis with EV-specific markers: EVs isolated from the conditioned medium of Renca cells were probed/immunoblotted with CD63 (D263-3, MBL) and CD81 (166029, Santa Cruz). Immuno-TEM against CD-81 antibody on Renca (**b**) EVs and (**c**) EV-FMSN. Scale bar 200 nm

2. This step can be performed at room temperature. Be careful not to disturb the pellet or discard any part of the pellet. To avoid this, do not leave the samples for too long after the rotor has stopped, to prevent part of the pellet from diffusing into the supernatant. Try to remove all the remaining supernatant on the tube wall.
3. The Renca EVs or EV-FMSN can be used immediately or stored in a tube wrapped with aluminum foil at −80 °C.
4. Before leaving the electrophorese, make sure the inner chamber of the western blot setup is not leaking. In case of leaking, uneven separation of bands occurs.
5. Do not use the blotting buffer more than twice. The blotting buffer should be ice-cold for the blotting.
6. Do not let the membrane dry at or after this point. Drying of the membrane leads to high background.
7. This work should be done in a ventilated fume hood.
8. All incubation and washing steps are performed by floating grid on top of solution droplets placed on parafilm in humidity chamber.
9. This work should be done in a ventilated fume hood. UA is radioactive and light sensitive. Avoid exposure to direct light sources. Use as little as possible of it.

References

1. Chen Y-P, Xu L, Tang T-W et al (2020) STING activator c-di-GMP-loaded mesoporous silica nanoparticles enhance immunotherapy against breast cancer. ACS Appl Mater Interfaces 12: 56741–56752
2. Chen Z-A, Wu S-H, Chen P et al (2019) Critical features for mesoporous silica nanoparticles encapsulated into erythrocytes. ACS Appl Mater Interfaces 11:4790–4798
3. Cheng B, Toh EKW, Chen K-H et al (2016) Biomimicking platelet–monocyte interactions as a novel targeting strategy for heart healing. Adv Healthc Mater 5:2686–2697
4. Elsharkasy OM, Nordin JZ, Hagey DW et al (2020) Extracellular vesicles as drug delivery systems: why and how? Adv Drug Deliv Rev 159:332–343
5. Han Z, Lv W, Li Y et al (2020) Improving tumor targeting of exosomal membrane-coated polymeric nanoparticles by conjugation with aptamers. ACS Appl Bio Mater 3:2666–2673
6. Huang S-S, Lee K-J, Chen H-C et al (2021) Immune cell shuttle for precise delivery of nanotherapeutics for heart disease and cancer. Science. Advances 7:eabf2400
7. Li Y, Raza F, Liu Y et al (2021) Clinical progress and advanced research of red blood cells based drug delivery system. Biomaterials 279: 121202
8. Liu T-P, Wu S-H, Chen Y-P et al (2015) Biosafety evaluations of well-dispersed mesoporous silica nanoparticles: towards in vivo-relevant conditions. Nanoscale 7:6471–6480
9. Midekessa G, Godakumara K, Ord J et al (2020) Zeta potential of extracellular vesicles: toward understanding the attributes that determine colloidal stability. ACS Omega 5:16701–16710
10. Pratiwi FW, Kuo CW, Wu S-H et al (2018) Chapter six – the bioimaging applications of mesoporous silica nanoparticles. In: Tamanoi F (ed) The enzymes. Academic Press, pp 123–153
11. Pratiwi FW, Peng C-C, Wu S-H et al (2021) Evaluation of nanoparticle penetration in the tumor spheroid using two-photon microscopy. Biomedicine 9:10
12. Sancho-Albero M, Encabo-Berzosa MDM, Beltrán-Visiedo M et al (2019) Efficient encapsulation of theranostic nanoparticles in cell-derived exosomes: leveraging the exosomal biogenesis pathway to obtain hollow gold nanoparticle-hybrids. Nanoscale 11:18825–18836
13. Théry C, Witwer KW, Aikawa E et al (2018) Minimal information for studies of extracellular vesicles 2018 (MISEV2018): a position statement of the International Society for Extracellular Vesicles and update of the MISEV2014 guidelines. J Extracell Vesicles 7:1535750
14. Van Deun J, Roux Q, Deville S et al (2020) Feasibility of mechanical extrusion to coat nanoparticles with extracellular vesicle membranes. Cell 9:1797
15. Wang F, Hou W, Xiao C et al (2021) Endothelial cell membrane-based biosurface for targeted delivery to acute injury: analysis of leukocyte-mediated nanoparticle transportation. Nanoscale 13:14636–14643
16. Wu S-H, Mou C-Y, Lin H-P (2013) Synthesis of mesoporous silica nanoparticles. Chem Soc Rev 42:3862–3875
17. Yan H, Li Y, Cheng S et al (2021) Advances in analytical technologies for extracellular vesicles. Anal Chem 93:4739–4774
18. Zhang J, Ji C, Zhang H et al (2022) Engineered neutrophil-derived exosome-like vesicles for targeted cancer therapy. Sci Adv 8:eabj8207
19. Zhang R-L, Pratiwi FW, Chen B-C et al (2020) Simultaneous single-particle tracking and dynamic pH sensing reveal lysosome-targetable mesoporous silica nanoparticle pathways. ACS Appl Mater Interfaces 12:42472–42484

Chapter 18

Magnetic Separation of Cell-Secreted Vesicles with Tailored Magnetic Particles and Downstream Applications

Mireia Bernuz, Arnau Pallarès-Rusiñol, Rosanna Rossi, Carolina Fernández-Senac, Mercè Martí, and María Isabel Pividori

Abstract

The analysis of the receptors on the surface of the cell-secreted vesicles provides valuable information of the cell signature and may also offer diagnosis and/or prognosis of a wide range of diseases, including cancer.

Due to their low concentration, conventional procedures for extracellular vesicle (EV) detection usually require relatively large sample volumes, involving preliminary purification or preconcentration steps from complex specimens. Here, we describe the separation and preconcentration in magnetic particles of extracellular vesicles obtained from cell culture supernatants from MCF7, MDA-MB-231, and SKBR3 breast cancer cell lines, human fetal osteoblastic cell line (hFOB), and human neuroblastoma SH-SY5Y cell line, as well as exosomes from human serum. The first approach involves the covalent immobilization for the exosomes directly on micro (4.5 μm)-sized magnetic particles. The second approach is based on tailored magnetic particles modified with antibodies for further immunomagnetic separation of the exosomes. In these instances, micro (4.5 μm)-sized magnetic particles are modified with different commercial antibodies against selected receptors, including the general tetraspanins CD9, CD63, and CD81 and the specific receptors (CD24, CD44, CD54, CD326, CD340, and CD171). The magnetic separation can be easily coupled with downstream characterization and quantification methods, including molecular biology techniques such as immunoassays, confocal microscopy, or flow cytometry.

Key words solid-phase preconcentration, immunomagnetic separation, magnetic particles, extracellular vesicles, breast cancer biomarker, antibodies, ELISA, confocal microscopy

1 Introduction

The most common methods for targeting exosomes to date typically involve purification followed by the specific characterization of their cargo [1]. The isolation of the exosomes is best performed with differential ultracentrifugation. Purification can also be done with precipitation, size-exclusion chromatography, or ultrafiltration [2]. Identification of membrane vesicles as exosomes also requires morphological analysis [1, 2]. Given their small size, exosomes can only be visualized with an electron microscope.

Seppo Vainio (ed.), *Cell-Secreted Vesicles: Methods and Protocols*, Methods in Molecular Biology, vol. 2668,
https://doi.org/10.1007/978-1-0716-3203-1_18,

Nanoparticle tracking analysis (NTA) is usually used to count the exosomes, followed by downstream processes for specific detection, including LC-MS/MS and Western blot for proteins and qPCR for genetic material.

Due to their low concentration, conventional procedures for exosome characterization and detection usually require relatively large sample volumes and involve a preliminary purification and preconcentration step. The isolation of the exosomes is best performed with differential ultracentrifugation, but it is a notoriously laborious, low-throughput approach and incompatible with emerging platforms for low-resource settings. Moreover, it can produce mechanical damage. Differential centrifugation was the first technique described [3] and later optimized [1] by increasing *g* values to up to 100,000 *g* (ultracentrifugation). The main goal is to eliminate interferences of dead cells, cell debris, and soluble proteins. The major drawback is the lack of specificity, since it separates the whole population of exosomes, regardless their cell origin, without taking advantage of the rich information contained within exosome subpopulations. These subpopulations could reveal data about processes happening at different areas of the body, such as disease signatures, which would otherwise fade into the background of a global analysis.

Therefore, there is a global demand for simple and robust exosome isolation methods from complex biological fluids amenable to point-of-care diagnosis. Novel developments that are needed involve solid-phase preconcentration procedures which can be easily integrated in emerging technologies. Ideally, exosomes should be specifically preconcentrated while the interfering matrix is removed at the same time, increasing also the sensitivity of the detection. Since the early reports on magnetic separation technology [4], magnetic particles (MPs) have been used as a powerful and versatile preconcentration tool in a variety of analytical and biotechnology applications [5] and in emerging technologies including microfluidic devices and biosensors for extracellular vesicles [6–8].

It was also shown that MPs can be easily functionalized with different molecular groups to be conjugated with a broad range of biomolecules for the specific interaction with a target in a complex sample [9, 10]. If the biorecognition element is an antibody which reacts with the target, this procedure is called immunomagnetic separation (IMS). In order to study the IMS of an exosome as a target, it is very important to understand their structure, especially the receptor in the membrane. For instance, an important feature of EVs is the presence of tetraspanins (CD81, CD63, CD9) in their outer membrane. Another important feature of IMS is the careful selection and rational immobilization of specific antibodies toward the exosomes in an oriented way on the MPs. In this chapter, the covalent immobilization of commercial antibodies on magnetic particles is described. Tosyl-activated magnetic microparticles (tosyl-MP) from a commercial source are reacted with amine groups present in the antibody structure. In order to quantify the total amount of antibody immobilized on the magnetic particles, ELISA can be performed or eventually a protein quantification method such as Bradford. Another approach that is presented is the direct covalent immobilization of exosomes on the magnetic particle. Downstream applications for exosomes attached on magnetic particles involve flow cytometry, magneto-immunoassay, or confocal microscopy, among others.

2 Materials

Prepare all solutions using Milli-Q and analytical grade reagents. Prepare and store all reagents at 4 °C. Diligently follow all waste disposal regulations when disposing waste materials. All the procedures involving the manipulation of potentially infectious materials should be performed following the safe handling and containment guidelines. According to these guidelines, place contaminated disposable pipet tips in conveniently located puncture resistant containers used for sharps disposal. Before final disposal, decontaminate by autoclaving all cultures, stocks, laboratory waste, laboratory glassware, and other potentially infectious materials. The ultimate disposal should be performed according to local regulations.

2.1 Covalent Immobilization of Antibodies on Tosylactivated Magnetic Particles

1. Eppendorf thermomixer for temperature-controlled incubations of the Eppendorf tubes.
2. Magnetic separator DynaMag™-2 Magnet (ref. 12321D, Thermo Fisher Scientific).
3. Specific antibodies anti-CDX to exosomes (being CDX either CD9, CD63, CD81, CD24, CD44, CD54, CD171, CD326, and CD340 biomarkers), for instance rabbit polyclonal antibody anti-CD81 (ref. HPA007234, Sigma-Aldrich) (as described in Table 1).
4. Dynabeads™ M-450 Tosylactivated (tosyl-MP, 4.5 μm, ref. 14013, Thermo Fisher Scientific).
5. Phosphate-buffered saline (1 X PBS) (0.137 mol L^{-1} NaCl, 0.0027 mol L^{-1} KCl, 0.01 mol L^{-1} Na_2HPO_4, 0.0018 mol L^{-1} KH_2PO_4, pH 7.4). Dissolve 8 g NaCl (MW 58.44), 0.200 g KCl (MW 74.55), 1.420 g of Na_2HPO_4 (MW 141.96), 0.245 g KH_2PO_4 (MW 136.09) in 0.800 L Milli-Q water. Mix thoroughly and adjust the pH to 7.4. Add Milli-Q water until 1 L.

Table 1
Summary of the antibodies covalently immobilized on Dynabeads® M450 Tosylactivated, n° 14,013, 4.5 μm diameter for the immunomagnetic isolation of exosomes

Antibody	Target	Clonality	Conjugate	Host	Reference	Commercial source
Anti-CD24	CD24	Monoclonal	No	Mouse	ab76514	Abcam
Anti-CD54	CD54	Monoclonal	No	Mouse	ab2213	Abcam
Anti-CD326(*)	CD326	Monoclonal	No	Mouse	ab7504	Abcam
Anti-CD340	CD340	Monoclonal	No	Mouse	Ab30	Abcam
Anti-CD9	CD9	Monoclonal	No	Mouse	10626D	Thermo fisher
Anti-CD63	CD63	Monoclonal	No	Mouse	10628D	Thermo fisher
Anti-CD81	CD81	Monoclonal	No	Mouse	10630D	Thermo fisher
Anti-CD171	CD171	Monoclonal	No	Mouse	14–1719-82	Thermo fisher
Anti-CD44	CD44	Monoclonal	No	Mouse	BMS113	eBioscience
Anti-CD81	CD81	Polyclonal	No	Rabbit	HPA007234	Sigma-Aldrich

Note (*) No isolation of exosomes was achieved for CD326 (Epcam during IMS, since upon immobilization on the MP, no binding of exosomes was observed, perhaps due to a bad orientation during covalent immobilization). However, this antibody was useful for indirect labeling (as primary antibody) in flow cytometry, ELISA, and electrochemical immunosensor
Adapted from Ref. [11]

6. Borate buffer (0.1 mol L^{-1}, pH 8.5). Dissolve 6.18 g H_3BO_3 in 600 mL Milli-Q water. Adjust pH to 8.5 using 5 mol L^{-1} NaOH and adjust the volume to up to 1 liter with Milli-Q water.
7. Ammonium sulfate buffer (3 mol L^{-1} in borate buffer). Dissolve 39.64 g $(NH_4)_2SO_4$ in 100 mL borate buffer.
8. Phosphate-glycine blocking buffer (1 X PBS, 0.5 mol L^{-1} glycine, pH 7.4). Add 37.54 g glycine (MW 75.07) to 0.8 L phosphate-buffered saline. Mix thoroughly and adjust volume to 1 L with the same buffer.
9. Phosphate storage buffer (1 X PBS, 0.1% w/v BSA, 0.01% w/v sodium azide, pH 7.4). Add 1 g BSA and 100 mg sodium azide (MW 65.01) to 0.8 L phosphate-buffered saline. Mix thoroughly and adjust volume to 1 L with the same buffer.

2.2 Determination of the Amount of Antibody Immobilized on Tailored Magnetic Particles by ELISA

1. Microplate reader (λ 450) (for instance, TECAN Infinite m200 PRO, Switzerland).
2. Multichannel micropipettes.
3. Micronic tubes.
4. Microtiter plate (Polystyrene MaxiSorp microplates, ref. 442404, Nunc, Thermo Fisher Scientific).

5. Specific antibody to exosomes. Any of the described in Table 1, for instance mouse monoclonal anti-CD63 antibody (ref. 10628D, Thermo Fisher Scientific).
6. Secondary antibody, for instance rabbit polyclonal anti-mouse IgG H&L HRP-conjugated antibody (ref. ab6728, Abcam).
7. Hydrogen peroxide and TMB (3,3′,5,5′-tetrametylbenzidine) (TMB Substrate Kit, Reference no. 34021, Thermo Fisher Scientific).
8. Stop solution. H_2SO_4 2 mol L^{-1}.
9. Adsorption buffer (1.25 mol L^{-1} ammonium sulfate in borate buffer, prepared as detailed in § 2.1).
10. Phosphate-buffered saline (1 X PBS) (0.137 mol L^{-1} NaCl, 0.0027 mol L^{-1} KCl, 0.01 mol L^{-1} Na_2HPO_4, 0.0018 mol L^{-1} KH_2PO_4, pH 7.4), prepared as detailed in § 2.1.
11. Phosphate blocking buffer 3% (1 X PBS, 3% w/v BSA, pH 7.4). Add 30 g BSA to 0.8 L phosphate-buffered saline. Mix thoroughly and adjust volume to 1 L with the same buffer.
12. Phosphate blocking buffer 0.5% (1 X PBS, 0.5% w/v BSA). Add 5 g BSA to 0.8 L phosphate-buffered saline. Mix thoroughly and adjust volume to 1 L with the same buffer.

2.3 Determination of the Amount of Antibody Immobilized on Tailored Magnetic Particles by Bradford

1. Microplate reader (λ 595) (for instance, TECAN Infinite m200 PRO, Switzerland).
2. Multichannel micropipettes.
3. Micronic tubes.
4. Microtiter plate (Polystyrene MaxiSorp microplates, ref. 442404, Nunc, Thermo Fisher Scientific).
5. Specific antibody to exosomes. Any of the described in Table 1, for instance mouse monoclonal anti-CD63 antibody (ref. 10628D, Thermo Fisher Scientific).
6. Antibody buffer (1.25 mol L^{-1} ammonium sulfate in borate buffer, prepared as detailed in § 2.1).
7. Pierce™ Coomassie (Bradford) Protein Assay kit (ref. 23200, Thermo Fisher Scientific).

2.4 Immuno-magnetic Separation of the Exosomes on Tailored Magnetic Particles

1. Magnetic separator DynaMag™-96 Side Skirted Magnet (ref. 12027, Thermo Fisher Scientific).
2. Multichannel micropipettes.
3. Microtiter plate (Corning® 96-well Clear Round Bottom Polypropylene Not Treated Microplate, ref. 3359, Corning, USA).

4. Magnetic particles modified with the antibody. Select from commercial sources (Dynabeads™ Epithelial Enrich, ref. 16,102, Thermo Fisher Scientific) or tailored anti-CDX-MPs, modified according to §3.1, being CDx any of CD9, CD24, CD44, CD54, CD63, CD81, CD171, CD326, or CD340 biomarkers).
5. Phosphate-buffered saline (1 X PBS) (0.137 mol L^{-1} NaCl, 0.0027 mol L^{-1} KCI, 0.01 mol L^{-1} Na_2HPO_4, 0.0018 mol L^{-1} KH_2PO_4, pH 7.4), prepared as detailed in § 2.1.
6. Phosphate blocking buffer 0.5% (1 X PBS, 0.5% w/v BSA), prepared as detailed in § 2.2.

2.5 Covalent Immobilization of Exosomes on Tosylactivated Magnetic Particles

1. Eppendorf thermomixer for temperature-controlled incubations of the Eppendorf tubes.
2. Magnetic separator DynaMag™-2 Magnet (ref. 12321D, Thermo Fisher Scientific).
3. Dynabeads™ M-450 Tosylactivated (tosyl-MP, 4.5 μm, ref. 14013, Thermo Fisher Scientific).
4. Phosphate-buffered saline (1 X PBS) (0.137 mol L^{-1} NaCl, 0.0027 mol L^{-1} KCI, 0.01 mol L^{-1} Na_2HPO_4, 0.0018 mol L^{-1} KH_2PO_4, pH 7.4). Dissolve 8 g NaCl (MW 58.44), 0.200 g KCI (MW 74.55), 1.420 g of Na_2HPO_4 (MW 141.96), 0.245 g KH_2PO_4 (MW 136.09) in 0.800 L Milli-Q water. Mix thoroughly and adjust the pH to 7.4. Add Milli-Q water until 1 L.
5. Borate buffer (0.1 mol L^{-1}, pH 8.5). Dissolve 6.18 g H_3BO_3 in 600 mL Milli-Q water. Adjust pH to 8.5 using 5 mol L^{-1} NaOH and adjust the volume to up to 1 liter with Milli-Q water.
6. Phosphate-glycine blocking buffer (1 X PBS, 0.5 mol L^{-1} glycine, pH 7.4). Add 37.54 g glycine (MW 75.07) to 0.8 L phosphate-buffered saline. Mix thoroughly and adjust volume to 1 L with the same buffer.

2.6 Characterization of the Exosomes by Flow Cytometry

1. Flow cytometer BD FACSCANTO II (BD Biosciences, San Jose, CA, USA).
2. Magnetic separator DynaMag™-96 Side Skirted Magnet (ref. 12027, Thermo Fisher Scientific).
3. Multichannel micropipettes.
4. Microtiter plate (Corning® 96-well Clear Round Bottom Polypropylene Not Treated Microplate, ref. 3359, Corning, USA).

5. Magnetic particles modified with the antibody, for instance, MPs modified with anti-CD81 antibody produced in rabbit (ref HPA007234, Sigma-Aldrich).
6. Magnetic particles covalently modified with the exosome, obtained according to § 3.5.
7. For direct labeling, CD81 monoclonal antibody, FITC (ref. A15753, Thermo Fisher Scientific).
8. For indirect labeling, primary antibody anti-CD81 monoclonal antibody (ref. 10630D, Thermo Fisher) or any of the mouse monoclonal CD9, CD24, CD44, CD54, CD63, CD171, CD326, or CD340 described in Table 1, which has been successfully used also as a primary antibody in the indirect labeling.
9. For indirect labeling, secondary antibody goat anti-mouse IgG H&L (Cy5 ®) (ref. ab97037, Abcam).
10. Phosphate-buffered saline (1 X PBS) (0.137 mol L^{-1} NaCl, 0.0027 mol L^{-1} KCl, 0.01 mol L^{-1} Na_2HPO_4, 0.0018 mol L^{-1} KH_2PO_4, pH 7.4), prepared as detailed in § 2.1.
11. Phosphate blocking buffer 0.5% (1 X PBS, 0.5% w/v BSA), prepared as detailed in § 2.2.

2.7 Characterization of the Exosomes by Confocal Microscopy

1. Confocal microscope Leica, TCS SP5.
2. Magnetic separator DynaMag™-96 Side Skirted Magnet (ref. 12027, Thermo Fisher Scientific).
3. Multichannel micropipettes.
4. Microtiter plate (Corning® 96-well Clear Round Bottom Polypropylene Not Treated Microplate, Catalogue no. 3359, Corning, USA).
5. Magnetic particles covalently modified with the exosome, obtained according to § 3.5.
6. For indirect labeling, primary antibody anti-CDX mouse monoclonal antibody (any of the mouse monoclonal CD9, CD24, CD44, CD54, CD63, CD171, CD326, or CD340 described in Table 1 can successfully work).
7. Secondary antibody goat anti-mouse IgG H&L (Cy5 ®) (ref. ab97037, Abcam).
8. Phosphate-buffered saline (1 X PBS) (0.137 mol L^{-1} NaCl, 0.0027 mol L^{-1} KCl, 0.01 mol L^{-1} Na_2HPO_4, 0.0018 mol L^{-1} KH_2PO_4, pH 7.4), prepared as detailed in § 2.1.
9. Phosphate blocking buffer 0.5% (1 X PBS, 0.5% w/v BSA), prepared as detailed in § 2.2.

2.8 Quantification of the Exosomes by Magneto-Actuated Immunoassay

1. Microplate reader (λ 450) ((for instance, TECAN Infinite m200 PRO, Switzerland).
2. Multichannel micropipettes.
3. Microtiter plate (Polystyrene MaxiSorp microplates, ref. 442404, Nunc, Thermo Fisher Scientific).
4. Microtiter plate (Corning® 96-well Clear Round Bottom Polypropylene Not Treated Microplate, Catalogue no. 3359, Corning, US).
5. Magnetic separator DynaMag™-96 Side Skirted Magnet (ref. 12027, Thermo Fisher Scientific).
6. Magnetic particles modified with the antibody. Select from commercial sources (Dynabeads™ Epithelial Enrich, ref. 16,102, Thermo Fisher) or tailored anti-CDX-MPs, modified according to §3.1, being CDx any of CD9, CD24, CD44, CD54, CD63, CD81, CD171, CD326, or CD340 biomarkers.
7. Magnetic particles covalently modified with the exosome, obtained according to § 3.5.
8. For direct labeling, the mouse monoclonal CD63 antibody (HRP) (anti-CD63-HRP) (ref. NBP2-42225H-100, Novus Biologicals, Bio-Techne R&D).
9. For indirect labeling, primary antibody anti-CD63 monoclonal antibody (ref. 10628D, Thermo Fisher) or any of the mouse monoclonal CD9, CD24, CD44, CD54, CD81, CD171, CD326, or CD340 described in Table 1, which has been successfully used also as a primary antibody in the indirect labeling.
10. For indirect labeling, secondary antibody anti-mouse-HRP, for instance, rabbit anti-mouse IgG H&L (HRP) (ref. ab6728, Abcam).
11. Hydrogen peroxide and TMB (3,3′,5,5′-tetrametylbenzidine) (Pierce TMB Substrate Kit, ref. 23227, Thermo Fisher Scientific).
12. Stop solution. H_2SO_4 2 mol L^{-1}.
13. Phosphate-buffered saline (1 X PBS) (0.137 mol L^{-1} NaCl, 0.0027 mol L^{-1} KCl, 0.01 mol L^{-1} Na_2HPO_4, 0.0018 mol L^{-1} KH_2PO_4, pH 7.4), prepared as detailed in § 2.1.
14. Phosphate blocking buffer 0.5% (1 X PBS, 0.5% w/v BSA), prepared as detailed in § 2.2.

3 Methods

The antibody specific for a surface receptor of the exosome can be covalently coupled to different moieties on magnetic particles, such as tosyl-MP (§ 3.1). The outline of the procedure is schematically represented in Fig. 1. After the immobilization, the supernatant is collected for the determination of the total amount of antibody immobilized on the magnetic particles by ELISA (§ 3.2) (or any other protein quantification method such as Bradford, § 3.3), as it will be further described. The modified magnetic particles are used for immunomagnetic separation (§ 3.4) of the exosomes. The other approach is based on the direct covalent immobilization of exosomes on tosyl-MP (§ 3.5), as depicted in Fig. 2. Downstream studies of the modified MP are flow cytometry (§ 3.6), confocal microscopy (§ 3.7), or magneto-immunoassay (§ 3.8). Magnetic separation by using the magnetic separator is performed after each incubation/washing step to separate the magnetic particles from the supernatant.

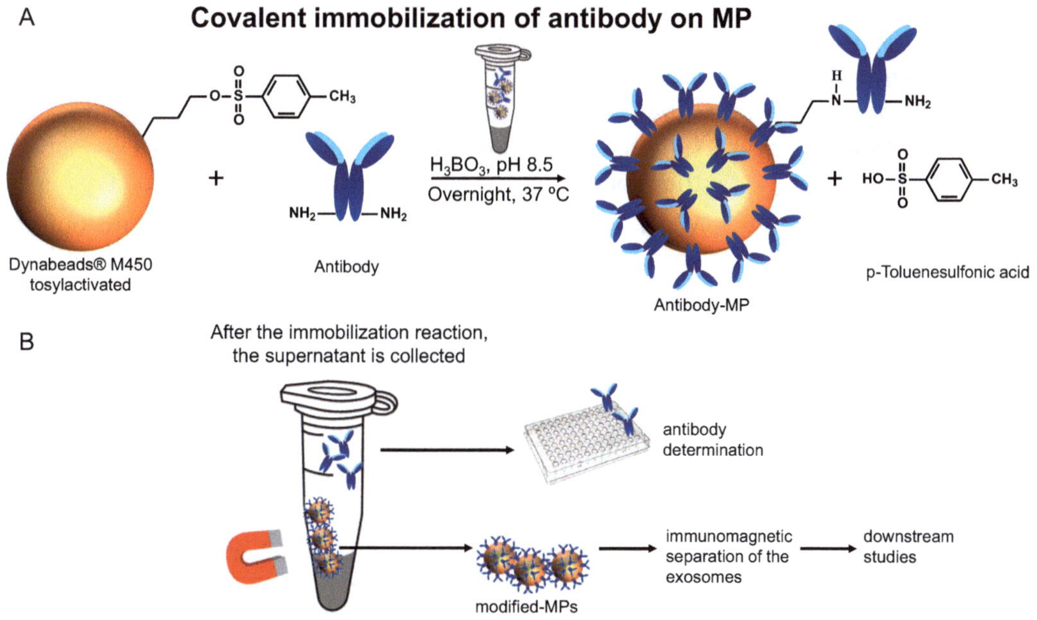

Fig. 1 Schematic representation of the tailored covalent immobilization of antibodies on magnetic particles (§ 3.1), followed by the determination of the coupling efficiency by determining the remaining antibodies on the supernatant by ELISA (§ 3.2) or any other total protein quantification method, as Bradford assay (§ 3.3) or BCA

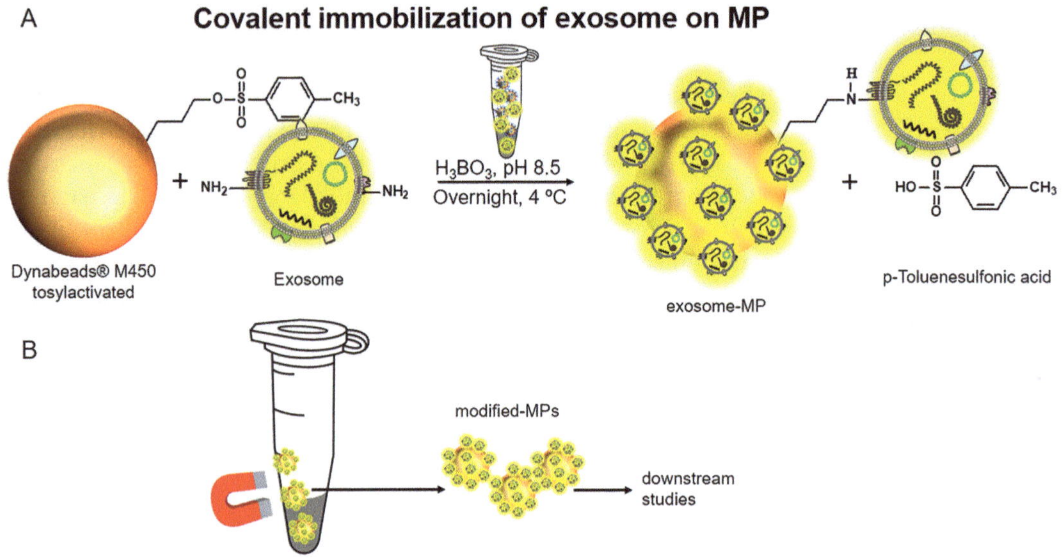

Fig. 2 Schematic representation of the covalent immobilization of exosomes on magnetic particles (§ 3.4), followed downstream studies

3.1 Covalent Immobilization of Antibodies on Tosylactivated Magnetic Particles

1. Place an Eppendorf tube containing 55 μL of tosylactivated magnetic particles (tosyl-MP 4.5 μm, 4 × 10^8 MP mL^{-1} in distilled water) on the magnetic separator.
2. Allow the MP to pellet completely until observing a clear supernatant and remove the supernatant. After elimination of the supernatant, remove the tube from the magnet.
3. Wash the tosyl-MP twice with 0.2 mL of borate buffer by vortexing or pipetting but avoiding foaming (*see* **Note 1**).
4. Afterward, add 7.5 μg of antibody in borate buffer to the tosyl-MP (a total volume of 150 μL is recommended) (*see* **Note 2**). All the antibodies specified in Table 1 has been successfully tested.
5. Add 100 μL of ammonium sulfate buffer to the previous solution containing the antibody and the tosyl-MP.
6. Incubate the MPs for a total reaction time of 20 h 37 °C at 750 rpm in the Eppendorf thermomixer (*see* **Note 3**).
7. After the incubation, collect the supernatant to perform quantification of the remaining antibodies by ELISA, as described in §3.3.
8. Add 1 mL of phosphate-glycine blocking buffer to the tosyl-MP and incubate under shaking for 2 h at 37 °C, in order to block the remaining tosyl groups.
9. Finally, wash the tailored magnetic particles and resuspend them in 220 μL phosphate storage buffer to reach a concentration of 10^5 MP per μL^{-1}.

10. Store the tailored MPs at 4 °C for further use.
11. Before each use, wash the tailored MPs twice and resuspend them in phosphate-buffered saline (*see* **Note 4**).

3.2 Determination of the Amount of Antibody Immobilized on Tailored Magnetic Particles by ELISA

1. Calibration plot. Prepare serial dilutions of the antibody used in the covalent immobilization (any of the described in Table 1, for instance mouse monoclonal anti-CD63 antibody, ref. 10628D, Thermo Fisher Scientific) on micronic tubes, ranging from 0 to 1 μg mL^{-1} and including a blank, in adsorption buffer.
2. Samples. Prepare in micronic tubes three serial dilutions of the supernatant collected during the immobilization (**step** 7 in § 3.1) to achieve 1/10, 1/50, 1/100 in a final volume of 180 μL.
3. From the micronic tubes, take 50 μL of each dilution (including calibration plot and samples) in triplicates to a Maxisorp microtiter plate.
4. Place the microplate (properly covered) in the fridge for overnight incubation at 4 °C.
5. Carefully remove the solutions of the calibration plot and samples. Make sure you switch tips in between, to prevent cross-contaminations (*see* **Note 5**).
6. Washing step. To each well, add 200 μL of phosphate-buffered saline and place the microplate on a microplate shaker for 5 min at 750 rpm and RT. Afterward, remove the solutions and tap the plate on absorbent paper.
7. Blocking step. Add 200 μL of phosphate blocking buffer 3% to each well for 2 h at 500 rpm and RT.
8. Discard the solutions and add 200 μL of phosphate-buffered saline for washing. Place the microplate on the microplate shaker for 5 min at 750 rpm.
9. Repeat the previous step. Afterward, remove the solutions and tap the plate on absorbent paper.
10. Prepare in advance a solution of secondary antibody (for instance rabbit polyclonal anti-mouse IgG H&L HRP-conjugated antibody, ref. ab6728, Abcam) diluted 1/25000 in phosphate blocking buffer 0.5% in micronic tubes.
11. Add to each well 100 μL of the secondary antibody solution.
12. Place again the microplate on the microplate shaker for 30 min at 750 rpm for the enzymatic labeling.
13. Discard the solution and add 200 μL of PBS for washing. Place on the microplate shaker for 5 min at 750 rpm.

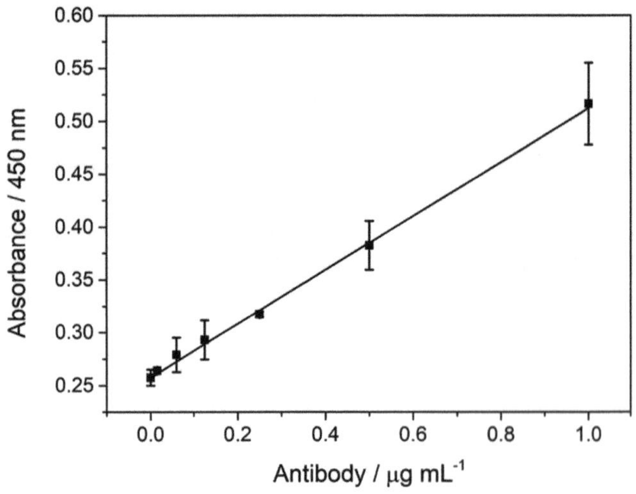

Fig. 3 Example of the calibration plot performed as described in §3.3, ranging from 0 to 1.0 μg mL^{-1} and $n = 3$. In this instance, the calibration plot was found to be absorbance = 0.2604 + 0.2307 [Antibody/μg mL^{-1}], with an $r^2 = 0.9976$

14. Repeat the previous step. Afterward, remove the solutions and tap the plate on absorbent paper.
15. Add to each well 100 μL of a solution of TMB/H_2O_2. Protect the microplate from the light and place it on the microplate shaker for 30 min at 750 rpm.
16. Add to each well 100 μL of the stop solution. Place the microplate on the microplate shaker for 1 min at 750 rpm.
17. Insert the microplate on the microplate reader. The absorbance should be measured at 450 nm.
18. Plot the concentration of the calibration curve vs the absorbance (450 nm) of antibody against exosome and obtain the concentration of the samples. An example is shown in Fig. 3.
19. Calculate the concentration of the antibody in the supernatant (μg mL^{-1}). To do that, fit the absorbance values obtained for each supernatant dilution in the dynamic range of the calibration curve (*see* **Note 6**).
20. Afterward, calculate the total immobilization in percentages (coupling efficiency) based on the following formula:

$$\text{Immobilization } (\%) = \frac{[\text{ab initial}] - [\text{ab supernatant}]}{[\text{ab initial}]} \quad (1)$$

Typically, by following the procedure, coupling efficiencies of around 99% for MMP are obtained.

3.3 Determination of the Amount of Antibody Immobilized on Tailored Magnetic Particles by Bradford

1. Calibration plot. Prepare serial dilutions of the antibody used in the covalent immobilization on micronic tubes, ranging from 0 to 0.1 μg mL^{-1} and including a blank, in antibody buffer.
2. Samples. Prepare in micronic tubes dilutions of the supernatant collected during the immobilization (**step** 7 in §3.1) to achieve 1/2, 1/4, in a final volume of 300 μL.
3. Transfer a volume of 150 μL of each dilution to the corresponding well.
4. Add 150 μL of the Coomassie dye solution.
5. Mix on the microplate shaker for 30 s.
6. Incubate for 10 min at room temperature.
7. Insert the microplate on the microplate reader. The absorbance should be measured at 595 nm.
8. Plot the concentration of the calibration curve vs the absorbance (595 nm) of antibody and obtain the concentration of the samples.
9. Calculate the concentration of the antibody in the supernatant (μg mL^{-1}). To do that, fit the absorbance values obtained for each supernatant dilution in the dynamic range of the calibration curve (*see* **Note 6**).
10. Afterward, calculate the total immobilization in percentages (coupling efficiency) based on Eq. 1.

3.4 Immunomagnetic Separation of the Exosomes on Tailored Magnetic Particles

The immunomagnetic separation (IMS) involves the reaction of the magnetic carriers with the exosomes, throughout an immunological reaction, performed in solution. It can be performed either in microplate or Eppendorf tubes. After the reaction, the separation and preconcentration of the exosomes are achieved under magnetic actuation, as schematically shown in Fig. 4. Upon magnetic separation of the exosomes, downstream evaluation methods can be easily performed.

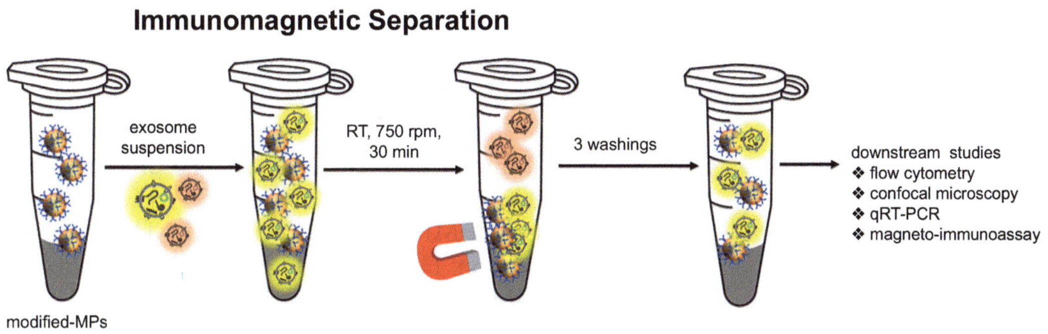

Fig. 4 Representation of the IMS procedure of the exosomes and downstream evaluation methods

1. Place in 96-well polypropylene microtiter plate a volume of 10 μL per well of anti-CDX-MPs (1 × 10^6 anti-CDX-MPs per well), including commercial (for instance Dynabeads™ Epithelial Enrich, ref. 16,102, Thermo Fisher Scientific) or tailored magnetic particles, modified according to §3.1.
2. Add 100 μL of exosome sample suspension (typically in the order of 10^9 exosomes per well) to the solution of the MPs and incubate 30 min at RT and 750 rpm (*see* **Note 7**).
3. Place the microplate on the magnetic separator.
4. Allow the MP to pellet completely until observing a clear supernatant and remove the supernatant. Remove the plate from the magnet.
5. Wash the MP with 100 μL of phosphate blocking buffer 0.5% (x3) under shaking for 1 min at RT and 750 rpm (*see* **Note 1**).
6. Collect the modified MPs in 10 μL of PBS to reach a concentration of 1 × 10^6 MP per 10 μL for downstream analysis.

3.5 Covalent Immobilization of Exosomes on Tosylactivated Magnetic Microparticles

1. Place an Eppendorf tube containing 40 μL of tosylactivated magnetic particles (tosyl-MP 4.5 μm, 4 × 10^8 MP mL^{-1} in distilled water) on the magnetic separator.
2. Allow the MP to pellet completely until observing a clear supernatant and remove the supernatant. After elimination of the supernatant, remove the tube from the magnet.
3. Wash the tosyl-MP twice with 0.2 mL of borate buffer by vortexing or pipetting but avoiding foaming (*see* **Note 1**).
4. Afterward, add 3.5 10^{10} exosomes to the tosyl-MP in borate buffer (a total volume of 250 μL is recommended) (*see* **Note 7**).
5. Incubate the MPs for a total reaction time of 20 h 4 °C at 750 rpm in the Eppendorf Thermomixer (*see* **Note 3**).
6. Add 1 mL of phosphate-glycine blocking buffer to the tosyl-MP and incubate under shaking for 2 h at 25 °C, in order to block the remaining tosyl groups.
7. Finally, wash the exosome-modified magnetic particles and resuspend them in 160 μL PBS to reach a concentration of 1 10^6 MP per 10 μL for downstream analysis.

3.6 Characterization of the Exosomes by Flow Cytometry

1. Place in 96-well microtiter plate a volume 10 μL per well of exosomes immunocaptured on anti-CD81-MP (1 × 10^6 MPs per well, for instance, MPs modified with anti-CD81 antibody produced in rabbit, ref. HPA007234, Sigma-Aldrich) obtained by immunomagnetic separation according to §3.4.

or, instead

1. Place in 96-well microtiter plate a volume 10 μL per well of exosome-MPs (1 × 10^6 exosome-MPs per well), obtained according to §3.5.
2. Indirect labeling. 2a) To the solution of the modified MPs, add 100 μL (5 μg mL^{-1}) of the antibodies anti-CDX (mouse) (for instance anti-CD81 monoclonal antibody, Ref 10630D, Thermo Fisher Scientific) and incubate for 30 min at 25 °C and 750 rpm. 2b) Wash the MP with 100 μL of Phosphate blocking buffer 0.5% under shaking for 1 min at RT and 750 rpm (*see* **Note 1**). Repeat this step twice. 2c) Afterward, add 100 μL (2 μg mL^{-1}) of, for instance, goat anti-mouse IgG H&L (Cy5 ®), ref. ab97037, Abcam, and incubate for 30 min at 25 °C and 750 rpm. 2d) Wash the MP with 100 μL of phosphate blocking buffer 0.5% (x3) under shaking for 1 min at RT and 750 rpm.

or, instead

2. Direct labeling. 2a) To the solution of the modified MPs, add 100 μL (5 μg mL^{-1}) of the antibody anti-CDX-Cy5 (for instance, CD81 monoclonal antibody, FITC, ref. A15753, Thermo Fisher Scientific), and incubate for 30 min at 25 °C and 750 rpm. 2b) Wash the MP with 100 μL of phosphate blocking buffer 0.5% (×3) under shaking for 1 min at RT and 750 rpm (*see* **Note 1**).
3. In any case, proceed with the readout of the resulting suspension of the magnetic particles modified with the antibodies using a cell cytometer.

Typically, by following the procedure, the histograms shown in Fig. 5 can be obtained.

3.7 Characterization of the Exosomes by Confocal Microscopy

1. Place in 96-well microtiter plate a volume 10 μL per well of exosome-MPs (4 × 10^9 exosomes in 1 × 10^6 exosome-MPs per well), obtained according to §3.5.
2. To the solution of the exosome-MPs, add 100 μL (5 μg mL − 1) of the antibodies anti-CDX (mouse) (for instance, any of the mouse monoclonal CD9, CD24, CD44, CD54, CD63, CD171, CD326, or CD340 described in Table 1 can be successfully used) and incubate for 30 min at 25 °C and 750 rpm.
3. Wash the modified MP with 100 μL of phosphate blocking buffer 0.5% under shaking for 1 min at RT and 750 rpm (*see* **Note 1**). Repeat this step twice.
4. Afterward, add 100 μL (2 μg mL^{-1}) of anti-mouse-Cy5 antibody, for instance, goat anti-mouse IgG H&L (Cy5 ®), (ref. ab97037, Abcam) and incubate for 30 min at 25 °C and 750 rpm.

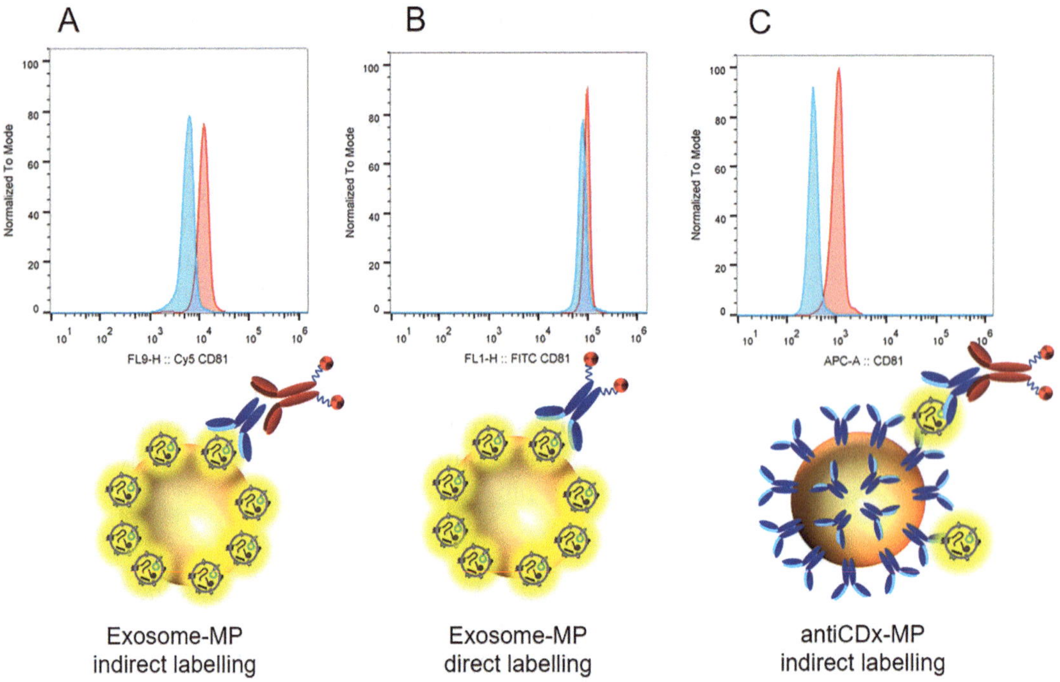

Fig. 5 Schematic representation of the covalent immobilization of exosomes on magnetic particles (§ 3.5), followed downstream studies by flow cytometry with indirect (panel A) or direct (B) label of exosomes derived from SKBr3 (ATCC® HTB-30™) breast cancer cell line. Panel C shows similar results for exosomes derived from human neuroblastoma SH-SY5Y cell line, immunocaptured on anti-CD81-MPs, followed by indirect label based on CD81 receptor, as described in § 3.6

5. Wash the MP with 100 μL of phosphate blocking buffer 0.5% (x3) under shaking for 1 min at RT and 750 rpm.
6. Finally, resuspended the MPs in 100 μL of phosphate blocking buffer 0.5%.
7. Collect the confocal images by selecting the laser lines for Cy5 (633 nm excitation, 650–785 nm emission).

Typically, by following the procedure, the images shown in Fig. 6 can be obtained.

3.8 Quantification of the Exosomes by Magneto-Actuated Immunoassay

1. Place in 96-well polypropylene microtiter plate a volume 10 μL per well of exosome-MPs (containing 1×10^6 MPs per well, with a number of exosomes covalently immobilized from 0 to up to 2×10^9 exosomes per well), obtained according to §3.5.

or, instead

1. Place in 96-well microtiter plate a volume 10 μL per well of immunocaptured exosomes (ranging from 0 to 2×10^9 exosomes) on anti-CDX-MP (1×10^6 MPs per well) for instance,

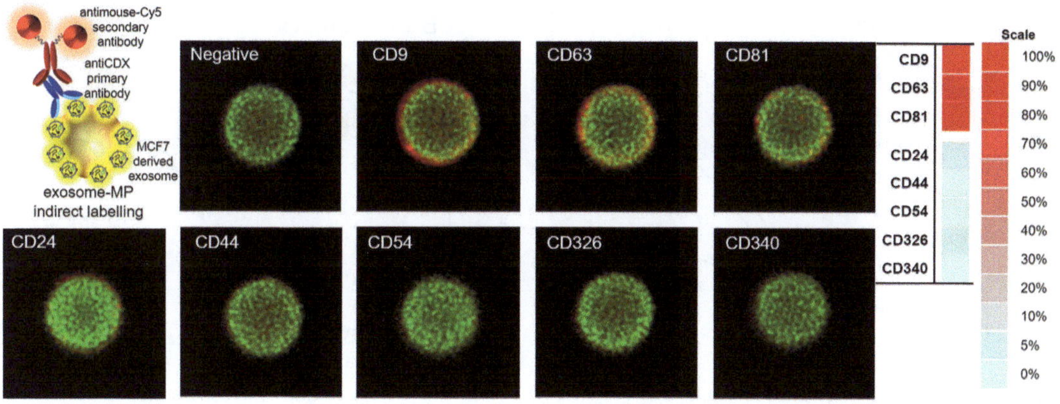

Fig. 6 Confocal microscopic characterization of MPs covalently modified with MCF7 breast cancer-derived exosomes, followed by indirect labeling with mouse anti-CDX (being CDX either CD9, CD24, CD44, CD54, CD63, CD81, CD326, and CD340 biomarkers) and anti-mouse-Cy5. The concentration of exosomes was set in 4×10^9 immobilized on 1×10^6 MPs per assay. The scale indicates the percentage of positive entities

MPs modified with mouse monoclonal anti-CD63 antibody (ref. 10628D, Thermo Fisher Scientific), obtained by immunomagnetic separation according to §3.4.

2. Direct labeling. 2a) To the solution of the modified MPs, add 100 μL (1.24 μg mL^{-1}) of the antibody anti-CDX-HRP (for instance, the mouse monoclonal CD63 antibody (HRP) (anti-CD63-HRP) (ref. NBP2-42225H-100, Novus Biologicals) and incubate for 30 min at 25 °C and 750 rpm.

or, instead

2. Indirect labeling. 2a) To the solution of the modified MPs, add 100 μL (0.5 μg mL^{-1}) of the antibodies anti-CDX (mouse) (for instance, anti-CD63 monoclonal antibody, ref. 10628D, Thermo Fisher Scientific) and incubate for 30 min at 25 °C and 750 rpm. 2b) Wash the MP with 100 μL of phosphate blocking buffer 0.5% under shaking for 3 min at RT and 750 rpm (*see* **Note 1**). Repeat this step twice. 2c) Afterward, add 100 μL (0.08 ng mL^{-1}) of anti-mouse-HRP, for instance, rabbit anti-mouse IgG H&L (HRP) (ref. ab6728, Abcam) and incubate for 30 min at 25 °C and 750 rpm.
3. Wash the MP with 100 μL of phosphate blocking buffer 0.5% (x3) under shaking for 3 min at RT and 750 rpm.
4. Add to each well 100 μL of a solution of TMB/H_2O_2. Protect the microplate from the light and place it on the microplate shaker for 30 min at 750 rpm.
5. Add to each well 100 μL of the stop solution. Place the microplate on the microplate shaker for 1 min at 750 rpm.
6. Place the microplate on the magnetic separator.

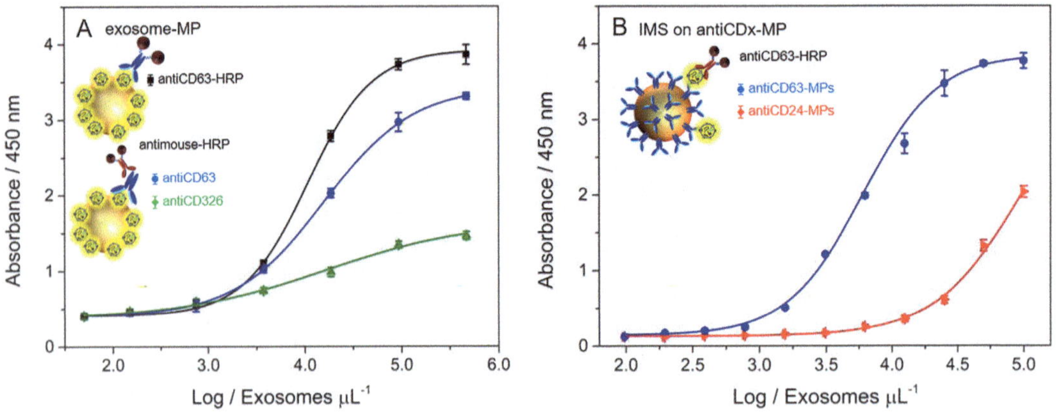

Fig. 7 Magneto-actuated immunoassay for detection of MCF7 exosomes ranging from 0 to 4.5×10^5 exosomes μL^{-1}. (**a**) covalently immobilization of the exosomes on MPs (exosomes-MPs), followed by indirect labeling with (●) anti-CD63 and (▲) anti-CD326, and by direct labeling with (■) anti-CD63- HRP antibody. (**b**) IMS of the exosomes on (▶) anti-CD24-MPs and (●) anti-CD63-MPs, followed by direct labeling with anti-CD63-HRP antibody. In all cases, the concentration of MPs was fixed in 1×10^6 MPs. The error bars show the standard deviation for $n = 3$. (Adapted from Ref [12])

7. Allow the MP to pellet completely until observing a clear supernatant and remove the supernatant. Remove the plate from the magnet.
8. Transfer the supernatant to polystyrene microplate using multichannel micropipettes.
9. Insert the microplate on the microplate reader. The absorbance should be measured at 450 nm.
10. Plot the concentration of the calibration curve vs the absorbance (450 nm) of exosome concentration and obtain the concentration of the samples. An example is shown in Fig. 7.
11. Calculate the concentration of the exosome in the sample. To do that, fit the absorbance values obtained for each sample in the dynamic range of the calibration curve (*see* **Note 6**).

Typically, by following the procedure, the calibration plots shown in Fig. 7 can be obtained.

4 Notes

1. Magnetic separation by using the magnetic separator is performed after each incubation/washing step in order to separate the magnetic particles from the supernatant. To do that, place the tube on the magnet, allow the MP to pellet, and remove (or eventually collect) the supernatant.

2. The antibody which will be used for the coating must be free of any other protein, sugar, or stabilizer to avoid competition during the immobilization. Do not use any buffer containing protein or amine groups (glycine, Tris buffer) for prewashing or coupling in any case.
3. Use a mixer that provides tilting and rotation of the tubes for incubation taking more than 1 hour in order to prevent the magnetic particles to settle the tubes.
4. Plating the modified magnetic particles in BHI agar plates is highly recommended to ensure possible microbiological contamination.
5. Make sure you switch tips in between, to avoid contaminations.
6. For a more accurate calculation, from the different dilution of the supernatant, take the value near the centroid of the calibration plot (in this example, absorbance value of approximately 0.75 AU).
7. The suspension of the exosomes can be obtained from cell culture supernatant (any of MCF7, MDA-MB-231, and SKBR3 breast cancer cell lines, human fetal osteoblastic cell line (hFOB), and human neuroblastoma SH-SY5Y cell line), separated from other particles including cells, or cell debris by any of the reported separation methods, differential and density gradient centrifugation, size-exclusion chromatography, filtration, and polymer-based precipitation are the most popular isolation methods for exosomes, or from clinical samples including blood samples centrifugated at 1500 *g*. The number of exosome can be achieved by nanoparticle tracking analysis (NTA).

Acknowledgments

This work was funded by Ministry of Science and Innovation (Project PID2019-106625RB-I00/AEI/10.13039/501100011033). Also, Ministry of Universities (Grant FPU16/01579, Grants Margarita Salas y María Zambrano) are gratefully acknowledged.

References

1. Théry C, Amigorena S, Raposo G, Claytonet A (2006) Isolation and characterization of exosomes from cell culture supernatants and biological fluids. Curr Protoc Cell Biol 30: 3.22.1-3.22.29
2. Patel GK, Khan MA, Zubair H, Srivastava SK, Khushman M, Singh S, Singh AP (2019) Comparative analysis of exosome isolation methods using culture supernatant for optimum yield, purity and downstream applications. Sci Rep 9:5335
3. Johnstone RM, Adam M, Hammond JR, Orr L, Turbideet C (1987) Vesicle formation during reticulocyte maturation. Association of

plasma membrane activities with released vesicles (exosomes). J Biol Chem 262:9412–9420

4. Rembaum A, Yen RCY, Kempner DH, Ugelstad J (1982) Cell labeling and magnetic separation by means of immunoreagents based on polyacrolein microspheres. J Immunol Methods 3:341–351
5. Reddy LH, Arias JL, Nicolas J, Couvreur P (2012) Magnetic nanoparticles: design and characterization, toxicity and biocompatibility, pharmaceutical and biomedical applications. Chem Rev 112:5818–5878
6. Moura SL, Pallarès-Rusiñol A, Sappia L, Martí M, Pividori MI (2022) The activity of alkaline phosphatase in breast cancer exosomes simplifies the biosensing design. Biosens Bioelectron 198:113826
7. Sanchez MA, Felice B, Sappia LD, Moura SL, Martí M, Pividori MI (2020) Osteoblastic exosomes. A non-destructive quantitative approach of alkaline phosphatase to assess osteoconductive nanomaterials. Mater Sci Eng C 115:110931
8. Moura SL, Martín CG, Martí M, Pividori MI (2022) Electrochemical immunosensing of nanovesicles as biomarkers for breast cancer. Biosens Bioelectron 150:111882
9. Brandão D, Liébana S, Pividori MI (2015) Multiplexed detection of foodborne pathogens based on magnetic particles. New Biotechnol 32:511–520
10. Carinelli S, Marti M, Alegret S, Pividori MI (2015) Biomarker detection of global infectious diseases based on magnetic particles. New Biotechnol 32:521–532
11. Moura SL, Martín CG, Martí M, Pividori MI (2020) Matrix effect in the isolation of breast cancer-derived Nanovesicles by Immunomagnetic separation and electrochemical Immunosensing—a comparative study. Sensors 20(4):965
12. Moura SL, Martín CG, Martí M, Pividori MI (2020) Multiplex detection and characterization of breast cancer exosomes by magneto-actuated immunoassay. Talanta 211:120657

Chapter 19

Cilia-Derived Extracellular Vesicles in *Caenorhabditis Elegans*: In Vivo Imaging and Quantification of Extracellular Vesicle Release and Capture

Adrià Razzauti, Teresa Lobo, and Patrick Laurent

Abstract

Caenorhabditis elegans is a microscopic model nematode characterized by body transparency and ease of genetic manipulation. Release of extracellular vesicles (EVs) is observed from different tissues; of particular interest are the EVs released by the cilia of sensory neurons. *C. elegans* ciliated sensory neurons produce EVs that are environmentally released and/or captured by neighboring glial cells. In this chapter, we describe a methodological approach to image the biogenesis, release, and capture of EVs by glial cells in anesthetized animals. This method will allow the experimenter to visualize and quantify the release of ciliary-derived EVs.

Key words Caenorhabditis elegans, Ectosomes, Ciliary-derived EVs, In vivo imaging, EV capture, Intercellular communication

1 Introduction

EVs comprise a variety of vesicles released by multiple cell types and by multiple biogenesis mechanisms. EVs are categorized as exosomes or ectosomes based on size and subcellular origin. The biogenesis of endosomal-derived exosomes includes several steps, from cargo selection and inward budding of the endosome to form intraluminal vesicles (ILVs), to plasma membrane fusion of multivesicular bodies (MVBs) and release of ILVs/exosomes [1]. The trafficking to endosomal membranes and cargo sorting to ILVs is mediated by 3 complementary pathways: the ESCRT-dependent [2], the ESCRT-independent pathway [3, 4], and a LAMP2A-dependent pathway [5]. Conversely, cargo selection and biogenesis of plasma membrane-derived ectosomes or microvesicles are poorly characterized.

Seppo Vainio (ed.), *Cell-Secreted Vesicles: Methods and Protocols*, Methods in Molecular Biology, vol. 2668,
https://doi.org/10.1007/978-1-0716-3203-1_19,

Most of the EV research is performed in mammalian cell cultures due to their accessibility to pharmacological agents, proficiency of manipulation, and ease of EV purification. These in vitro studies lack a comprehensive understanding of how EVs behave within a tissue or at an organism-scale in physiological or pathological conditions. Thus, model organisms where EVs can be tracked from origin to their destination confer an advantage over classical in vitro approaches. *C. elegans* has proven to be an excellent platform for the study of ectosomes during development [6–8]. Additionally, the nematode is recurrently used as a model to explore the biogenesis of EVs released from cilia, sorting of ciliary EV cargo, and the role of ciliary EVs in inter-animal communication [9, 10]. Several advantages determine the success of EV research using *C. elegans* as a model organism: its stereotypical cell placement, the availability of mutants, the ease of transgenesis, and its transparency, which allows in vivo imaging of cells expressing fluorescently labeled EV cargoes (at endogenous levels). All these reasons make *C. elegans* an attractive model to observe and characterize EV biogenesis and their fate in an organism.

We and others have used the ciliated sensory neurons of *C. elegans* as a model to explore the mechanisms of EV production [9, 11, 12]. Sensory organs—also called sensilla—in *C. elegans* are developed by the organized bundling of a group of ciliated neurons and their supporting glial cells. Seven sensilla mediate the different sensory modalities of the hermaphrodite *C. elegans*. The amphid sensilla is formed by 12 sensory neurons and 2 glial cells: the amphid sheath glia (AMsh) and the amphid socket (AMso) (*see* Fig. 1a). AMsh ensheathes the nerve receptive endings of 4 amphid neurons and together with the AMso glia forms a channel that houses the cilia of the remaining 8 neurons, this channel is also known as the amphid pore (*see* Fig. 1b). We have previously shown how ciliated neurons within the amphid sensilla are capable of releasing EVs through ectocytosis [11]. Ectocytosis consists in the outward budding of the plasma membrane and subsequent excision, resulting in the release of ectosomes. We demonstrated how this process occurs at different locations along the cilium and their differential fates: (1) ectosomes that are shed from the ciliary tip compartment (distal cilium) are environmentally released. (2) Ectosomes produced from the base of the cilium (proximal cilium) are subsequently endocytosed by the surrounding AMsh glia (*see* Fig. 1c) [11]. We take advantage of the body transparency, the simple amphid sensilla anatomy and the availability of fluorescently labeled reporter strains to perform live imaging. We developed a setup to observe EV release events to the environment, or alternatively image the EVs captured by the AMsh glia. This methodology has the potential to be combined with forward genetic or candidate

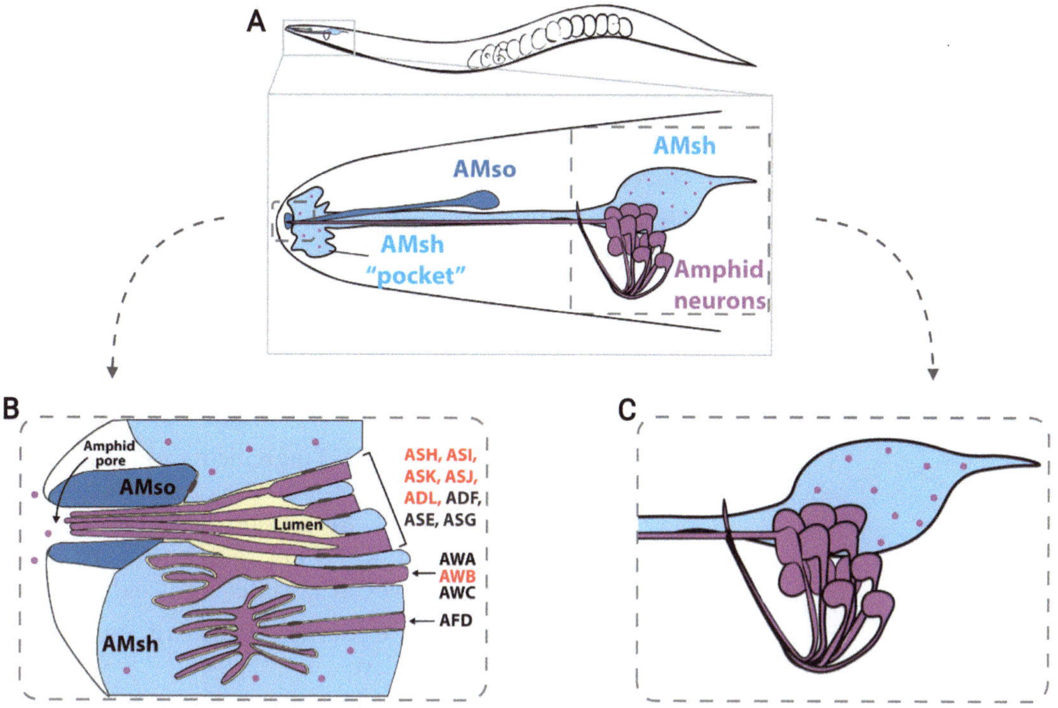

Fig. 1 *C. elegans* amphid ciliated sensory neurons shed ectosomes, displayed as purple dots. (**a**) *C. elegans*' amphid sensilla is located in the head and it is composed of 12 neurons ciliated sensory neurons and 2 glial cells, the AMsh and AMso. (**b**) Amphid neurons shed ectosomes at least from 2 distinct ciliary locations: at the tip of the cilium—these are released to the external environment; and at the base of the cilium—the ones that are uptaken by AMsh glia. (**c**) Ectosomes captured by AMsh glia accumulate in its cell body, located posteriorly to the cell bodies of amphid neurons

gene screening approaches as a platform to find genes participating in EV biogenesis, EV cargo selection, and EV capture by the glial cell. In this chapter, we define a step-by-step protocol that we routinely use in the lab to acquire live images during the EV formation and export or still images of the resulting EV intercellular transfer, together with the methods to quantify such events. We provide instructions to follow this protocol to achieve a simple and reproductive quantification of EVs derived from ciliated sensory neurons in *C. elegans*. We describe how to rear and synchronize animals for imaging EVs, methods to anesthetize the animals, sample preparation, and how to do in vivo imaging. We also list the available strains up to date to track and study ciliary-derived EVs. Additionally, we show the pipeline that we used for post-imaging analysis of acquired data using open-source software. This protocol can be adapted to perform live imaging of fluorescent proteins in any region of interest of the animal.

2 Materials

- Incubator (set at 20 °C).
- Bunsen burner.
- Sterile Petri dishes (55 mm diameter and 90 mm diameter).
- *E. coli* strain OP50: obtained from Caenorhabditis Genetics Center (cgc.umn.edu/strain/OP50). *E. coli* bacteria are used as the standard food source for *C. elegans* [13].
- Inoculating rod.
- Glycerol.
- Parafilm M tape.
- Nematode growth medium (NGM) ingredients and stock solutions: NaCl, Agar, Bacto™ peptone, cholesterol in ethanol (5 mg/mL concentration), 1 M phosphate buffer pH 6.0 (108.3 g KH2PO4, 35.6 g K2HPO4, H2O to 1 L), 1 M MgSO4 (filter-sterilized), 1 M CaCl2 (filter-sterilized).
- 2xYT Liquid medium: 16 g tryptone, 10 g yeast extract, 5 g NaCl, fill to 1 L H2O. Autoclave.
- 2xYT Agar medium: 16 g tryptone, 10 g yeast extract, 5 g NaCl, 15 g Agar, fill to 1 L H2O. Autoclave and dispense in 90 mm culture plates.
- Pasteur pipette (glass and plastic).
- Rubber suction bulb.
- 10 mL sterile plastic pipettes.
- Platinum wire (90% platinum, 10% iridium).
- Worm pick: Homemade pick using 2–3 cm of platinum wire. We modify the end of the platinum wire to generate a flat and rounded edge (shaped like a duck beak) using a hammer and with the help of heat and pliers. Break off the elongated tip of a glass Pasteur pipette and melt the broken glass tip with the Bunsen burner. Using tweezers, immediately attach the platinum wire leaving the flattened end exposed and the sharp end fused to the melting glass. Let the glass cool down. The worm pick allows to transfer worms between NGM plates or to pick the worms from NGM plates to mount them for imaging.
- M9 Buffer: 3 g KH2PO4, 6 g Na2HPO4, 5 g NaCl, fill to 1 L H2O. Autoclave. Only when the solution has cooled down, add 1 mL 1 M MgSO4 (filter-sterilized) (*see* **Note 1**).
- Heating block at 55–60 °C (with adaptor for glass tubes).
- Borosilicate glass tubes.
- Vortex mixer.
- Precision tweezers.

- Sodium azide anesthetic solution (25 mM in M9 buffer). A 1 M sodium azide is prepared with H2O mq and stored at RT, in a cool and well-ventilated space, out of heat and light sources (*see* **Note 2**).
- (−)-Tetramisole hydrochloride (Sigma) anesthetic solution (10 mM in M9 buffer). Make several 100 mM stocks and keep them frozen at −20 °C. Aliquots are thawed and diluted with M9 buffer on the day of the experiment.
- FluoroDish (35 mm diameter, 10 mm well, WPI, #FD3510).
- Microscope glass slides.
- Coverslips (24 × 24 mm, Thickness: 1).
- Vaseline (in a 3 mL syringe).
- Lab tape.
- Cryovials.
- Transgenic *C. elegans* strain bearing the ciliary EV marker of interest. For this protocol, the strain OQ369 *osm-3(p802)*; PHX4122 (tsp-6(*syb4122*[tsp-6::wrmScarlet]); Ex[pF16F9.3::CFP] was used as example (*see* **Note 3**).
- Laser scanning confocal microscope (LSM 780 or later models).
- Computer with FIJI installed [14].

3 Methods

3.1 Preparation of NGM Plates

1. The following amounts are used to prepare 1 L of NGM solution. Add 3 g NaCl, 17 g agar, 2.5 g Bacto™ peptone in a 1 L autoclavable bottle, fill with Milli-Q H2O to 975 mL, put a cap but make sure not to fully screw it.
2. Mix well and autoclave the bottle for 1 h.
3. Allow the bottle to cool to approximately 55 °C and then add 25 mL phosphate buffer 1 M pH 6.0 (autoclaved), 1 mL 1 M CaCl2 (filter-sterilized), 1 mL 1 M MgSO4 (filter-sterilized), and 1 mL cholesterol 5 mg/mL. Mix well (*see* **Note 1**).
4. Dispense 10.5 mL of NGM agar in each 55 mm plate with the aid of a peristaltic pump under sterile procedures (*see* **Note 4**).
5. Leave plates at room temperature for 2–7 days before using them (waiting time will depend on the laboratory temperature and ventilation), this allows the drying of excess moisture and the detection of contaminants. Plates can be turned upside down once the excess of moisture is evaporated; this will slow down the drying process and keep them at an optimal moisture content for a longer period.

3.2 OP50 Bacterial Stock, OP50 Liquid Culture, and Seeding Plates

3.2.1 OP50 Bacterial Stock

1. Using sterile procedures, streak the ordered OP50 bacteria into a culture plate with no antibiotics and grow overnight at 37 °C.
2. Using an inoculating rod, select an isolated colony and grow an overnight liquid culture in 5 mL of 2xYT medium.
3. Add 500 μL of the overnight culture to 500 μL of 50% glycerol in a 2 mL cryovial and mix.
4. Freeze the glycerol stock tube at −80 °C. Subsequent freeze and thaw cycles can reduce shelf life. We recommend doing several stocks.
5. To recover bacteria from your glycerol stock, using an inoculating rod (or a sterile pipette tip), scrape some of the frozen bacteria off the surface and streak the bacteria onto a 2xYT agar plate.
6. Grow overnight. Plates with individual OP50 colonies can be sealed with Parafilm tape and stored at 4 °C for up to 2 months. Individual colonies will be used to start OP50 liquid cultures.

3.2.2 OP50 Liquid Culture

1. Pick a single colony from a bacterial plate with individual OP50 colonies.
2. Using sterile procedures, inoculate a 100–250 mL bottle of sterile 2xYT liquid medium.
3. Grow overnight at 37 °C. OP50 liquid cultures can be stored at 4 °C for up to 2 months (*see* **Note 5**).

3.2.3 Seeding Plates

1. Dispense and spread 100–200 μL of OP50 liquid culture onto each NGM place, trying to distribute the liquid at the center of the plate and avoiding the edges (*see* **Note 6**).
2. Allow the seeded bacteria to dry and grow overnight at RT. The dried bacterial will grow during the following days and the resulting bacterial lawn will be used as a food source by the nematodes. If needed, bacteria can be grown overnight at 37 °C, plates should be allowed to cool down at the *C. elegans* cultivation temperature before using to avoid thermal stress.

3.3 Synchronizing Worm Populations by Egg-Laying Window

To consistently get a population of *C. elegans* of a desired developmental stage, we perform egg-laying windows. Animals are allowed to lay eggs on a plate for a limited amount of time. Afterward, the animals are removed from the plate leaving the eggs on the plate, which will hatch and lead to a population of age-matched animals.

1. Transfer 20–25 animals of a desired genotype (i.e., transgenic worms bearing a fluorescent EV reporter) to a seeded plate with OP50 bacterial lawn. Label the plates according to their genotype name, date, and hour.
2. Let the animals, lay eggs on the plate for 1 h 30 min.

3. Remove all the animals from the plate such that only the eggs remain (*see* **Note** 7).
4. Place the plates in the incubator at 20 °C.
5. Wait 3 days to obtain a synchronized population of Day 1 animals. (Alternatively, wait 2 days to obtain a L4-staged synchronized population.)

3.4 Imaging of EV Release from Ciliated Neurons in Living Animals

As described in the introduction, EVs can be released from at least 2 different locations along the cilium of ciliated sensory neurons. To image and quantify such release events, we can take 2 different approaches. The first involves performing in vivo imaging of the cilia of the amphid sensory neurons at the tip of *C. elegans* head for a given amount of time. The second involves acquiring confocal images of the immobilized animals to quantify the EVs captured by AMsh. (For the second approach, see sect. 3.5.)

3.4.1 Sample Preparation

1. Prepare 4% agarose solution by melting 0.16 g of agarose in 4 mL of M9 buffer in a borosilicate glass tube. Melt the agarose solution by placing the solution in a heating block set at 95 °C or by repeatedly placing the tube over the flame in short intervals. Mix the solution using the vortex until agarose has completely dissolved (*see* **Note 8**).
2. Once the solution has cooled to 55 °C in the heat block, add 40 μL of 1 M (−)-tetramisole hydrochloride solution.
3. With a plastic Pasteur pipette, dispense 2–3 drops of the melted agarose solution to completely cover the hollow part of a FluoroDish. This agarose plug will be used as a lid to cover the worms and avoid dehydration (*see* Fig. 2a).
   (a) With the aid of some sharp tweezers, remove the agarose plug and set it aside (*see* Fig. 2b).
   (b) Pipette a 2–3 μL drop of 10 mM (−)-tetramisole hydrochloride anesthetic solution at the center of the FluoroDish.
   (c) Transfer a small number [10–20] of synchronized animals within the drop with the help of a worm pick. Wait for the animals to be anesthetized, it should take around 15–20 minutes (*see* Fig. 2a).
   (d) Once the animals have been anesthetized, gently place the agarose plug on top of the anesthetized animals. Gently place back the agarose plug. To do so, place it in an angle to avoid air bubbles to be trapped within the sample (*see* Fig. 2a-b).
   (e) Worms will be ready to be imaged at that stage (*see* Fig. 2a-c).

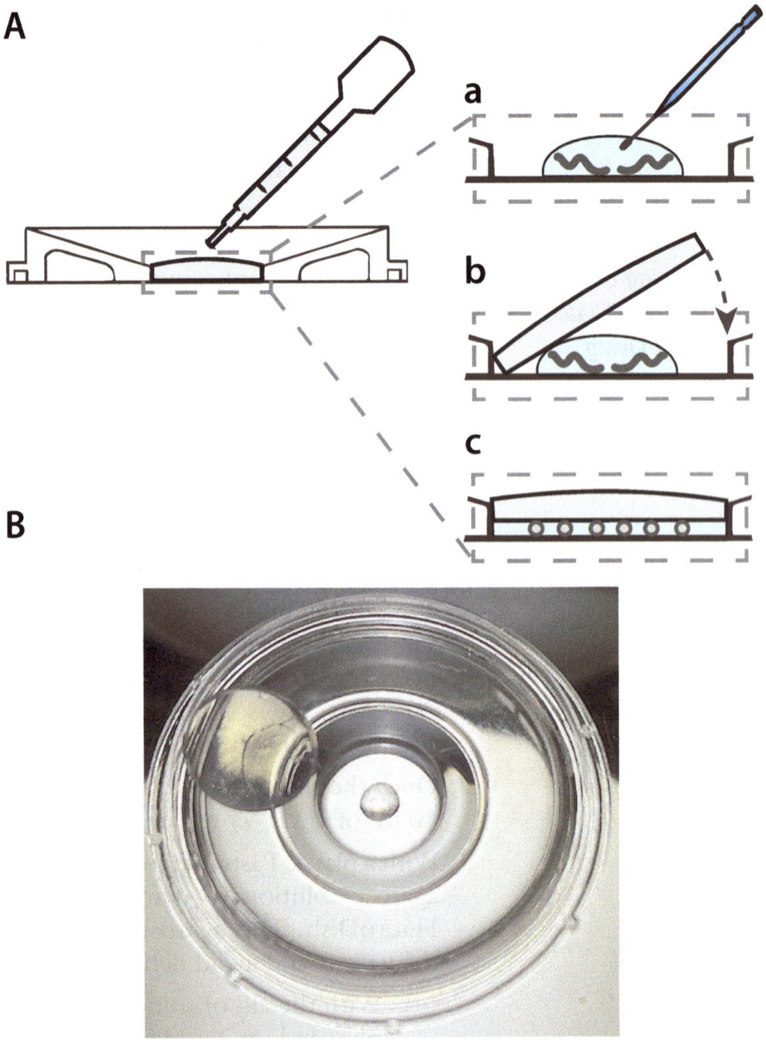

Fig. 2 Mounting the animals for live imaging in a FluoroDish. (**a**) Scheme depicting in a stepwise manner (**a-c**) the procedure to mount *C. elegans* for live imaging. First a drop of 4% agarose is placed to cover the hollow side of a FluoroDish and left 1 minute to solidify. (**a**) Remove the agarose plug created beforehand with tweezers. Then, place a 2–3 μL drop of the anesthetic 10 mM (−)-Tetramisole solution at the center of the FluoroDish. Pick 10–20 animals using the worm pick avoiding the transfer of OP50 bacteria, allow the worms to anesthetize with the plastic lead covered. Animals should become anesthetized in the time range of 15–20 min. (**b**) Under a stereodissecting scope or microscope with a high magnification (40X), check whether the animals' heads have stopped twitching, when the heads have ceased moving, use the tweezers to gently place the agarose plug on top of the drop. (**c**) The worms should now be immobilized and ready to image. If the animals are heavily displaced to the margins of the plate and liquid overflows after placing the lid, repeat the process and try reducing the volume of anesthetic used during step (**a**). (**b**) Picture of the mounting step in the laboratory. This image corresponds to the transition between the aforementioned steps (**a-b**)

3.4.2 Time-Lapse Imaging Using Confocal Microscopy

Amphid cilia are easy to image due to their location within the body, they reside within an opening at the nose of the animal. Since the animal has bilateral amphid sensilla, we always focus on the sensillum that lies closer to the objective once mounted. Doing so will minimize light scattering and improve the signal-to-noise ratio. One of the limitations of live imaging recordings is photobleaching. To prevent it, we use settings that allow fast acquisition with low laser excitation combined with sensitive detectors. Acquisition of time series is possible with the use of conventional confocal microscopes or microscopes with super-resolution features. However, the use of spinning-disk microscopes with good objectives might be advantageous to minimize photobleaching during long acquisition times.

1. Locate the sample using a low magnification objective (10× or 20×) (*see* **Note 9**).
2. Once located, change to a higher magnification objective (40× or 63× or 100×) with high NA (*see* **Note 10**).
3. Set the laser excitation power to a range where the fluorescence at the tip or ciliary axoneme is visible. Avoid using high laser powers to minimize photobleaching. Instead, increase detector gain settings when possible.
4. Select a focal plane that captures the area of the cilium (Setting the pinhole at 1 AU) (*see* **Note 11**).
5. Using your default confocal software, select an option that allows acquisition of time-lapse images (i.e. Zeiss ZEN 2012 –> Check Time series option).
6. EV release is a relatively fast event (within the range of seconds). Select time intervals between acquisitions of 1–3 sec. We recommend setting a 5–10 min total acquisition time to capture several release events.
7. Before starting recording, verify that the ciliary tip is still in focus and begin the acquisition (*see* **Note 12**).
8. Once the acquisition has finished, move the stage to the next visible animal in range and repeat the procedure.
9. *Optional*: We have seen that even after anesthesia samples can slightly move. If this movement is noticeable but does not affect the focal plane where the cilium is focused, it can be corrected by image post-processing. We use the FIJI plugin StackReg [15] to align a series of consecutive images and correct for slight XY head movements occurring during acquisition, for that we use rigid body as transformation option.

3.4.3 Quantification of the Number of Released EVs/Time

Using FIJI, we analyze the time series frame by frame to quantify the number of released EVs from the ciliary tip in an elapsed time. This measurement can be used as an indicator of relative EV release when comparing a given mutant genotype to control animals.

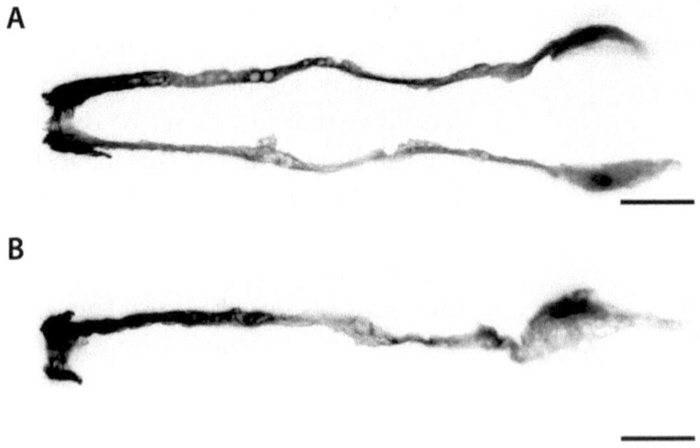

Fig. 3 Animals that are mounted for imaging can display different orientations within the preparation. (**a**) Animals that show the orientation in which both AMsh are visible should be avoided. (**b**) Instead, preference should be given to the orientation in which only one of the AMsh can be observed. Scale bars are 20 μm

3.5 Imaging and Quantification of Ciliary-Derived EVs in Their Capturing Tissue

Imaging and quantifying EVs can also be performed through an indirect approach: vesicles that are released from the base of the cilium are captured by their surrounding AMsh glia. We consider that quantifying the number of puncta/vesicles within AMsh cytoplasm serves as a proxy of the amount of EVs released by the donor cell. In the interest of experimental reproducibility, when comparing between different genotypes (i.e., Control vs. Mutant increasing EV secretion), it is imperative to be consistent and take into account several factors like: always synchronize *C. elegans* populations equally and use proper controls and imaging with the same acquisition parameters, tissue depth, and image animals that are similarly oriented (*see* **Note 13** and Fig. 3).

3.5.1 Sample Preparation

1. Prepare 4% agarose solution by melting 0.16 g of agarose in 4 mL of M9 buffer in a borosilicate glass tube. Melt the agarose solution by placing the solution in a heating block set at 95 °C or by repeatedly placing the tube over the flame in short intervals. Mix the solution using the vortex until agarose has completely dissolved (*see* **Note 8**).
2. Once the solution has cooled to 55 °C in the heat block, add 40 μL of sodium azide 1 M in M9.
3. Align 3 slides in parallel and place lab tape on both sides of the ones on the extremities (*see* Fig. 4a).
4. With a plastic Pasteur pipette, place a drop of melted agarose at the center of a glass slide (*see* Fig. 4b).
5. Proceed fast while the agarose is still hot: Place another glass slide orthogonally to flatten the melted agarose into a pad shape. We do this step by placing the glass slide at an angle as depicted in Fig. 4c (*see* **Note 14**).

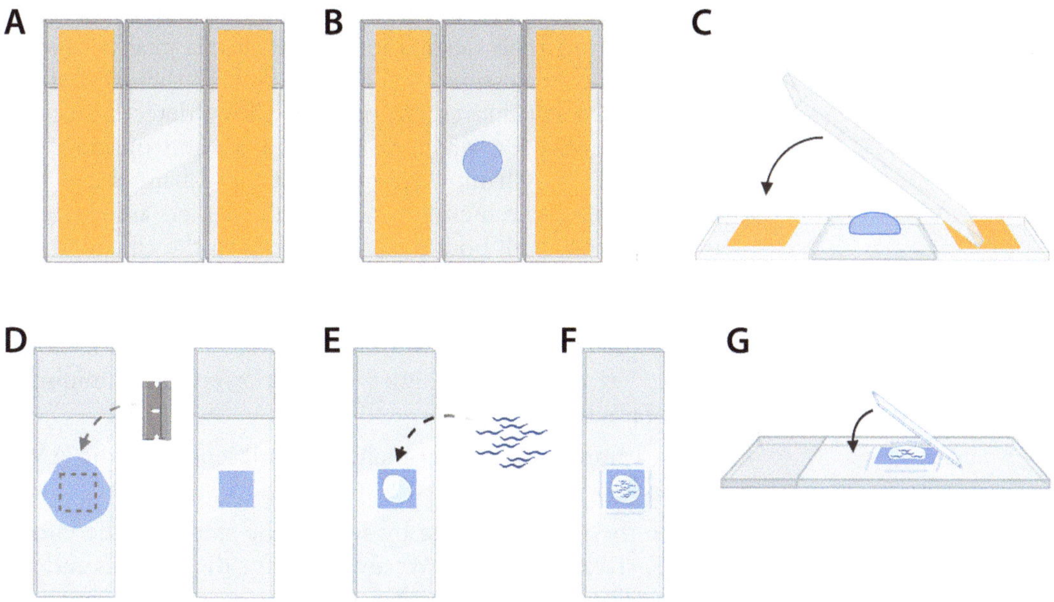

Fig. 4 Mounting animals with agar pads for confocal imaging. (**a**) Align three glass slides and place lap tape on the ones on the extremities. (**b**) Deliver a drop of melted agarose in the center of the middle glass slide. (**c**) Spread the agarose to form a pad by lowering another glass slide diagonally. (**d**) Reshape the pad using a scalpel blade. (**e**) Deliver a drop of the solution sodium azide/M9 on the pad then transfer the animals. (**f**) Flank the sides of the pad with lines of Vaseline. (**g**) Place the coverslip in a tilted manner to avoid air bubbles to be trapped in between the agarose and the coverslip

6. Allow the pad to solidify before separating the 2 glass slides. Gently slide the upper glass slide to expose the agar pad. Using a blade, reshape the agar pad: cut the edges so it forms a square/rectangle (*see* Fig. 4d).
7. Place a drop of 25 mM sodium azide diluted in M9 in the center of the agar pad.
8. Using the worm pick, select 20–30 animals outside of the bacterial lawn and transfer them to the anesthetic drop (*see* Fig. 4e).
9. Using the syringe loaded with Vaseline, deliver small lines 3–5 mm away from the agar pad (*see* Fig. 4f) (*see* **Note 15**).
10. Wait 5–10 min until the movements of the animals are significantly reduced then place the coverslip (*see* Fig. 4g) (*see* **Note 16**).
11. Worms will be ready to image at that stage.

3.5.2 Imaging (Z-Stacks) Using Confocal Microscopy

1. Locate the sample with a low magnification objective (10× or 20×) (*see* **Note 9**).
2. Switch to a higher magnification 40× objective.
3. To later facilitate the quantification of the EVs within the AMsh, we use 2 complementary fluorescent proteins: the

fluorescently tagged EV marker allows us to visualize EVs from their origin to their destination and we use a glial co-marker (cytoplasmic expression of CFP) to delimit the area of the AMsh glia. Each fluorescent protein will be detected using a specific channel for which you should set and optimize the parameters (i.e. Channel 1:wrmScarlet and Channel 2: CFP). It is important to remember that the EV markers are produced at the neuronal cell bodies and trafficked to the cilium; therefore, fluorescence will be observed in the neuronal cell bodies too. Avoid signal saturation when setting the acquisition parameters, as these images will later be used for quantification. Lastly, avoid using high laser powers to minimize photobleaching.

4. Select a frame size that contains the full length of the AMsh cell body.
5. Define the imaging range by setting the start and the end of your Z-stack acquisition. We always image the neurons and glia that lie closer to the objective.
6. Start the acquisition.
7. Once the first animal has been imaged, move the stage keeping in mind the location of the animals that have been previously imaged to avoid repeating acquisitions of the same sample (*see* **Note 17**).

3.5.3 Quantification of EV Fluorescence in a ROI

Using FIJI, one can quantify the fluorescence of the EVs that have been transferred into the targeting tissue to determine the amount of transferred material from the donor to the acceptor tissue. In our setup, we determine the fluorescence of the ciliary-derived EVs in a selected region of interest within the AMsh glial cell. Please keep in mind two things before starting your analysis:

- After a maximum intensity projection of the Z-stack in 2D, the cell body of AMsh partially overlaps with the cell bodies of the amphid neurons. This overlapping region should be avoided when quantifying, as fluorescence in that region corresponds to the fluorescent proteins within the ER-Golgi of the donor cell.
- Background fluorescence measurements should be performed for each picture; this will be important at later steps for background subtraction.

1. Open the image to be analyzed with FIJI. A tab named "Bio-formats Import Options" will open. By default, leave all the import options unchecked.
2. Split the channels by clicking on "Image" on the toolbar then "Color" and "Split Channels." Two different images will open, corresponding to each of the acquired channels.

3. Make a maximum intensity projection. To do that, go to "Image" on the toolbar, then "Stacks" and "Z project...". We use the Maximum intensity option as a Projection type. This will render a volumetric 3D dataset into a 2D dimension (*see* **Note 18**).
4. Merge the channels by clicking on "Image" on the toolbar then "Color" and "Merge Channels." Select the colors to display the signal. This step is necessary to display both channels on the same window to analyze the EV fluorescence within AMsh limits (*see* **Note 19**).
5. To determine which parameters will be evaluated by FIJI, click "Analyze" on the toolbar, then "Set measurements." A new tab will open, for the EV signal the following criteria should be selected for EV signal quantification: "Area," "Integrated density" and "Mean gray value." Click "ok" (*see* Fig. 5a).
6. Click on the icon of the preferred selection tool (i.e. Rectangle tool) to draw an area where to measure the fluorescence of the background. Make sure the channel that corresponds to the EVs signal is displayed and, with your cursor, draw a uniform shape on the background region. Once the region of interest is drawn, click "Analyze" on the toolbar then "Measure." A new tab "Results" will appear containing the measurements (*see* Fig. 5b) (*see* **Note 20**).
7. Using the polygon tool, outline the cell body of AMsh as depicted in Fig. 5c. As previously mentioned, we avoid outlining the region where AMsh overlaps with the neuronal cell bodies.
8. Using a spreadsheet, save all the fluorescence and area values for the background and AMsh for each measured individual.
9. Calculate the average value of the mean fluorescence of the background readings of all the pictures (*see* Fig. 5d).
10. Using a spreadsheet, determine the corrected total cell fluorescence (CTCF) using the following formula (*see* Fig. 5e):

$$\text{Integrated density (IntDen) of AMsh} - \text{AMsh area} \\ * \text{Mean fluorescence of the background readings}$$

(*see* **Note 21**).

11. Using statistical software, perform an ANOVA analysis to compare the mean fluorescence intensity of the control to the mean fluorescence intensity of the tested strains (*see* Fig. 6). (*see* **Note 22**)

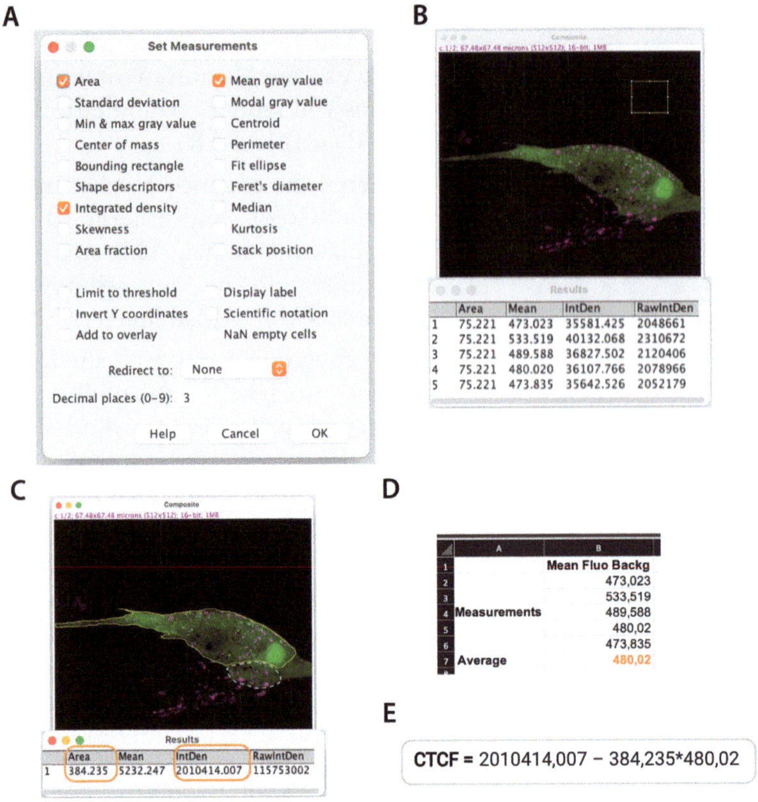

Fig. 5 Detection of the fluorescence intensity in a ROI. (**a**) Select the input parameters required to calculate fluorescence intensity: area, integrated density, and mean gray value. (**b**) Select a ROI for the background, this should be located outside the area where EVs are located. Examples show the values for the background of different images. (**c**) Select the ROI to evaluate the fluorescence intensity of EVs by outlining the AMsh (yellow line) and excluding the region where the neuronal cell bodies are located (white dashed line). Example shows the fluorescence intensity values (Integrated Density) of AMsh ROI of one picture. (**d**) Determine the average of the mean fluorescence of the background readings using a spreadsheet. (**e**) Calculate the CTCF using the values depicted in orange

3.5.4 Quantification of EV Number in a ROI Using ComDet

Using the ComDet v0.5.5 plugin built for FIJI, one can detect the number of fluorescent particles inside a Region of Interest (ROI). We recommend reading the plugin's documentation before you start the experiment, you can find it under the following link: https://FIJI.net/plugins/spots-colocalization-comdet

1. Download the plugin by going to FIJI's menu, then "Help" and "Update," and press the "Manage update sites" button. Press on the "Add update site" button and introduce the following link: "https://sites.FIJI.net/Ekatrukha/." See documentation page for a manual installation.

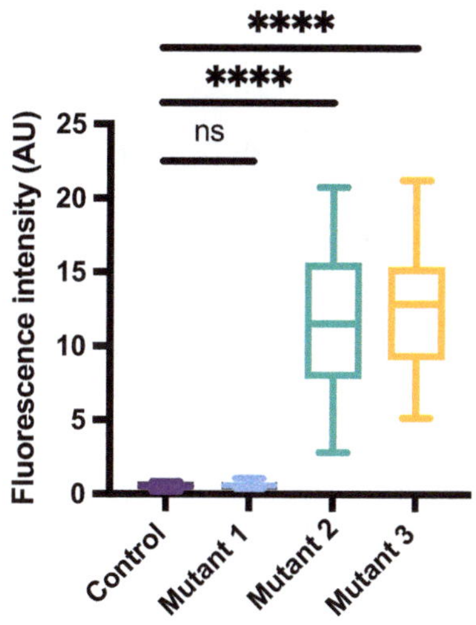

Fig. 6 Example of fluorescence quantification in AMsh cell body. Comparison of the mean fluorescence of several mutant strains compared to the control using one-way ANOVA ($p < 0,0001$)

2. Open the image to be analyzed with FIJI. A tab named "Bio-formats Import Options" will open. By default, leave all the import options unchecked.
3. If the acquired image is a 2-channel image, split the channels by clicking on "Image" on the toolbar then "Color" and "Split Channels." Two different images will open, corresponding to each of the acquired channels.
4. Merge the channels back by clicking on "Image" on the toolbar then "Color" and "Merge Channels," select option "Create composite." This step allows to display both channels within the same window, allowing the experimenter to visualize the limits of the AMsh glia and the presence of transferred EVs contained within the glia (*see* **Note 19**).
5. Make a Maximum intensity projection. To do that, go to "Image" on the toolbar, then "Stacks" and "Z project..." We use the Maximum intensity option as a Projection type. The channels back by clicking on "Image" on the toolbar then "Color" and "Merge Channels." This will render a volumetric 3D dataset into a 2D dimension (*see* **Note 18**).
6. Click on the icon of the Polygon Selection tool to draw a ROI outlining the area of the glial cell containing the fluorescent foci (*see* Fig. 7a) (*see* **Note 23**).
7. With the EV marker channel selected and the ROI drawn, duplicate the image of the selected channel. Go to "Image," then "Duplicate" and then write the Channel number (i.e., Select 1 instead of 1–2). A new window should appear with a duplicated image of the selected ROI for only the selected channel.

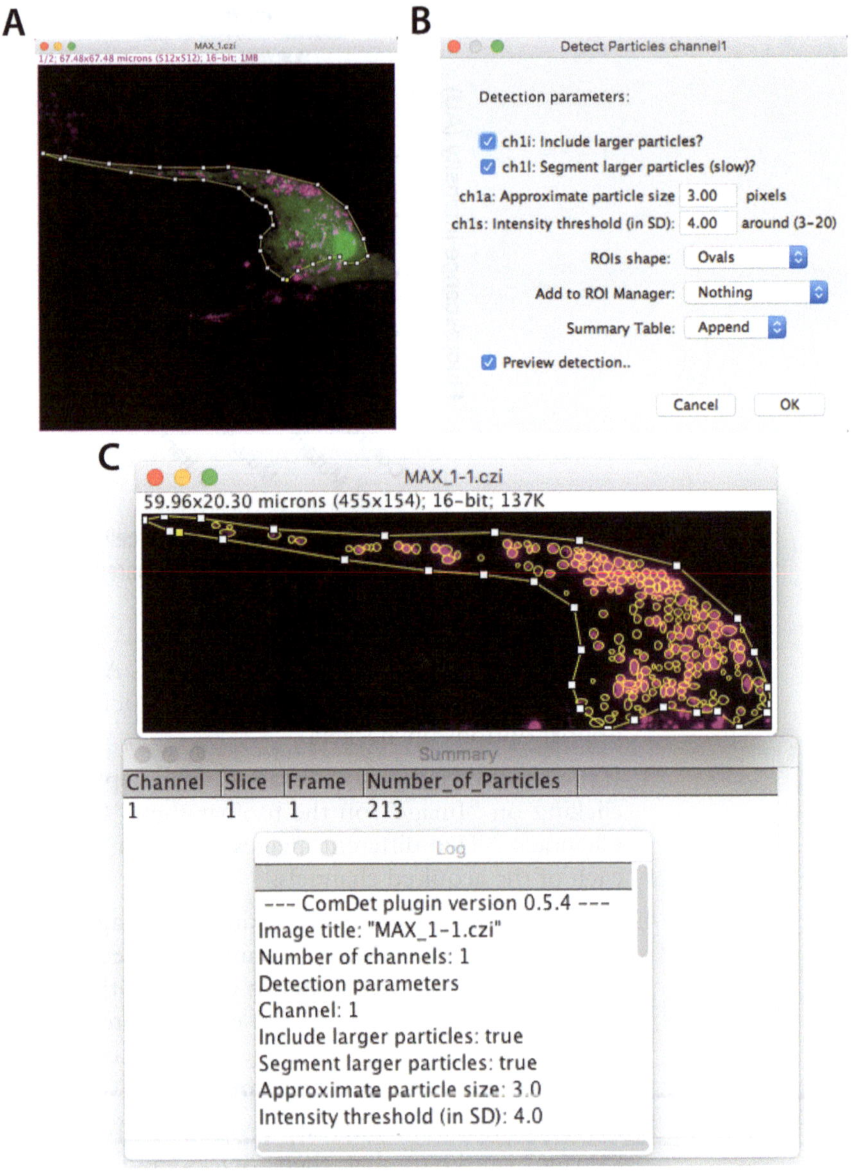

Fig. 7 ComDet is a FIJI plugin that allows particle detection in a given image/ROI. (**a**) We first outline the glial cell using the Polygon selection tool. Avoid outlining the amphid neurons cell bodies as they can partially overlap with the cell body of the AMsh glia; in this image, the cell bodies appear to contain brighter foci and they are normally right below the AMsh cell nuclei. (**b**) We have selected these parameters as an input for the ComDet plugin. These parameters might change according to the imaging settings, objectives, and microscope used for image acquisition. Each user should optimize these parameters according to their data. (**c**) After running the plugin, 3 windows should appear: (Top) one with the selected ROI and the displayed number of calculated foci by the ComDet plugin, appearing as oval shapes drawn on the original image. (Middle) The summary window which tells you the total amount of detected particles. (Bottom) The log file with the used parameters

8. On the toolbar, click on "Plugins," then "ComDet v.0.5.5" and "Detect particles." A new tab "Detect particles Channel 1" will open. (Check at this point if the EVs signal is displayed on channel 1 or 2.) A new tab "Detect Particles Channel1" will open, if your EVs signal is displayed on channel 1. Select the parameters as depicted in Fig. 7b.
9. A "Summary" tab will open, exhibiting the number of particles contained in your select area in both channels (*see* Fig. 7c). Additionally, a "Results" tab will open with a list of each detected foci and their relative X-Y positions, with their fluorescence measurements displayed as integrated density.
10. Using a spreadsheet, save all the particle counts for each individual measured AMsh area.
11. Repeat this process 25–50 times with different experimental replicates and compare the results across all your different samples (i.e., different mutant strains, different Extracellular markers, different treatment conditions) using the appropriate statistical tests where needed. Results can be displayed as an average number of transferred EVs within the receiving tissue.

3.6 Markers for the Study of Ciliary EVs in C. Elegans

A wide range of ciliary membrane proteins, but not all, can be used to study the production of ciliary-derived EVs from *C. elegans* sensory neurons. We and other labs have established the use of ciliary EV markers such as sensory G-protein coupled receptors (GPCRs) or guanylate cyclases (GCYs), tetraspanins (TSPs), or polycystin (PKD) proteins [9–11]. These EV markers are endogenously expressed in one or several neurons in hermaphrodites and/or males.

We have used two strategies to generate strains to visualize EV production:

1. We have generated strains overexpressing EV markers under cell-specific promoters. Cell-specific promoters exist for almost every ciliated neuron of *C. elegans* [16]. The advantage of this strategy is that we can cell-specifically express such EV cargoes in a single ciliated neuron or a subset of neurons of interest, which facilitates the imaging of EVs in scenarios where otherwise would be difficult to resolve. Multiple copies of the transgene as well as strong promoters will generate overexpression of the EV marker in the neuron of interest. We observed that overexpression of several of these ciliary proteins promotes generation of EVs [11]. This approach is useful to study production of EVs from a single neuron or how EVs are generated in stress conditions (i.e., proteostatic challenges, cilia trafficking disruption).
2. We have generated strains expressing EV markers at endogenous levels by knocking in a fluorescent protein at the

endogenous locus via CRISPR-Cas9 gene editing. Native expression levels of EV markers allow avoiding overexpression artifacts. However, one disadvantage of this strategy is that the level of expression of the marker relies on the strength of the promoter and its unmodifiable expression pattern.

Here, we show Table 1 with the different markers that have been used by our lab and others.

4 Notes

1. Adding salts (specially MgSO4) to hot NGM medium or M9 buffer mixture will make them precipitate. It is important to allow the solutions to cool down to 55–60 °C before the addition of salts to avoid formation of precipitates.
2. Sodium azide is highly toxic through skin contact, inhalation, and ingestion. When manipulating sodium azide, ensure good ventilation of the workstation and wear personal protective equipment. Do not get in eyes, on skin, or on clothing.
3. Strain description: The strain carries an EV marker TSP-6 tagged with a red fluorescent protein (wrmScarlet) at endogenous expression levels generated by CRISPR-Cas9 gene editing. The strain also carries a cytoplasmic cyan fluorescent protein (CFP) expressed in the AMsh glial cell to visualize the limits of the glial cell.
4. *Optional*: When filling many plates, it is advisable to use an automated culture media dispenser (we use the Automatic dispenser from Mediajet, Integra). This is the most time-effective manner to fill the NGM plates with a precise volume. We prepare 4 L of NGM media at once; this yields approximately 340 plates.
5. *Caution*: OP50 liquid cultures stored in the fridge are susceptible to contamination by fungi spores, yeast, and other bacteria. Sterile laboratory procedures must be followed at all times when working with bacterial cultures to avoid contamination. Renew the OP50 liquid culture if contamination appears on the seeded plates.
6. When seeding dozens of plates with OP50 bacteria, a time-effective method is to stack the plates in groups of 4–6 plates and dispense the OP50 using a 10 mL sterile plastic pipette attached to a rubber suction bulb.
7. *Caution*: Count the number of worms that are transferred into a plate for a given egg-laying window. This number should be the same when removing them after the egg-laying window has concluded. Leaving any adult animals on the plate will result in

Table 1
List of ciliary proteins used for EV tracking in hermaphrodite and male *C. elegans*. The expression pattern is reporter using CenGEN or as reported in the literature. The expression level indicates how these proteins have been used in previous publications, either by overexpression or by expression at endogenous levels by single copy insertion using the Mos system or CRISPR-Cas9 genome editing

Gene name	Ciliary protein type	Expression pattern	Expression level	Reference
GCY-22	Guanylate cyclase	ASER neuron	Multicopy Oex. / Knock-in	[11, 17]
GCY-8	Guanylate cyclase	AFD neuron	Multicopy Oex.	[11]
SRTX-1	G-protein-coupled receptor	AFD neuron	Multicopy Oex.	[11]
TSP-6	Tetraspanin (CD9 ortholog)	Ciliary neuron subset: ASK, ADL, ASH, ASJ, ADF, ASG, AWC, AWA,ASI, ASER, ASEL, AQR, PQR, URX	Multicopy Oex. / Knock-in	[11]
TSP-7	Tetraspanin (CD63 ortholog)	Cholinergic neurons, GABAergic neurons, oxygen-sensing neurons, AWA, AWC (ON/OFF), ASEL, IL2 LR	Multicopy Oex.	[11]
LOV-1	TRP channel Polycystin receptor 1	Male IL2s, CEM, HOB, ray neurons	Multicopy Oex.	[9]
PKD-2	TRP channel Polycystin receptor 2	Male IL2s, CEM, HOB, ray neurons	Multicopy Oex. / Single copy insertion (Mos system) / Knock-in	[9]
ASIC-1	Acid-sensing sodium channel	IL2 neurons	Multicopy Oex.	[18]
EGAS-1	Acid-sensing sodium channel	IL2 quadrant (IL2Q) neurons	Multicopy Oex.	[18]
CIL-7	Myristoylated coiled-coil protein	Male IL2s, CEM, HOB, ray neurons	Multicopy Oex. / Knock-in	[19, 20]
SID-2	Transmembrane dsRNA transporter	IL2s, CEM, HOB, ray neurons	Knock-in	[10]
ENPP-1	Phosphodiesterase	IL2s, male CEM, and ray neurons	Multicopy Oex.	[10]
MCM-3	DNA replication licensing factor	IL2s, male CEM, HOB, and ray neurons	Multicopy Oex.	[10]
CLHM-1	Ion channel	IL2, ASE, ASG, ASI, ASJ, ASK, ADE, PHA, PHB neurons	Single copy insertion (Mos system)	[12]

(continued)

Table 1 (continued)

Gene name	Ciliary protein type	Expression pattern	Expression level	Reference
TBB-4	β-Tubulin	Male IL2s, CEM, HOB, ray neurons	Multicopy Oex.	[20]
NPHP-1	Nephrocystin-4	Ciliated sensory neurons	Multicopy Oex.	[20]
CWP-1	Ciliary protein coexpressed with Polycystins	Male IL2s, CEM, HOB, ray neurons	Multicopy Oex.	[9]

an unsynchronized population of worms that are younger than desired. It is also important not to make holes on the surface of the NGM agar plates; *C. elegans* will crawl onto these holes, reducing the number of available experimental subjects.

8. *Danger*: Overheating the agarose solution by directly putting it on a flame could cause the solution to flash boil, resulting in splashes or burn injuries to the hands, arms, and face. Flame heating and vortexing generate air bubbles within the mixture. Once the agarose has been melted, the solution should be placed on a heat block set at 55 °C. This allows air bubbles to dissipate.
9. We have noticed that not all animals release EVs continuously, the best way to capture such events is finding animals where cilia have bulged ends. This is an indicator that their cilia are at the initial phases of EV formation, where the tip of the cilia begins to swell and later on is excised and released.
10. We used the Zeiss LSM780-NLO confocal microscope with the LD C-Apochromat 40X/1.1 W Korr M27 objective or the alpha Plan Apochromat 63X/1.46 Oil Korr M27 objective. Also, virtual zooming is recommended to focus onto the cilia endings where EVs shed from.
11. In conditions where cilia are not positioned in a straight manner or bent in their trajectories, we recommend slightly increasing the pinhole diameter to collect more out of focus light, slightly improving detection. However, this increase in detection outside of the focal comes at the expense of a reduction in the signal-to-noise ratio.
12. Recording can be manually stopped at any time. Stop the recording if the animals make sudden head movements that render the cilium out of the focal plane.
13. Bear in mind that anesthetized animals will be arranged in different positions within the preparation, this will lead to only a fraction of the animals being correctly oriented. When

locating the animals for imaging, one should target animals in which the glial cell is displayed as in Fig. 3b, this allows to better discriminate neurons from glia. It is also recommended to mount several samples of the same genotype to increase the number of animals correctly oriented.

14. Positioning the glass slide in an angle (as in Fig. 4c) avoids the formation of bubbles which can act as sinkholes that trap the animals during the mounting step. The agar pad needs to have the even thickness all throughout to prevent the mounted animals from being at different focal planes.
15. In our experience, Vaseline seems to be toxic to the animals so beware to keep a distance between the agar pad and the Vaseline sealant; otherwise, it will spread into the pad when you place the coverslip. Vaseline is autofluorescent; therefore, it can interfere with the imaging.
16. The coverslip should be placed before the animals are completely anesthetized, which takes 10–15 min, such that they are immobilized in their anatomical orientation. Animals will become anesthetized even if they are covered with the coverslip.
17. Mounting several animals on the same slide can lead to mistakes by imaging the same sample twice. To prevent that, it is helpful to draw schemes and notes of the preparation before and during acquisition. Additionally, when locating the samples, we always start from the top left to top right, and then repeat this in serpentine fashion (from top to bottom), this can help prevent this issue. Some acquisition software offers the possibility to mark the stage position, which can help prevent the imaging of an already acquired sample.
18. Our volumetric datasets comprise the whole volume of the glial cell. We only collect the images from a volume that targets only one of the bilateral amphid sensilla (neurons and glia) that lies closer to the objective. This provides consistency across all the dataset and faster acquisition times. We image at approximately 0.4 μm step size with 25–35 steps in total, which is equivalent to 10–15 μm Z-stacks; this allows us to capture the whole volume of AMsh glia cell body.
19. The color of the channel can be modified by pseudo-coloring the images. You can change the color associated with each channel by placing the slider bar on the channel you want to change, clicking on the toolbar "Image," "Lookup tables," and selecting the color of your choice. We recommend using in color-blind color combinations (i.e. magenta-green or cyan-orange) to make final figures accessible to readers with color-blindness.

20. We usually keep the same Region of Interest (ROI) to measure the fluorescence of the background. To do so, click "Analyze" on the toolbar, "Tools" and "ROI manager." A new tab named "ROI manager" will open. With your cursor, click on the preferred shape and draw the ROI. Once the ROI is displayed on the image in yellow, go to the "ROI manager" tab and click "Add" then "Save." This allows the experimenter to re-use the same ROI on other images. To re-open this ROI on another image, simply click on the ROI on the "ROI manager" tab and it will appear on your image. To move it, select the shape by clicking on the center of the ROI to displace it.
21. The parameters provided in the results tab are:
    - Mean (mean gray value): Average of the pixel values in the selection.
    - IntDen (integrated density): Area * Mean gray value.
    - RawInt (raw integrated density): Sum of the pixel values in the selection.
22. Several options are available to display your data. You can also normalize your data to a Control group and display the difference in fold change.
23. Make sure to delimit the outline of the AMsh glial cell in the channel that corresponds to the EVs signal. Be aware that the amphid neurons cell bodies overlap with that of the amphid sheath glia. When delimiting the ROI of the glial cell try avoiding the cell bodies of the amphid neurons. A trick to do that is to temporarily adjust the Brightness/Contrast of the image to saturate the fluorescent pixels coming from the neuronal cell bodies, this allows for a better discrimination of the amphid neuronal bodies position regarding the position of AMsh. Fluorescence from neurons usually appears brighter and with bigger foci.

Acknowledgments

The research and authors were supported by UNI-ULB and FRIA-FNRS grant 40004232 (A.R), FRIA-FNRS grant 40009291 (T.L) and FNRS fund 22445636 (P.L). Figures were created with the help of Biorender.com.

References

1. van Niel G, D'Angelo G, Raposo G (2018) Shedding light on the cell biology of extracellular vesicles. Nat Rev Mol Cell Biol 19(4): 213–228
2. Hurley JH (2008) ESCRT complexes and the biogenesis of multivesicular bodies. Curr Opin Cell Biol 20(1):4–11

3. Trajkovic K, Hsu C, Chiantia S et al (2008) Ceramide triggers budding of exosome vesicles into multivesicular endosomes. Science 319: 1244–1247
4. Stuffers S, Sem Wegner C, Stenmark H et al (2009) Multivesicular endosome biogenesis in the absence of ESCRTs. Traffic (Copenhagen, Denmark) 10(7):925–937
5. Ferreira JV, da Rosa Soares A, Ramalho J, Máximo Carvalho C, Cardoso MH, Pintado P, Carvalho AS, Beck HC, Matthiesen R, Zuzarte M, Girão H, van Niel G, Pereira P (2022) LAMP2A regulates the loading of proteins into exosomes. Sci Adv 8(12):eabm1140
6. Wehman AM, Poggioli C, Schweinsberg P, Grant BD, Nance J (2011) The P4-ATPase TAT-5 inhibits the budding of extracellular vesicles in C. elegans embryos. Curr Biol 21(23):1951–1959
7. Beer KB, Wehman AM (2017) Mechanisms and functions of extracellular vesicle release in vivo-what we can learn from flies and worms. Cell Adhes Migr 11(2):135–150
8. Beer KB, Rivas-Castillo J, Kuhn K, Fazeli G, Karmann B, Nance JF, Stigloher C, Wehman AM (2018) Extracellular vesicle budding is inhibited by redundant regulators of TAT-5 flippase localization and phospholipid asymmetry. Proc Natl Acad Sci U S A 115(6):E1127–E1136
9. Wang J, Silva M, Haas LA et al (2014) C. elegans ciliated sensory neurons release extracellular vesicles that function in animal communication. Curr Biol 24:519–52510
10. Nikonorova IA, Wang J, Cope AL et al (2022) Isolation, profiling, and tracking of extracellular vesicle cargo in Caenorhabditis elegans. Curr Biol 32(9):1924–1936.e6
11. Razzauti A, Laurent P (2021) Ectocytosis prevents accumulation of ciliary cargo in C. elegans sensory neurons. elife 10:e67670
12. Clupper M, Gill R, Elsayyid M, Touroutine D, Caplan JL, Tanis JE (2021) Kinesin-II motors differentially impact biogenesis of distinct extracellular vesicle subpopulations shed from C. elegans sensory cilia. bioRxiv 12(19): 473369
13. Brenner S (1974) The genetics of Caenorhabditis elegans. Genetics 77(1):71–94
14. Schindelin J, Arganda-Carreras I, Frise E et al (2012) Fiji: an open-source platform for biological-image analysis. Nat Methods 9: 676–682
15. Thevenaz P, Ruttimann UE, Unser M (1998) A pyramid approach to subpixel registration based on intensity. IEEE Trans Image Process 7:27–41
16. Lorenzo R, Onizuka M, Defrance M, Laurent P (2020) Combining single-cell RNA-sequencing with a molecular atlas unveils new markers for Caenorhabditis elegans neuron classes. Nucleic Acids Res 48(13):7119–7134
17. van der Burght SN, Rademakers S, Johnson J-L et al (2020) Ciliary tip signaling compartment is formed and maintained by intraflagellar transport. Curr Biol 30:4299–4306.e5
18. Wang J, Kaletsky R, Silva M, Williams A, Haas LA, Androwski RJ, Landis JN, Patrick C, Rashid A, Santiago-Martinez D, Gravato-Nobre M, Hodgkin J, Hall DH, Murphy CT, Barr MM (2015) Cell-specific transcriptional profiling of ciliated sensory neurons reveals regulators of behavior and extracellular vesicle biogenesis. Curr Biol CB 25(24):3232–3238
19. Maguire JE, Silva M, Nguyen KC, Hellen E, Kern AD, Hall DH, Barr MM (2015) Myristoylated CIL-7 regulates ciliary extracellular vesicle biogenesis. Mol Biol Cell 26(15): 2823–2832
20. Wang J, Nikonorova IA, Silva M, Walsh JD, Tilton PE, Gu A, Akella JS, Barr MM (2021) Sensory cilia act as a specialized venue for regulated extracellular vesicle biogenesis and signaling. Curr Biol CB 31(17):3943–3951.e3

Chapter 20

Exosome-Based COVID-19 Vaccine

Jaeyoung Kim and Nikita Thapa

Abstract

Extracellular vesicles (EVs) enable cell-to-cell communication and, by delivering antigens, can stimulate the immune response strongly. Approved in use SARS-CoV-2 vaccine, candidates immunize with the viral spike protein delivered via viral vectors, translated by injected mRNAs, or as a pure protein. Here, we outline a novel methodological approach for generating SARS-CoV-2 vaccine using exosome that delivers antigens from the SARS-CoV-2 structural proteins. Engineered EVs can be loaded with viral antigens, thus acting as antigens presenting EVs, eliciting strong and targeted CD8(+) T cell and B cell, offering a unique approach to vaccine development. Engineered EVs thus portray a safe, adaptable, and effective approach for a virus-free vaccine development.

Key words Extracellular vesicles, Exosomes, Ultracentrifugation, Exosome-based vaccine

1 Introduction

SARS-CoV-2 (CoV-2) has made its presence known globally as a highly contagious virus that has reported to spread rapidly and resulted in severe deaths [1]. Although currently approved vaccines have shown to be effective at preventing severe infection, vaccine hesitancy, the quick emergence of variants of concern (VOCs) that could escape pre-existing immunity, and the potential for the endemic establishment of SARS-CoV2, all indicate that COVID-19 cases will continue to be reported in the future. As a result, novel strategies that directly target and restrict SARS-CoV-2 during infection will need to be developed.

Extracellular vesicles (EVs) are lipid bilayer membrane vesicles derived from endosomes (multivesicular bodies), which can be produced by all cell types, including prokaryotic or eukaryotic and healthy or malignant cells. As per the guidelines stated by the International Society of Extracellular Vesicles, EVs can be categorized into three types: microvesicles (100–1000 nm), apoptotic bodies (100–5000 nm), and exosomes (50–200 nm). They are categorized based on size, biogenesis pathway, and content [2].

Seppo Vainio (ed.), *Cell-Secreted Vesicles: Methods and Protocols*, Methods in Molecular Biology, vol. 2668, https://doi.org/10.1007/978-1-0716-3203-1_20,

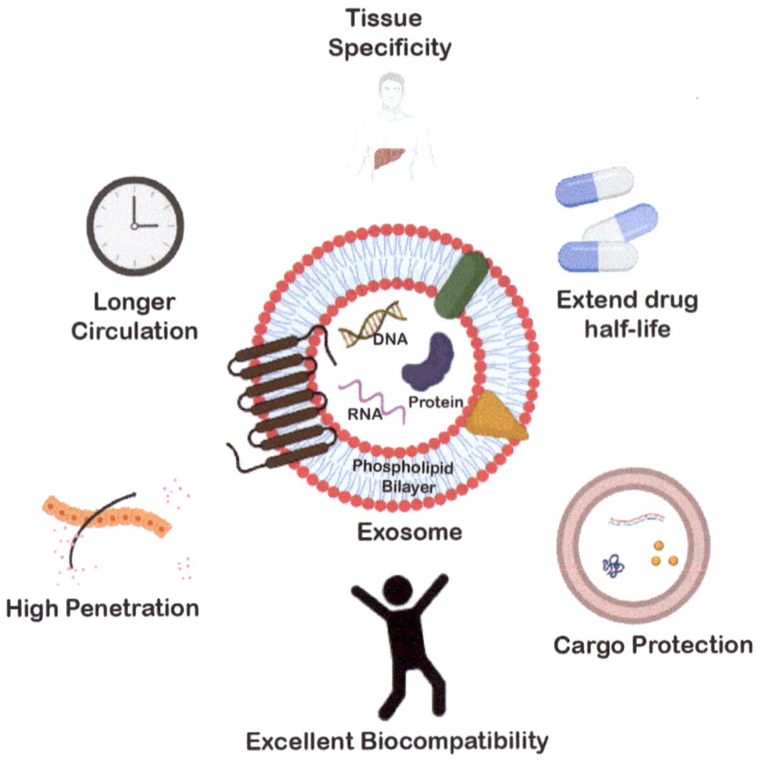

Fig. 1 Structure and advantages of exosome-based drug delivery

EV-based therapeutics is a promising drug delivery system which has the property of excellent penetration and carrying heterogenous cargo (DNA, proteins, lipid) enclosed within lipid bilayer [3, 4]. Exosomes in particular offer numerous benefits as drug delivery vehicles which includes small size, low cytotoxicity, long half-life in the circulation, and the ability to load various cargoes with high biocompatibility (*See* Fig. 1). Exosome-based vaccines present one such option that present an exceptional platform for delivering antigen effectively owing to their involvement in disease progression and their role in inhibiting viral infection and triggering host immune response.

The key features of EV-based vaccines include their ability to induce poor immunogenicity with zero toxicity, meaning that EVs can be safely and efficiently used in vaccine development. The ability of EVs to preserve naïve antigen conformation and access to all organs via bodily fluids give an added advantage compared with other delivery agents, such as lipid-based nanoparticles (LNPs) or viral vectors [5]. Therefore, engineered EVs fulfill the criteria for an efficacious vaccine due to their efficient antigen-presenting system and high biosafety.

The development of EV-based vaccines requires engineering of EV, or the artificial incorporation of target antigens into these vesicles to transform them into antigen-presenting EVs. Two major approaches, direct manipulation of EVs following cell isolation (e.g., by electroporation), or the engineering of the EV donor cells, were proposed. Modification at the level of donor cells is the most common approach as it enables continuous manufacturing of tailored EVs [6].

To use exosome technology in clinical settings for human use, it is imperative to follow standardized technique for its isolation. The lack of standardized isolation methodologies is one of the major drawbacks in the use of EVs for diagnostic and therapeutic applications. Existing isolation techniques for producing exosomes are not scalable with low quantity production, which led to restricting limited usage to assess exosome efficacy in the preclinical stage of animal testing. Furthermore, exosome, because of its scanty secretion from cell, poses a major roadblock to effective, large-scale exosome manufacturing, consequently leading to its limited application [7].

This chapter presents the methodology for isolating ultracentrifugation-based high yield exosome from HEK293 Cells (*See* Fig. 2). Ultracentrifugation is the most utilized and supposedly the gold standard technique for exosome isolation. Here, stepwise procedure for the development of a non-adjuvanted, virus-free vaccine inducing long-lasting immunity using EV-based platform has been described.

2 Materials

Prepare all solutions using ultrapure water and analytical grade reagents. Prepare and store all reagents at room temperature (unless indicated otherwise). Diligently follow all waste disposal regulations when disposing of waste materials.

2.1 Cell Culture and Lentiviral Transduction

1. Dulbecco's modified Eagle's medium (DMEM).
2. Fetal bovine serum (FBS), penicillin, and streptomycin.
3. Phosphate-buffered saline (PBS)—Measure and transfer prepare 800 mL of distilled water in a suitable container. Weigh 20.214 g of sodium phosphate dibasic heptahydrate and 3.394 g of potassium phosphate monobasic monohydrate and transfer to container. Adjust solution to desired pH (typically pH $\approx$ 6–7). Add distilled water until the volume is 1 L and store at room temperature.
4. Trypsin, Petri dish, sterile pipette.
5. 5% CO2 incubator, microscope.
6. Hartman solution, falcon tubes.

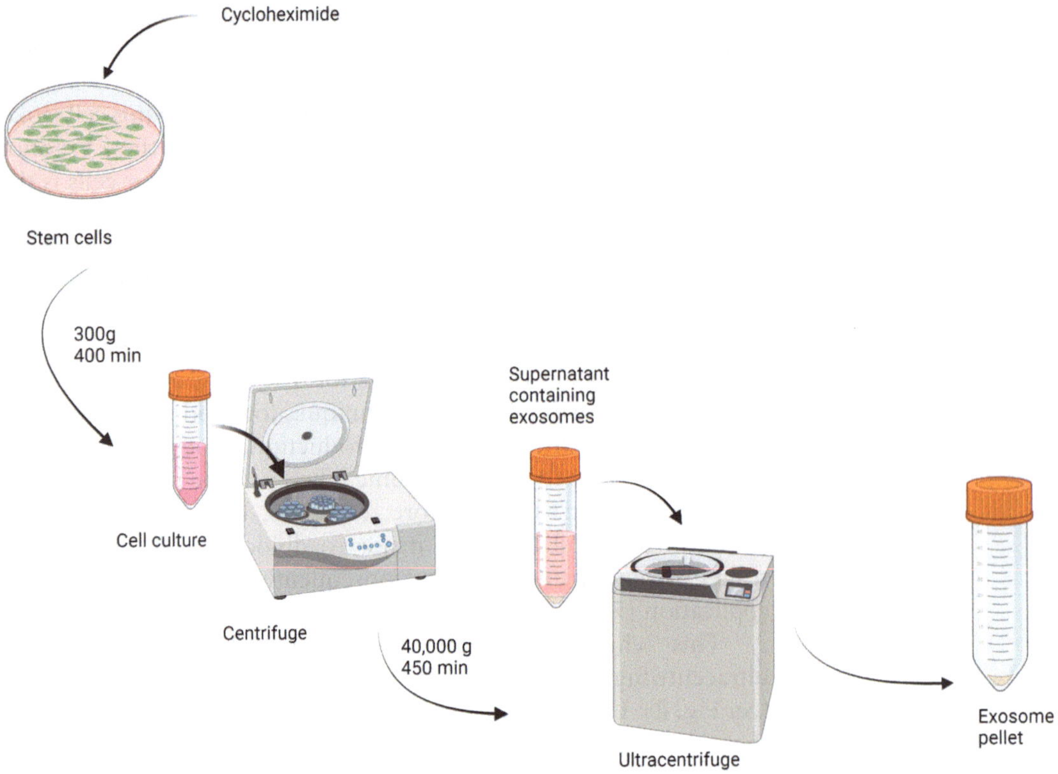

Fig. 2 Diagrammatic representation of ultracentrifugation-based exosome isolation from mesenchymal stem cell

2.2 Ultra-centrifugation

1. Centrifuge.
2. Ultracentrifuge (Beckman Coulter).
3. Ice-cold Hartman solution.
4. 50-mL Falcon tubes.
5. 25 mL sterile pipette.
6. Weighing machine.
7. Sonicator.
8. 150 mm petri dish.
9. Pipette tips, PCR tubes, 4-degree refrigerator.

2.3 Immunoblotting

1. RIPA lysis buffer—25 mM Tris–HCl, pH 7.6, 150 mM NaCl, 1% NP-40, 1% sodium deoxycholate, 0.1% SDS.
2. Pierce BCA Protein Assay Kit.
3. Tris-glycine/SDS electrophoresis buffer—10x premixed electrophoresis buffer, contains 25 mM Tris, 192 mM glycine, 0.1% SDS, pH 8.3 following dilution to 1× with water.
4. 4–20% Mini-PROTEAN TGX Precast Protein Gels.
5. Western blot transfer buffer: 0.025 M Tris–HCl, 0.192 M glycine, 20% methanol.

6. Nitrocellulose membranes.
7. Tris-buffered saline (TBS; 10×): 1.5 M NaCl, 0.1 M Tris–HCl, pH 7.4.
8. TBS containing 0.05% Tween-20 (TBS-T).
9. Blocking buffer: 2% BSA in TBS-T. Store at 4 °C.

2.4 Transmission Electron Microscopy (TEM)

1. Phosphate buffer pH 7.4 with 2.5 percent glutaraldehyde.
2. Cold PBS, OsO4 buffer, graded acetone, Durcupan ACM resin.
3. Uranyl acetate and lead citrate.
4. Hitachi H-7600 TEM.
5. 2% paraformaldehyde.
6. 0.4 percent methylcellulose.

3 Methods

Here, we present the methodology for the development of an exosome-based SARS-CoV-2 vaccine candidate, CKV-21. The type is an EV display vaccine consisting of human HEK293 cells transduced with lentiviral vectors for the expression of the SARS-CoV-2 proteins (Spike, Membrane, and Envelope). In turn, the released EVs will carry all viral antigens in their native context and conformation. Prior research has shown that vaccination with various protein enables the modification of the immune response's amplitude and character, in terms of cytokine production and Th1 or Th2 activation [8].

3.1 Cell Culture and Lentiviral Transduction

1. Grow HEK293 cells in DMEM supplemented with 10% fetal bovine serum (FBS) and 1% pen/strep solution at 37 °C in a 5% CO2-humidified incubator. Routinely test HEK293T cells for MycoAlert™ mycoplasma detection.
2. For the preparation of the exosomal vaccines, seed HEK293T cell lines on 10-cm diameter plates for 24 h to reach 70–80% confluency following lentiviral transduction of gene of interest as described below.
3. Clone cDNA encoding SARS-CoV-2 Spike RM (Spike RM is more stable form of spike protein which is created by introducing point mutation) membrane and envelope into eukaryotic cell expression vector pCDH-CMV.
4. Transfect HEK293T cells with psPAX2 (packaging plasmid), pMD2 (envelope plasmid), and pCDH-CMV-MCS-EF1-Puro encoding SARS-CoV-2 S glycoprotein (Spike RM), M, and E using calcium phosphate transfection.

5. After 17 h, replace the culture media with complete media. Harvest the supernatant containing viral particles every 24 h for up to 72 h. Centrifuge media at 500 g for 5 min and discard cell pellet.
6. Centrifuge harvested supernatant at 2000 *g* for 20 min to remove cell debris, and then filter through a 0.45 μm PVDF (low protein attachment) membrane.
7. Finally, apply centrifugation at 19,000 *g* for 1:30 h to concentrate viral particles and measure multiplicity of infection (MOI) (number of viral particles per target cell) for HEK293T cell transduction.
8. Seed mesenchymal stem cell (MSC) in a 24-well plate at a cell density of 1–2 × 10^4 /well. Add viral particles with MOI of 100 and protamine sulfate to each well and change the media after 24 h. Repeat the transduction procedure 3 times to increase efficiency.
9. Select positively transduced cells by supplementing puromycin or neomycin with periodic passaging until the cell lines returned to >90% viability, at which point they can be cryopreserved.

3.2 Ultra-centrifugation-Based Exosome Isolation

Differential ultracentrifugation-based technique for the extraction of exosomes from mesenchymal stem cells has been explained below. All steps described in this section must be carried out in a biosafety cabinet under sterile conditions. All centrifugations should be performed at 4-degree C.

1. Culture mesenchymal stem cells in DMEM media supplemented with penicillin and streptomycin and incubate in a 5% CO2 incubator at 37 °C.
2. Once cells are fully confluent, it is necessary to replace media with FBS-free media. This should be done 100 h before isolating exosomes from the cell culture media (*See* **Note 1**).
3. Provide stress-induced conditions such as addition of drugs such as cycloheximide to fully confluent cells after replacing with FBS-free media, 100 h prior to exosome isolation.
4. After 100 h, check the cells under the microscope and detach cells using sterile pipette. The cell suspension will be transferred to 50 mL falcon tube.
5. Perform centrifugation of cell suspension at 300 *g* for 400 min. Repeat this step twice. Low-speed centrifugation is used to separate the samples from the conditioned medium and remove dead cells, microvesicles, apoptotic bodies, and cell debris (*See* **Note 2**). Discard the pellet and transfer the supernatant to a new collection tube.

6. Ultracentrifugation is used to isolate the exosomes from the filtrate. Transfer the cell supernatant to ultracentrifuge tube. Weigh the ultracentrifuge tubes containing medium and normalize their weight using sterile PBS on weighing machine (*See* **Note 3**).
7. Make the pellets of collected supernatants using ultracentrifugation at 40,000× *g* for 450 min to harvest the isolated extracellular vesicles accumulated at the bottom. Aspirate the supernatant without touching the pellet (*see* **Note 4**).
8. Wash the exosome pellet by resuspending it in 30 mL of ice-cold sterile PBS (*see* **Note 3**). Weigh the ultracentrifuge tubes again and normalize their weight using sterile PBS. Centrifuge the samples at 100,000 Xg, 4 degree C for 20 min.
9. Discard supernatant again to get rid of contaminating proteins without touching pellets. This pellet is concentrated EV accumulated at the bottom of the ultracentrifuge tube.
10. Resuspend the pellet by scrapping the pellet with a pipet tip in phosphate-buffered saline (PBS) based on the size of pellet (20–100 μL) of fresh PBS and collect into PCR tubes. Perform sonication under very low speed for few seconds following vertexing the tube to achieve even distribution. Harvested exosomes could be stored in −80 degree (*see* **Note 5**).

3.3 Characterization of Exosomes

The exosomal pellets need to be resuspended into PBS, and the suspension should be aliquoted to check characteristic properties of isolated exosome using techniques such as transmission electron microscope (TEM), Western blot, and nanoparticle tracking analysis (NTA).

3.3.1 Immunoblotting

1. Determine the protein concentrations of the exosome lysates using the Bradford assay and normalize their concentrations. For Western blotting, common EV markers such as CD63, CD9, CD81, S1PR1, S1PR3, and phosphatidylserine can be used to confirm the successful isolation of EV (*See* Fig. 3a) (*see* **Note 6**).
2. Lyse HEK293 EVs and producer cells in RIPA lysis buffer, and the protein concentration was determined by a Pierce BCA Protein Assay Kit (Thermo Fisher Scientific) according to manufacturer protocols.
3. Load equal amounts of protein (10 ug) from each sample onto 4–20% Mini-PROTEAN TGX Precast Protein Gels (Bio-Rad, Hercules, CA, USA) and run in Tris-glycine/SDS electrophoresis buffer (Bio-Rad).
4. Transfer proteins onto nitrocellulose membrane overnight at 4 degree C in Tris-glycine buffer (Bio-Rad). Rinse the membrane and block using 2% BSA in TBS-T blocking buffer and probe using the following primary antibodies.

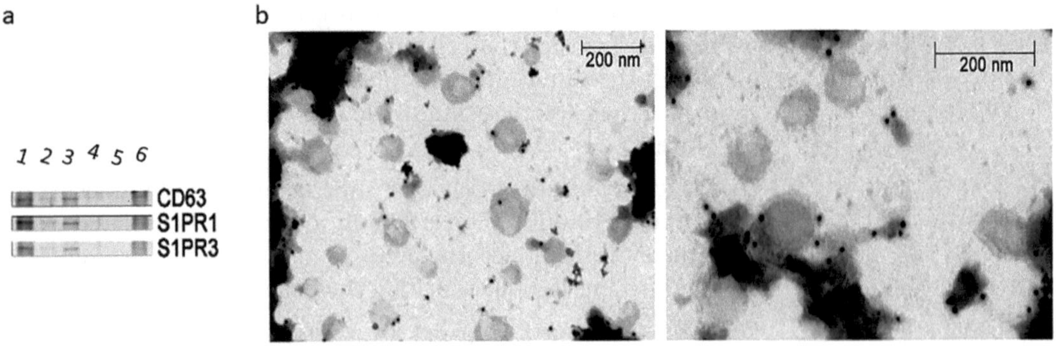

Fig. 3 Characterization of isolated exosomes from 293T-conditioned medium. (**a**) Six different samples of isolated exosomes were characterized using exosome marker CD63, S1PR1, and S1PR3 where sample 1 and 3 showed strong band against anti-CD63, anti-S1PR1, and anti-S1PR3 primary antibody (**b**) Transmission electronic micrograph (TEM) of HEK293-derived exosomes depicting spherical morphology with 200 nm size. The 293 cell-derived exosomes were stained with gold-particle-tagged anti-CD63 antibody and then negatively stained with uranyl acetate

5. Bind primary antibodies with a horseradish peroxidase (HRP)-conjugated a-mouse IgG goat antibody or Immun-Star goat anti-rabbit (GAR)-HRP conjugate and expose using a Pierce SuperSignal West Pico PLUS Chemiluminescent Substrate (Thermo Fisher Scientific).
6. Detect the signal on the Bio-Rad Chemidoc Touch Gel-Imaging System and analyze on Bio-Rad Image Lab software v.6.0.0.

3.3.2 Immunoblotting: Characterization of S, M, E Exosomal Vaccines

1. For the preparation of the exosomal vaccines, the expression of SARS-CoV-2 protein S, M, and E proteins needs to be analyzed using immunoblotting. To detect CoV-2 S, M, and E proteins, pellet the cells and then lyse as described above (3.3.1) and then load onto a Bio-Rad Mini-Protean TGX 7.5% gel and run in TGS running buffer.
2. Transfer protein to nitrocellulose membranes using Bio-Rad's TransBlot Turbo Transfer system. Block the membrane for 1 h in 1X TBS-T with 5% milk.
3. Wash membranes and then incubate overnight at 4 degree C with primary antibody, i.e. anti-Spike 3RM, anti-envelope, and anti-membrane diluted 1:1000 in blocking buffer.
4. Wash the membranes three times for 2 min each before incubation for 30 min at room temperature with the HRP-conjugated secondary antibodies.
5. Expose membranes with Pierce ECL Plus Western Blotting Substrate (Thermo Fisher Scientific) before imaging on the Bio-Rad Chemidoc Touch Gel-Imaging system. Analyze images on Bio-Rad Image Lab software v.6.0.0. Fig. 4a.

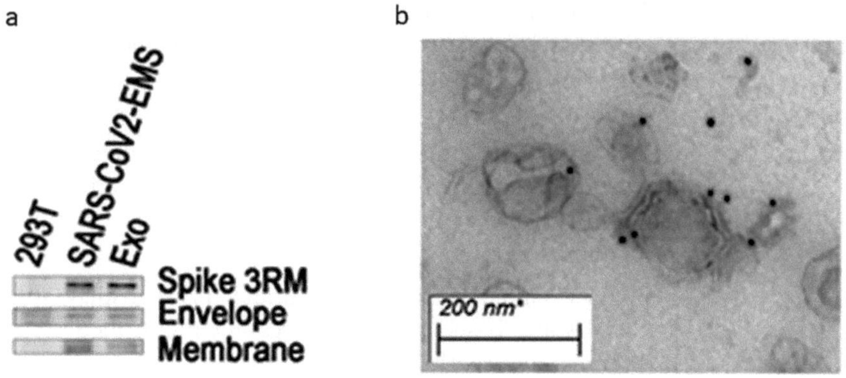

Fig. 4 Characterization of the SME exosomal vaccines. (**a**) Incorporation of S, M, and E proteins in exosomes. 293T cells were transduced with plasmids expressing the gene encoding S, M, and E proteins and vesicles pelleted from the conditioned media were analyzed by Western blot for the presence of the S, M, and E primary antibody. 293T cells are negative control, SARS-Covid-2 EMS is positive control, and Exo is exosome isolated from SME-transduced 293T cells. The samples obtained were analyzed for the presence of the S, M, and E protein by Western blot using the S, M, and E antiserum. (**b**) Transmission electronic micrograph (TEM) of exosome isolated from SME-transduced HEK293 cells depicting the appearance of CKV-21 (exosome vaccine) expressing structural proteins

3.3.3 Immuno-Transmission Electron Microscopy (TEM)

1. Transmitted electron microscopy (TEM) analysis can be applied to show the spherical morphology of the HEK293-derived exosomes with 200 nm size. TEM is the only system that confirms in terms of purity, the presence of contaminants, and the yield and size of the isolated exosome (*See* Fig. 3b).
2. Wash cell pellets twice in PBS. Use phosphate buffer pH 7.4 with 2.5 percent glutaraldehyde to fix the samples for 30 min at 4 °C. After washing twice with cold PBS, post-fix the pellets in buffered OsO4, dehydrate in graded acetone, and embedded in Durcupan ACM resin. Produce ultrathin sections and mount in copper grids, with uranyl acetate and lead citrate used as counterstains.
3. Use A Hitachi H-7600 TEM to examine the specimens at an 80 kV voltage. Resuspend the exosomal fractions in PBS, deposit onto formvar carbon-coated nickel grids for 60 min, rinse in PBS, and fix for 10 min with 2% paraformaldehyde.
4. Wash exosome-coated grids in PBS, transfer to a drop of antibody, and then immunogold label with anti-CD63 antibody for 40 min. Wash grids in 0.1 percent BSA/PBS, incubate for 40 min with gold-particle-tagged anti-CD63 antibody, rinse in PBS, and then post-fix for 10 min in 2.5 percent glutaraldehyde.
5. Stain the grids with 2 percent uranyl acetate for 15 min and 0.13 percent uranyl acetate 3 and 0.4 percent methylcellulose for 10 min, air-dried for 5 min, then examine with transmission electron microscopy at a dilution of 1:10,000 with PBS after being washed in deionized water.

3.3.4 Transmission Electron Microscopy

1. Perform transmission electron microscopy (TEM) on S, M, E displaying exosome placed on 200-mesh nickel formvar carbon-coated grids and leave to adhere for 20 min as stated above.
2. Incubate the grids with 2.5% glutaraldehyde containing 2% sucrose and, after washings in distilled water, negatively stain the EVs with NanoVan and observe under electron microscope.
3. EVs loaded with S, M, E under TEM display appearance of exosome-based vaccine expressing mentioned structural proteins. Figure 4b.

4 Notes

1. It is important to culture cells in FBS-free media for isolating EV. The reason being FBS, itself consist of abundant number of vesicles, if left within media, will be mixed with EVs derived from the parent cells. It is also important the cells should be full confluent when replacing media with FBS-free media and during addition of drugs in order to achieve enhanced exosome secretion.
2. The transfer of medium between tubes (i.e., conical and ultracentrifuge tubes) must be carried out in a sterile tissue culture hood. Also, we must be careful of spillage of cell suspension. In case of any spill of any cell suspension, this needs to be cleaned with 70% ethanol to avoid any sort of contamination.
3. Wash ultracentrifuge tubes and their lids thoroughly with sterile deionized and distilled water. Always use sterile ultracentrifuge tube in order to avoid cross-contamination. It is also advisable to use sterile ice-cold PBS for pellet washing.
4. At this point, an off-white to transparent pellet will be visible on the lower wall of the marked side of the ultracentrifuge tube. This pellet consists of EV and contaminating protein.
5. The exosomes resuspended in PBS can be frozen at −80 degree C and used for characterization assay next day. A small aliquot of this sample can also be subjected to NTA to quantify the exosomes.
6. Different antibodies can be employed as distinct markers to separate each of these various fractions. This method can also be used to evaluate the expression levels of different proteins of interest in EVs (*See* Figs. 3 and 4).

References

1. Ji Y, Ma Z, Peppelenbosch MP, Pan Q (2020) Potential association between COVID-19 mortality and health-care resource availability. Lancet Glob Health 8:e480
2. Thery C, Witwer KW, Aikawa E et al (2018) Minimal information for studies of extracellular vesicles 2018 (MISEV2018): a position statement of the International Society for Extracellular Vesicles and update of the MISEV2014 guidelines. J Extracell Vesicles 7:1535750
3. Lee Y, El Andaloussi S, Wood MJ (2012) Exosomes and microvesicles: extracellular vesicles for genetic information transfer and gene therapy. Hum Mol Genet 21:R125–R134
4. Lou G, Chen Z, Zheng M, Liu Y (2017) Mesenchymal stem cell-derived exosomes as a new therapeutic strategy for liver diseases. Exp Mol Med 49:e346
5. Walker S, Busatto S, Pham A et al (2019) Extracellular vesicle-based drug delivery systems for cancer treatment. Theranostics 9:8001–8017
6. Jafari D, Shajari S, Jafari R et al (2020) Designer exosomes: a new platform for biotechnology therapeutics. BioDrugs 34:567–586
7. Colao IL, Corteling R, Bracewell D, Wall I (2018) Manufacturing exosomes: a promising therapeutic platform. Trends Mol Med 24: 242–256
8. Morel PA, Falkner D, Plowey J, Larregina AT, Falo LD (2004) DNA immunisation: altering the cellular localisation of expressed protein and the immunisation route allows manipulation of the immune response. Vaccine 22:447–456

Index

A

Adhesion 7, 11, 92, 99, 146, 155, 169, 183
Advanced Therapy Medicinal Product (ATMP) 70, 71
Antibodies (Ab) 3, 4, 6, 8, 9, 12, 36, 39, 41, 42, 72, 81, 83, 85, 87, 99, 100, 102–104, 107, 129, 135, 137, 143, 147, 150, 151, 156, 162, 166–168, 172, 177, 178, 180, 182, 183, 202, 216, 219–224, 235, 244, 251, 252, 254, 255, 258–271, 273–275, 307–310
Artificial microRNAs 191–206
Asymmetrical flow field-flow fractionation 105–107

B

Bacterial extracellular vesicles (BEVs) 212, 218, 219, 224
Barcoded exosomal microRNAs (bEXOmiRs) 192–206
Biodistribution 24, 29, 31
Biogenesis 33, 121, 122, 145, 192, 247, 277–279, 301
Bioluminescence resonance energy transfer (BRET) 23–31

C

Caenorhabditis elegans 278, 279, 281–283, 286, 291, 296
Capto Core 700 multimodal chromatography (MMC) 46
Cardiac progenitors cells (CPC) 70, 71, 73, 76, 86, 90, 92, 94
Cargo delivery 160, 180, 241
Cell 15, 23, 34, 45, 57, 70, 111, 122, 133, 145, 159, 191, 212
Cell–cell interactions 111–118
Cell sample 20
Characterization 3–12, 18, 21, 23, 25, 28, 33–42, 59, 61, 65, 66, 70, 80, 82, 99, 121–131, 135, 137, 139, 140, 144, 149–150, 160, 202, 212, 219, 223–224, 247, 257, 277
Ciliary-derived EVs 279, 286–291
Conditioned medium (CM) 6, 7, 10, 26, 27, 45, 48, 50, 52, 54, 55, 70–97, 253, 255, 306
Confocal microscopy 66, 134, 135, 137, 170, 179, 187, 259, 263, 265, 285, 287

D

Density gradient ultracentrifugation 211–226

E

Ectosomes 277–279
ELISA 73, 75, 78, 81, 83, 84, 87, 90, 93, 259, 260, 265, 266
Encapsulation 121–131, 254
Endothelial cells 20, 21, 111, 112
EV capture 279
EV isolation 18, 25–27, 34, 45–52, 59–61, 70, 104, 142, 160, 194, 199, 205, 227
Exosome 6, 7, 10, 11, 15–21, 59, 73, 95, 99, 102, 111–118, 121–124, 126, 128, 129, 131, 140, 145–156, 159–188, 191–193, 201, 206, 224, 227, 228, 234–236, 257–266, 268–275, 277, 301–304, 306–310
Exosome-based vaccine 301
Extracellular vesicle (EV) 3–12, 23–31, 33–42, 45, 46, 49–54, 57–66, 70–97, 99–107, 121, 122, 133–145, 149, 154, 159–188, 191–206, 212, 213, 218, 219, 221–224, 227, 228, 231, 233, 236, 237, 241–243, 247, 249, 250, 252–254, 258, 278, 279, 281–286, 288, 289, 291, 294–296, 301, 303, 305, 307, 310

G

Good Manufacturing Practices (GMP) 70, 71, 75, 80–82, 92

H

Homing 145–156, 169, 183
Human feces 212

Seppo Vainio (ed.), *Cell-Secreted Vesicles: Methods and Protocols*, Methods in Molecular Biology, vol. 2668, https://doi.org/10.1007/978-1-0716-3203-1,

I

Immunoaffinity chromatography (IAC)............100, 102, 104–107
Immunoblot..201, 202
Immunogold labelling.. 37
Immunolabelling Electron Microscopy (IEM)............. 33 34, 36, 41
Immunomagnetic separation (IMS)..................258, 260, 265, 269, 270, 273, 274
Integrin... 99, 146, 147, 149–153, 155, 156
Integrity...39, 46, 79, 122, 123, 130, 228, 242
Intercellular communication 99, 133, 159, 227, 241
In vivo imaging 24, 26, 278, 279, 283
Isolation...4–7, 10, 18, 25–27, 31, 34, 45–52, 59, 60, 70, 77, 99–107, 117, 122, 138–140, 142, 146, 149, 160, 192, 194, 199, 205, 212, 213, 216–219, 223, 224, 228, 229, 231, 232, 234, 246, 248, 257, 258, 260, 275, 303, 304, 306, 307

L

Label-free...15, 99
Ligand...99, 100, 148, 151–153, 156, 159, 166
Liquid Chromatography–Mass Spectrometry (LC-MS) 57–66, 258
Localization...134, 169, 183–184
Luciferase..................................... 24, 134–136, 138–139

M

Magnetic particles...................... 258–266, 269–273, 275
Mass spectrometry 60, 63
Master cell bank (MCB).......................70, 71, 76, 77, 93
Medium/large EVs (m/lEVs) 23, 27–30
Mesoporous silica nanoparticles......................... 242–246
Metabolites.................................. 58, 59, 61–63, 66, 122
Metabolomics.......................................58, 61–63, 65, 66
Microvesicles 10, 11, 113, 114, 121, 128, 145, 149, 193, 227, 228, 234–236, 277, 301
Microwave Plasma Atomic Emission Spectroscopy (MP-AES)..123, 124, 126, 127, 129
Molecular analysis 15

N

Nanoparticle.. 15, 16, 21, 82, 86, 100, 122, 130, 212, 214, 238, 241, 242, 244, 249, 250, 254, 302
Nanoparticle tracking analysis (NTA).......................7, 18, 28, 48, 53, 73, 81, 86, 100, 124, 129, 138–140, 147, 149, 150, 212, 219, 223, 229, 230, 232–234, 247, 249, 253, 254, 257, 275, 307, 310
Non-lipidic staining 133–144
Nuclear magnetic resonance (NMR)......................58–63, 65, 66

O

On-line coupling...102

P

Plasmon-enhanced fluorescence...................................... 5
Primary cells 24, 228
Production methods 69–97, 138, 305
Protein 4, 11, 15, 16, 18, 20, 21, 23–31, 36, 38, 39, 41, 42, 46, 49, 51, 53, 54, 57, 58, 65, 70, 72, 78–81, 85, 87, 89, 93, 95, 99, 111, 117, 118, 122, 124, 128–131, 134, 136, 138–140, 143, 145–147, 149–151, 156, 159–168, 170–183, 192, 193, 206, 212, 216, 219, 221–223, 228, 231, 232, 242, 244, 247, 250–254, 258, 259, 261, 265, 275, 279, 287, 288, 291, 294–296, 302, 304–310
Purification 45–52, 54, 55, 70, 80, 84, 122, 127, 160, 162, 163, 171, 173, 174, 178, 187, 188, 192, 194, 196, 199, 212, 228, 253, 257, 258, 278

Q

Quality Control methods (QC) 70
Quantitative... 31, 73, 80, 84, 88, 116, 160, 202, 203

R

Raman scattering...15–21

S

Sequential filtration (SF) 228, 232, 235, 236
Silver nanoparticles 15–21
Size...7, 9–12, 18, 23, 34, 39, 41, 46, 51–54, 62, 64, 70, 71, 81, 86, 100, 102, 103, 105, 113, 121, 129, 130, 138, 140, 141, 149, 150, 156, 159, 173, 175, 177, 186, 198, 212, 214, 216, 219, 221, 223, 227, 228, 232–236, 238, 244, 247, 249, 253, 254, 257, 277, 288, 297, 301, 302, 307–309
Small EVs (sEVs) 23, 24, 27–30, 228, 232, 236
Solid-phase preconcentration......................................258
Stem-loop RT-PCR 199, 202

Surface-enhanced Raman scattering (SERS)..........15–21
Surface plasmon resonances 4

T

Tangential-flow filtration (TFF)............................ 46, 47, 49–53, 70, 72, 78–80, 94
T cells 146, 227, 228, 231–237
Tetraspanins................................ 99, 131, 134, 150–151, 212, 221, 224, 235, 258, 291, 295
Transmission electron microscopy (TEM) 17–19, 21, 33–39, 54, 219, 223, 226, 230, 232, 234–236, 247, 251, 253, 254, 305, 307–310
Tumor cells.................................. 24, 111, 113, 117, 227

U

Ultracentrifugation 18, 25, 34, 45, 58–60, 70, 137, 138, 140, 142, 149, 154, 155, 201, 224, 225, 253, 257, 258, 303, 307
Uptake 133–144, 159, 169–170, 180, 184–187, 242, 252, 253